System Safety Engineering and Risk Assessment

A Practical Approach

Second Edition

System Safety Engineering and Risk Assessment

A Practical Approach

Second Edition

Nicholas J. Bahr

CRC Press
Taylor & Francis Group
Boca Raton London New York

CRC Press is an imprint of the
Taylor & Francis Group, an **informa** business

CRC Press
Taylor & Francis Group
6000 Broken Sound Parkway NW, Suite 300
Boca Raton, FL 33487-2742

First issued in paperback 2017

ISBN-13: 978-1-4665-5160-2 (hbk)
ISBN-13: 978-1-138-89336-8 (pbk)

Library of Congress Cataloging-in-Publication Data

Bahr, Nicholas J., author.
    System safety engineering and risk assessment : a practical approach / Nicholas J. Bahr. -- Second edition.
        pages cm
    Includes bibliographical references and index.
    Summary: "We all know that safety should be an integral part of the systems that we build and operate. The public demands that they are protected from accidents, yet the two constituencies--industry and government--don't always know how to reach this common goal"-- Provided by publisher.
        ISBN 978-1-4665-5160-2 (hardback)
        1. Industrial safety. 2. System safety. 3. Risk assessment. I. Title.

T55.B23 2014
620.8'6--dc23   2014025988

Visit the Taylor & Francis Web site at
http://www.taylorandfrancis.com

and the CRC Press Web site at
http://www.crcpress.com

This book is dedicated to the memory of my wife, Carolina, and to my children, David and Julian.

Contents

Foreword

The engineering field is now so broad that we can become an expert in only a small part of it. Nevertheless, it is a mistake to restrict our reading to a specialized branch. We can learn much by reading widely and learning how others approach their problems. Safety engineers, in particular, can learn from each other. I have learned much from reading reports of railway and aircraft accidents and risk assessments in the nuclear industry.

Unfortunately, the books on other branches of engineering are often difficult to understand as we lack the specialized knowledge that the authors take for granted. Nicholas Bahr writes with such readers in mind. Even if you do not work in the process industry you can learn from this book how process safety experts identify and assess hazards. Their techniques may be useful elsewhere or they may suggest ways you can improve your own techniques. Similarly, process engineers can learn from others.

The emphasis of the book is on a systems approach, illustrated by case histories. Plants are so complex today that we cannot identify the hazards and operating problems just by looking at a drawing. A structured, systematic approach is essential. However, as the author points out, all a system can do is make sure that the knowledge and experience of the people involved are fully utilized. If they lack knowledge and experience, the system is just an empty shell. Knowledge and experience without a system achieve less than their potential; a system without knowledge and experience achieves nothing. This should be obvious, but is sometimes overlooked in the enthusiasm for downsizing.

The book makes it abundantly clear that safety is not a coat of paint added by a safety expert at the end of design, but an integral part of design. The safety expert can advise, monitor, and assist, but full involvement by the design team is essential. I hope, therefore, that this book will be widely read, not just by safety engineers but by all involved in the design and operation of complex systems. It is well written and easy to read.

Trevor Kletz, DSc, FEng

Preface to the Second Edition

Among the many objects to which a wise and free people find it necessary to direct their attention, that of providing for their safety seems to be the first. The safety of the people doubtless has relation to a great variety of circumstances and considerations, and consequently affords great latitude to those who wish to define it precisely and comprehensively.

Federalist #2
The Federalist Papers

Concerning Dangers from Foreign Force and Influence, 1787
John Jay

We all know that safety should be an integral part of the systems that we build and operate. The public demands that they are protected from accidents, yet the two constituencies—industry and government—do not always know how to reach this common goal. The reason I wrote this updated edition is that engineers and managers in companies need a succinct and practical way to effectively design safety into their systems and products within ever-increasing cost and time constraints. Government regulators need to better understand how to oversee those industries in such a way that protects the public, spurs innovation, and yet does not negatively impact the economy or competitiveness. This balance is not easy. This book will give both industry and government the practical insight and concrete success stories to make this happen.

The purpose of this book is to give engineers and managers, working in companies and governments around the world, a pragmatic and reasonable approach to system safety and risk assessment techniques and to design viable safety management systems. It is written in easy-to-understand language that gives you the tools to implement tested solutions immediately. You can pick it up and use it at once by following the numerous worked examples taken from real-life engineering problems. There are practical tips and best practices that tell you how to prevent accidents but also how to put safety into your systems at a sensible price. There are numerous case studies from real disasters that go into detail describing what went wrong and the lessons learned. And there are case studies of where safety was done correctly and are best practices.

Because our work is becoming increasingly global, examples and lessons learned are taken from around the world to explain how they are used in industries such as manufacturing, consumer products, chemical process, oil and gas, aviation, mass transit, military and space, and commercial nuclear power. No one country or industry has got it all figured out and there is much to learn from each other. Each chapter also includes *Notes from Nick's File*, actual experiences that I've lived through in applying the techniques discussed—successes and also my own failures.

Like the first edition, the book is aimed at working engineers who know that they need to build safe systems, but aren't sure where to start. They don't want to waste a lot of time sorting through the mountain of safety books that are more theoretical than practical or too narrowly focused. This book is for those looking for a single, comprehensive, but pragmatic reference, not a shelf full. It also makes a good companion text to university engineering design courses and can be the basis of a system safety course.

The first three chapters lay the necessary foundation for understanding the concepts and their applications in many different industries in system safety engineering, risk assessment, and safety management systems. It discusses important definitions and concepts and illustrates their application and helps you understand the system safety process, what constitutes a best-in-class safety management system, and how to actually apply the concepts in reducing real hazards that will be demonstrated through the tools discussed later in the book. A safety maturity model helps you determine where your company fits on the safety continuum, and lagging and leading safety performance indicators help you measure success and progress. Practical suggestions of where to find good national and international safety standards will make your search for the best standards worldwide easier.

Many books describe safety management systems almost as a laundry list of safety activities, but Chapter 4 gives an easy-to-understand and realistic approach that fits logically together for designing, implementing, and auditing safety management systems. Numerous examples and best practices illustrate what you should and shouldn't do in designing your safety management system. Safety culture must be embedded into your safety management system and this chapter gives you practical ways to do it.

Chapters 5 through 9 describe the different safety analysis tools available. Hazard Analysis, HAZOP, What-If, Fault Tree Analysis, Failure Modes, and Effects Analysis, Human Factors, Software Safety, and other safety tools are described with realistic worked examples. The chapters detail how to use them, give examples, describe common mistakes in using them, and also provide best practices and tips of how to apply them judiciously.

Chapter 10 discusses practical ways to find safety data and gives suggestions for creating your own safety knowledge management system to help you link all this information together. It also explains how to set up a safety training program and gives a sample safety training course outline used in a microprocessor production plant.

Unfortunately, accidents do happen. Chapter 11 helps you understand what to do when you must investigate an accident at your facility. It details how to set up the investigation team, investigate the accident, and document the results, lessons learned, and corrective actions. Companies and governments are learning the hard way the importance of how to communicate an accident to the public. This chapter tells you how to develop a crisis communication plan, common mistakes, and the dos and don'ts of communicating with the public during an accident.

Not enough has been written about how to develop balanced safety government oversight bodies and regulations. Chapter 12 focuses on how to set up a safety regulatory oversight body, its functions, and governance structure and how to implement them. Numerous examples from around the world are discussed, and a case study describes how one country developed their first safety oversight body ever. It also details another case study with the U.S. Federal Aviation Administration and explains how they've taken a strong and mature safety oversight program and made it even better by incorporating more advanced system safety techniques.

Chapter 13 describes the power of risk assessments. It discusses how risk is defined and how it is perceived differently. It takes these concepts and details a quantitative risk assessment methodology and explains how to use it appropriately. But results need to be communicated to the public and others the right way. Suggestions for effectively communicating complicated risk information are also given.

Chapter 14 gives a detailed example of conducting a probabilistic risk assessment of launching a payload into space. Accident scenario generation, event trees, consequence determination, and uncertainty are described and worked through. It also discusses how this information can be used to determine safety costs.

The book includes appendices of useful lists to help you apply the system safety engineering and risk assessment tools and safety management system program described in the book. Typical energy sources, generic hazard checklists, and facility safety checklists are included. The final appendix gives you some useful website addresses for more information.

Accident rates are dropping worldwide; companies are more proactive in preventing accidents and disasters; governments seem to be using a more effective safety oversight process. Despite the overall statistics, we are still seeing shocking accidents worldwide. Why?

In the last few decades, we have seen a significant increase in operational complexity—rapid implementation of new and advanced technologies, just-in-time operations, increased complexity of quality assurance, and organizations themselves becoming more complex and global. At the same time, we are seeing much greater interdependencies in our systems, between systems and their operational environments, and these interdependencies are dynamic, not static.

Our companies and governments are creating more international alliances and unfortunately accidents are more transnational than before. Supply chain networks are very complex and business interruptions are less tolerable than earlier. Our global markets seem to be more unstable and very fluid. Corporations are facing not just increased competition from around the world and at home but also tougher internal financial constraints, controls, and oversight. Insurance costs continue to rise and labor movement now happens on a global scale. Natural disasters are a more significant threat than ever before.

On top of all these complexities, companies and governments are also facing a public that is much less tolerant to risk. The public perception of risk to a company's brand and a government's reputation has increased significantly in the last 20 years, and many have lost it overnight. It takes a long time to build a strong brand but it can be lost very quickly. It doesn't take much for the public to quickly lose confidence in a company or government.

We still need to improve how we assess risks and manage their impacts. But we also have to do it in a cost-conscious way, balancing risk management with cost efficiencies—they are not mutually exclusive. Governments must also play a more proactive role in the safety oversight process, but still allow the free market to find their own solutions. Industry and government must work collaboratively to find the best solutions. Hopefully, this book will help take us closer to that goal. And, I also hope that this book demystifies safety from risk, shows its power, and proves that it can even be fun.

Preface to the First Edition

This book came about when I tried to find a comprehensive but inexpensive book that really shows an engineer how to design and build equipment that is safe. I was looking for something that I could give to working engineers who know nothing about safety, but who are asked to build their products safely. I couldn't find anything that succinctly demonstrates the most important aspects of safety analysis and risk assessment. I wanted something written by an engineer, for engineers. Most books on the market describe what system safety is—not how to apply it.

The purpose of this book is to give engineers a comprehensive, practical guide on how to build safety into their products and industrial processes. It is for those who are concerned about safety but who have no idea where to start, nor wish to spend a lot of precious time trying to find out. This is a book for any engineer who has been told to build safe products and who wants one reference book, not a shelf full, that will show how to apply the concepts immediately, without wasting time on unnecessary things.

This book is very pragmatic—you can pick it up and use it at once by following the numerous worked examples of real-life engineering problems. There are practical tips that tell you how to avoid common mistakes engineers make. Many suggestions explain not only how to prevent disasters, but also how to put safety into your system for a reasonable price.

Most of all, key safety and risk methods are clearly discussed, with useful examples that show an engineer how to apply them to a job. Real-life examples are worked through, explaining why one system is safe and another is not.

Many features of system safety and risk assessment are common to various industrial situations, and one industry can learn from the experience of another. This book crosses various industries so that you can learn the best techniques from the chemical, nuclear, aerospace and military, manufacturing, and mass transit industries.

Though occupational safety has long been taught at universities, system safety is still not an integral part of the college engineering program. This book makes a very good companion text for engineering design courses.

The first three chapters describe why safety is important, what it is, and how different industries use it. Safety and risk concepts are briefly discussed so that the reader has the necessary foundation to understand, and use appropriately, the safety and risk techniques in later chapters.

Chapter 4 illustrates how to implement a cost-effective safety management organization quickly and efficiently. Examples demonstrate what not to do in a safety management program, including how others have failed and the mistakes that engineers typically make. Actual, successful system safety program and audit plans are included. Ideas are given to help you sell safety to your management.

The heart of the book are Chapters 6 through 9, where best system safety techniques from different industries are presented. Various kinds of hazard analyses are detailed, with actual engineering examples of a laser and a hazardous waste storage facility, demonstrating how to do the analysis.

HAZOP and *what-if*/safety checklists, two of the most common safety methods in the chemical industry, are explained. Sample process problems, which engineers face every day at work, are shown. Other safety tools, such as fault tree analysis, failure modes and effects analysis, human factors safety analysis, and software safety, are explained. Examples of the use of these tools are also presented.

Chapter 10 provides useful information about how to create and maintain the necessary data management system to keep a system safety program running smoothly. The chapter also shows practical examples and suggests how to set up a safety awareness and training program.

No one likes to deal with accidents, yet they are inevitable. Chapter 11 provides useful information about how to set up your own investigation board, create a closed-loop reporting system, and learn from the accident.

Many engineers are unaware of the power of risk assessment. This is a very cost-effective technique that cannot only help you make your system safer, but also can help you decide how to allocate resources to do so in the most efficient way possible. Chapter 12 explains what risk assessment is. It discusses how the public perceives risk and offers suggestions about how best to communicate engineering risks to the public.

Chapter 13 details how to conduct a risk evaluation. The necessary models are developed and explained. A risk assessment example of launching a payload into space, describing how to decide which design or operational changes will make the system safe, and which will increase the risk, is given.

At the end of the book, appendices give the reader useful checklists to help in identifying hazards. Typical energy sources, hazard checklists, and facility safety checklists are included. Several Internet sources are listed so that you can get the best, most up-to-date safety and risk information available.

As technological systems become more and more complex, it becomes increasingly difficult to identify safety hazards and control their impact. The cost is measured not only in dollars lost due to accidents, but also in lawsuits by employees injured on the job, degradation of the environment, loss of market share, and even ruined reputations. Engineers are finding that safety and risk touch upon every aspect of the engineering system design, operation, and disposal life cycle.

It is obvious that many of our current system safety techniques have come about as the result of horrendous accidents. A lot of pain, suffering, and economic loss have been endured before we, as a society, decided to take safety more seriously. At the same time, however, through this difficult learning process we have found that making systems safer is not just something we should do because it is ethical and moral, but also because it makes very good business sense.

Engineers are making an honest effort to design, build, and operate their systems safely. Many engineers, however, just don't have the necessary tools to do the job right. This book demonstrates that the way engineers produce safety in one industry can be used in another with few changes.

It is very important that safety is designed into the system or process. Failure to do so will eventually result in an accident, with the accompanying downtime, lost production, injuries, lawsuits, and possible loss of business.

System safety and risk assessment does not have to be an expensive part of designing and building technological products. If it is done early and efficiently, it will more than pay for itself.

The important point to remember is to take the various system safety tools discussed in this book and apply them *as you see fit*. The system safety analyses and programs are described and demonstrated in detail so that you can take these well-established methods and tailor them to *your needs*. What is critical is that the system safety process be comprehensive. It is much better to use a shortened safety analysis than none at all.

This book is intended to be used as a tool by practicing engineers in all disciplines, to help identify, control, and mitigate safety issues before they become serious problems. I also hope it demonstrates that there is absolutely no reason that safety has to be difficult, a problem, or mysterious. In fact, it can even be fun.

Acknowledgments

I thank Bill Crittenden and John Rauscher for their incisive comments and support in developing the HAZOP example. Albert Powell contributed significantly with his ideas for the facility hazard analysis section. Special thanks to Adrian Rad for supplying information for the laser example. Because their examples are so useful, I've reproduced them for this edition.

I also thank Mark Davis, Khalil Allen, and Jason Sergent for our numerous discussions and debates about how to best apply system safety. Special thanks to Len Neist for his insights and thoughts on how best to regulate system safety.

And of course a very special thanks to my children, David and Julian, for keeping me on task.

Thanks also to Phillip Johnson for help with graphics and Colin Holmes for the Waterfall Rail Accident photographs.

It is also a pleasure to express gratitude to the fine people at Taylor & Francis Group, for their support and enthusiasm in developing this book.

Author

Nicholas J. Bahr is an internationally recognized expert in system safety, risk assessment, and enterprise risk management systems and has over 25 years of professional experience working around the world. He has set up safety management systems for companies and helped governments improve their safety oversight programs. Over his career, Mr. Bahr has conducted programs for commercial and government clients, detailed technical risk assessments, implemented enterprise risk management business processes, and developed regulatory oversight programs throughout the United States, United Kingdom, Europe, South America, Australia, the Middle East, and North Africa. His diverse experience and background cover many industries including aerospace, utilities, oil and gas, manufacturing, and transportation.

After a high-profile rail accident in Australia, Mr. Bahr was asked to lead an international team conducting a safety management systems audit of both the regulator and the railway. The audit methodology is now considered the *new international gold standard* for safety management systems. His client engagements range from risk strategy for senior government and commercial executives, to detailed risk assessments for front-line management. He has helped CEOs, senior VPs, and senior government officials realize tangible and sustainable benefits from their safety and risk management programs. Mr. Bahr is a past U.S. delegate to various standards-writing bodies. Currently, Mr. Bahr is a principal at Booz Allen Hamilton and is the regional manager for the Middle East and North Africa.

1 Introduction

Better safe than sorry.

Nineteenth-century proverb

The way to be safe is never to be secure.

Gnomologia, 1732
Thomas Fuller

Appearances often are deceiving.

The Wolf in Sheep's Clothing, c. 550 BCE
Aesop

1.1 WHY DO WE NEED SAFETY ENGINEERING?

It is difficult to go on the Internet, open a newspaper, or turn on the television and not be reminded of how dangerous our world is. Both large-scale natural and man-made disasters seem to occur on an almost daily basis and seemingly never ending. An accident at a plant in Bhopal, India, in 1984, killed over 2500 people. A magnitude 9.0 earthquake and resulting tsunami in 2011 triggered a series of fires and explosions at a commercial nuclear power plant in Japan, resulting in three of the six reactors melting down and over 100,000 residents permanently evacuated.

Though there is a downward trend in fatal accidents in the United States and many other countries, recent high-profile accidents still command headlines worldwide. In 2011, a Chinese high-speed train collided into another killing 38 people. In the Gulf of Mexico, the oil company at the center of the worst oil spill in history put aside $41 billion in 2010 to pay for damages from the spill. An automobile airbag manufacturing plant exploded, killing one worker, after it had had over 21 fire emergencies in 1 year. Swarms of helicopters with television cameras were drawn to the plant after every call, creating a public relations nightmare and forcing the government to shut down the plant temporarily.

An airliner crashed into an apartment building in downtown Sao Paolo, Brazil, killing all on board and many in the apartment building. The Air France Concorde went from a 27-year record of zero crashes to a single crash in July 2000, killing 100 passengers and 9 crew members, becoming one of the worst aircraft-type safety records (due to the low frequency of flights). In June 2009, the new Airbus A330 flying from South America to Europe experienced an aerodynamic stall, caused by inconsistent sensor readings and inadequate pilot response, crashed into the Atlantic Ocean killing all 228 on board, resulting in the highest death toll of any aircraft type worldwide. Although it was the first crash while in commercial passenger flight, it was the second fatal accident of the new design. After hitting a flock of geese,

1

a commercial airliner miraculously landed safely, without loss of life, in the Hudson River right off of Manhattan in New York City. The April 2010 crash of the Polish Air Force Tu-154, likely due to human error, killed the current president and his wife, the national Polish bank president, chief of Polish general staff, deputy foreign minister, 15 members of parliament, and other political notables.

In 1995, the Air Route Traffic Control Center, Fremont, California, lost power, causing radar screens covering Northern California, Western Nevada, and 18 million square miles of Pacific Ocean to go dark for 34 min while 70 planes were in the air, almost resulting in two separate midair collisions. In another incident, a worker in downtown Chicago cut into a cable and brought down the entire Air Route Traffic Control System for thousands of square miles.

But it is not just builders, manufacturers, and operators that significantly impact accident rates—governments do too. An independent U.S. government panel (U.S. Department of Labor, 2012) found that government mine safety regulators and their leadership failed to heed warning signs or implement and enforce their own safety regulations that allowed a coal mine to operate unsafely, resulting in 29 deaths from an explosion and fire. A 2005 Special Commission of Inquiry (McInerney, 2005) in Australia found that both the railway operator and the regulator failed to carry out their safety duties adequately. The rail accident that killed seven followed on the heels of a previous rail accident in which government oversight was not sufficiently strengthened, resulting in the special commissioner to request the stand-up of an oversight board to ensure that both the rail regulator and the operator implemented safety improvements and strengthened oversight programs.

The commercial nuclear accident in Japan in 2011 followed a much more devastating nuclear power plant accident in Ukraine in 1986; the reactor explosion burned out of control, sending a radioactive cloud to over 20 countries, severely affecting its immediate neighbors' livestock and farming. The Ukrainian disaster forced countries to rethink reactor safety. Government regulators worldwide instituted changes to their oversight regimes. But, just as many were starting to feel comfortable with commercial nuclear energy again, in 2011, 2 months after the Fukushima nuclear accident in Japan, Germany announced that it was shutting down all of its nuclear power plants by 2022. Germany gets 25% of its energy from commercial nuclear power plants.

Some of these accidents occurred many years ago. Some of them occurred recently. Many of the accidents crossed international borders and affected millions of people in other countries. Many more did not extend beyond national borders but still affected a great number of people. And some of the accidents did not kill anyone.

We all know how quickly technology is changing; as engineers, it is difficult just to keep up. As technology advances by leaps and bounds, and business competition heats up with the globalization of the economy, turnaround time from product design to market launch is shrinking dramatically. The problem quickly becomes evident: *How do we build products with high quality, cheaply, quickly, and still safely?* But also, we have to ask, *how do governments protect the public and regulate industry without negatively impacting competitiveness or the national economy?*

An American Society of Mechanical Engineers' national survey (Main and Ward, 1992) found that most design engineers were very aware of the importance of safety and product liability in designs but did not know how to use the system safety tools available. In fact, most of the engineers who responded said that the only safety analyses they used were the application of safety factors in design, safety checklists, and the use of compliance standards. Almost 80% of the engineers had never taken a safety course in college, and more than 60% had never taken a short course in safety through work. Also, 80% had never attended a safety conference, and 70% had never attended a safety lecture. A complementary study produced for the UK Health and Safety Executive (Lee, 1999) found that undergraduate students still did not have a good grasp of safety and risk concepts. It further states that continuing to rely on codes and standards as the primary safety education method was insufficient in today's complex world. Unfortunately, the report states many professors in universities do not have a good understanding of hazard identification and risk reduction, let alone teach it. Although safety engineering is still not a core course in most engineering education programs, there are many safety engineering degree programs offered at universities around the world. There is hope that engineering programs will start to integrate safety into the engineering curriculum. The UK Health and Safety Laboratory and University of Liverpool (Schleyer et al., 2006) have collaborated to propose a safety and risk curriculum to be included in the European Union's project to mainstream occupational health and safety into the educational system.

So how do engineers design, build, and operate systems safely if they have never really been prepared for it? And, to make matters worse, engineers are now more frequently called to testify in court about failures in their designs. And how do governments protect the public without damaging the economy or national competitiveness? It almost seems that these ends are mutually exclusive. But they are not.

Like most engineering problems, this one does have a solution. And the solution is not that difficult to implement nor costly. What it does entail is considerable forethought, systematic engineering analysis, and a methodical approach to managing risk. It also requires government and industry to collaborate to improve safety as a joint project—not independently. As you will see in reading this book, system safety engineering is not difficult to apply—in fact, it is almost easy.

NOTES FROM NICK'S FILE

One of my first jobs out of engineering school was building river dredging equipment in South America. Because of the prohibitive cost of importing equipment, we reconfigured a lot of used and old equipment. Using a systems approach, coupled with hazard analyses, helped us to understand the implications of our contraptions on the human operator (our customers). It was easy and fast to apply.

1.2   WHAT IS SAFETY ANALYSIS?

Safety analysis is a generic term for study of the system, identification of dangerous aspects of the system, and correction of them. *System safety* is the formal name for a comprehensive and systematic examination of an engineering design or mature operation and control of any particular hazards that could injure people or damage equipment.

System safety engineering is a compilation of engineering analyses and management practices that control dangerous situations, specifically

- Identify the hazards in a system
- Determine the underlying causes of those hazards
- Develop engineering or management controls to either eliminate the hazards or mitigate their consequences
- Verify that the controls are adequate and in place
- Monitor the system after it has been changed and modify further as needed

1.3   SYSTEM SAFETY AND RISK ASSESSMENT

Many engineers confuse system safety with risk assessment and use the terms interchangeably. *System safety is the assurance and management that the system is safe for all people, environment, and equipment.* Risk assessment, like system safety engineering, can be used to determine how safe something is, but it also can be used to determine the various trade-off alternatives to lower the risk in a system. Risk in this case does not have to be related to safety; it could just mean the risk of losing market share or delivering a product late. A question many of us face is how do I balance the business risks with the safety risks?

In this book, two concepts of system safety engineering and risk assessment are combined. System safety engineering is considered a working part of the risk assessment process. Engineers must use system safety engineering analyses to truly understand what causes hazards and how they should be controlled. Risk assessment takes that information and helps the engineer weigh the options and decide which is the most cost-effective.

At first glance, it seems that every industry performs safety and risk assessment differently. On closer look, however, the fundamental precepts are the same: The methods are systematic and comprehensive. An industry may favor one method over another, but in most cases, this is mostly out of tradition. Now is a good time to review the way different industries apply safety and try to learn from each other. In most cases, you can literally lift the safety method from one industry and apply it directly to another.

The remainder of this book will go deeply into the two concepts of system safety and risk assessment. Proven tools and techniques are discussed, and actual engineering examples are shown. It will address safety from the operator and manufacturer's point of view, as well as from government regulations and oversight. But most importantly, it will help you understand how to manage the safety of your systems.

1.4 GOVERNMENT SAFETY REGULATIONS VERSUS SAFETY FROM INDUSTRY'S POINT OF VIEW

The perennial fight in governments, occurring in some more than others around the world, is how much should government regulate industry and how does it impact a company's bottom line and the nation's competitiveness. The natural societal tension is for governments to push safety regulations, while industry wants to minimize undue intrusion. Media stories regularly report the over or under safety regulation of industry, bemusing the public to understand which side to support. One side screams that safety regulation is unduly onerous and is strangling the economy and ultimately eliminating jobs. The other screams with equal vigor that companies are not valuing human life and focusing only on making money. Historically, we see an increase in government oversight after a major disaster, especially with significant loss of life and despoiling of the environment. The question is what is too much or too little?

NOTES FROM NICK'S FILE

A few years ago, I was helping a government agency that had lots of accidents design a safety oversight regime. We debated quite a bit on how far the regulations should go and how free industry should be to meet the regulations (prescriptive vs. performance based). In the end, we agreed to use a prescriptive regulatory process initially to quickly help industry get its safety performance up but that would migrate over time to performance based as the industry and regulatory regime matured, as measured by safety performance indicators.

The debate between the cost of safety regulation and the actual risk reduction realized is fierce. The Cato Institute (Viscusi and Taylor, 2002) highlights the cost of safety regulations, strongly questioning the realized benefits to a safer public. Some have argued that regulations do not add costs but merely move them from one group to another. Of course the various safety associations and regulators themselves have numerous studies on the benefits of risk reduction across industries. Yet countries, such as Great Britain, are actively trying to balance the two. The UK Treasury issued a report (HM Treasury, 1996) that details their roadmap at attempting to balance the protection of the public from safety hazards with the reduction of governmental red tape. Their intent is to move an industry's cost of compliance to a more proactive safety approach that prevents hazards without unduly impacting the bottom line.

Certainly the solution is the delicate balance between government safety regulations that truly do protect the public and industry's ability to be innovative, cost competitive, and have a sustainable business. But the question is "What is the correct balance?" An easy answer is to recognize that it is okay to have a healthy tension and debate between the two ends of the spectrum, but it must focus on how to support industry competitiveness while protecting the public. Leaving either side to independently meet the challenge will result in a lopsided emphasis on one or the other. And, of course, endless debate equals no action, leaving the public vulnerable and industry in limbo on how to predict and manage costs. The answer is for government and industry to jointly develop the solution. This book discusses both sides of the equation with practical, real, and actionable steps that each can take to correctly balance the two. But first, it is illustrative to briefly review the history of safety and how we arrived to where we are today.

1.5 BRIEF HISTORY OF SAFETY

Of course, the need for safety has always been with us. One of the earliest written references to safety is from the Code of Hammurabi, around 1750 BCE. His code stated that if a house was built and it fell due to poor construction, killing the owner, then the builder himself would be put to death. The first laws covering compensation for injuries were codified in the Middle Ages. The Great Fire of London in 1667 resulted in the first English fire insurance laws to be created.

Some of the first maritime safety regulations, came about around 1255 in Venice, stated that a ship's draught could not be exceeded and must be verified by visual inspection. The Comité Maritime International, established in Antwerp in 1897, codified the need for ongoing maritime regulations by bringing together numerous maritime law associations.

Around 1834, Lloyd's Register of British and Foreign Shipping was created, institutionalizing the concept of safety and risk analysis. In response to the sinking of the Titanic, the International Convention for the Safety of Life at Sea treaty was passed in 1914, stipulating that the number of lifeboats and other safety equipment must be commensurate with the number of passengers on board the ship. The German safety certification company, TUV Rheinland, was founded in 1872, providing technical safety certification services. In 1877, the U.S. Commonwealth of Massachusetts passed a law to safeguard machinery and also created employers' liability laws. Underwriters Laboratory was founded in Illinois, United States, in 1894, creating one of the most recognized product testing, certification, and standard bodies in the world.

At the end of the nineteenth century, a rash of boilers exploding gave urgency and impetus to the American Society of Mechanical Engineers to create the boiler and pressure vessel design codes and standards. Beginning in 1911, the United States saw safety groups forming, with one of the most important as the American Society of Safety Engineers. The National Safety Council was founded in 1913.

Around the 1920s, private companies started to create formalized safety programs. The early 1930s was the beginning of the implementation of accident prevention programs across the United States. By the end of the decade, the American National Standards Institute had published hundreds of industrial manuals.

Most of the current safety techniques and concepts we use today were born at the end of the World War II. Operations research led the way, suggesting that the scientific method could be applied to the safety profession. In fact, operations research gave some legitimacy to the use of quantitative analysis in predicting accidents. One of the earliest concept definitions for system safety (looking at safety from a system perspective) first appeared at the *Fourteenth Annual Meeting of the Institute of Aeronautical Sciences* in New York City in January 1946. "The Organization of an Aircraft Manufacturer's Air Safety Program" emphasized looking at safety from a holistic perspective: design safety into the system, detailed analysis of systems, and taking proactive steps to prevent accidents from occurring in the first place.

However, the system safety concept and profession really started during the American military missile and nuclear programs in the 1950s and 1960s. Liquid-propellant missiles exploded frequently and unexpectedly. During that period, the Atlas and Titan programs saw many missiles blow up in their silos during practice operations. Some of the accident investigations found that these failures were due to design problems, operations deficiencies, and poor management decisions.

Because of the loss of thousands of aircraft and pilots during the same time frame, the U.S. Air Force started to pull together the concepts of system safety and, in April 1962, published BSD Exhibit 62-41, "System Safety Engineering: Military Specification for the Development of Air Force Ballistic Missiles."

Safety was also starting to enter the public mind. Ralph Nader publicized safety concerns during the mid-1960s and started making people aware of how danger-ous cars really were with his book, *Unsafe at Any Speed* (published in 1965). He continued being a powerful voice to the U.S. Congress to bring automobile design under federal control and to regulate consumer protection. In 1959, Volvo was the first to introduce three-point seat belts, and General Motors introduced airbags in the late 1960s.

In the United Kingdom in the early 1960s, Imperial Chemical Industries started developing the concept of the HAZOP study (a chemical industry safety analysis). In 1974, it was presented at an American Institute of Chemical Engineers conference on loss prevention.

The U.S. National Aeronautics and Space Administration (NASA) sponsored government-industry conferences in the late 1960s and early 1970s to address system safety. Part of this was safety technology transfer from the *man-rating* program—to develop ballistic missiles safe enough to carry humans into space—of the Mercury program.

One of the earliest comprehensive probabilistic risk assessments conducted was borne out of the Apollo 1 fire in 1967 that killed all three astronauts on board. For the following 20 years, NASA vacillated conducting probabilistic risk assessments for human space flight, at times concerned that the numbers showed that human space flight was too risky. However, the 1986 Challenger space shuttle accident returned NASA to regular probabilistic risk assessments.

Also in the early 1960s, Pillsbury Company, United States, collaborated with the U.S. Army to produce food for astronauts on NASA missions and created the Hazard Analysis and Critical Control Point (HACCP) methodology. It is a systematic approach to food and pharmaceutical safety that identifies physical, chemical, and biological hazards during the entire supply chain—especially during production—that can cause the product to be unsafe for humans. In 1993, it became a regulation for all Europe community countries and, in 2005, was incorporated into the ISO 22000, *Food safety management system-requirements for any organization in the food chain.*

In 1970, the U.S. Occupational Safety and Health Administration (OSHA) pub-lished industrial safety requirements. Later in the decade, the U.S. military published

Mil-Std-882, "Requirements for System Safety Program for Systems and Associated Subsystems and Equipment." This document is still considered the cornerstone of the system safety profession. It is one of the most cited requirements in procurement contracts. Most of the safety analysis techniques described in Chapters 5 through 9 were created during the heady days of safety from the 1950s to the 1980s. Mil-Std-882C was released in January of 1993, integrating the analysis of software systems with hardware and giving strong emphasis in studying how the software/hardware interface impacts safety. Software systems now control almost all industrial operations and are an integral part of all technology systems. Version E was released in 2012, reinstating safety task descriptions deleted from the prior version and strengthening the integration of functional disciplines into systems engineering.

A key development in system safety is the ALARP principle that states that the residual risk of a system shall be *as low as reasonably practicable* (ALARP principle) and was codified through the UK Health and Safety at Work Act of 1974. The concept asserts that safety-critical systems and operations should be safe as far as reasonably practicable without risks to health and safety. This is important because it forces the overt decision to balance the realized safety benefits to the actual costs to implement; in other words, residual risks are tolerable and thus do not need further mitigations.

OSHA published a process safety standard for hazardous materials in 1992. This is one of the strongest cross-fertilizations of system safety techniques taken from various industries and applied to the chemical industry.

Government safety regulation traditionally has been reactive and very prescriptive. Most safety regulation was borne out of a specific accident or series of accidents in an industry. Though the famous Reactor Safety Study WASH-1400 for the commercial nuclear power industry was written in 1975, it did anticipate the failure scenarios of the near failure of the reactor core at Three Mile Island in 1979. The scenarios were correct, but the probability of human failure was underestimated.

And today, even relatively young countries have made tremendous progress in safety practices. The United Arab Emirates became a country in 1971. Abu Dhabi, the capital, created the Environment, Health, and Safety Center in 2010. In a few short years, a series of safety standards and codes of practices have started being developed across many industries including transportation, health, tourism, water and electricity, waste management, education, food industry, building and construction, and industrial and commercial activity sectors.

It is obvious that the system safety engineering profession, like all professions, has evolved over time. In most cases, out of necessity—an unacceptable number of deaths, accidents, and loss of revenue—engineers have been forced to take a more serious approach to designing safety into both systems and products.

REFERENCES

HM Treasury. 1996, June 28. The setting of safety standards: A report. London, U.K.: HM Treasury.
Lee, J. F. 1999. *Education of Undergraduate Engineers in Risk Concepts: Scoping Study.* Great Britain: Health and Safety Executive.

Main, B. W. and Ward, A. 1992, August. What do design engineers really know about safety? *Mechanical Engineering*, 114:44–51.

McInerney, P. 2005, January. Special commission of inquiry into the waterfall rail accident, final report, vol. 2. Sydney, New South Wales, Australia.

Schleyer, G., Duan, F. R., Stacety, N., and Williamson, J. 2006. Educating engineers in risk concepts. *International Conference on Innovation, Good Practice and Research in Engineering Education*. July 24–26, 2006. Liverpool, U.K.: University of Liverpool, pp. 496–501.

U.S. Department of Labor. 2012, March 22. An independent panel assessment of an internal review of MSHA enforcement actions at the upper big branch mine south. Washington, DC: U.S. Department of Labor.

Viscusi, K. W. and Gayer, T. 2002. Safety at any price. *Regulation*, 25(3):54–63. www.cato.org/pubs/regulation/regv25n3/v25n3-12.pdf, downloaded May 9, 2014.

FURTHER READING

Aldrich, M. 2001, August 14. History of workplace safety in the United States, 1880–1970. In Whaples, R. (ed.), *Economic History Association, EH.Net Encyclopedia*. http://eh.net/encyclopedia/history-of-workplace-safety-in-the-united-states-1880-1970, downloaded May 9, 2014.

Bernstein, P. 1998. *Against the Gods: The Remarkable Story of Risk*. New York: Wiley.

Institute of Aeronautical Sciences. 1948. Aeronautical engineering review. *The Organization of an Aircraft Manufacturer's Air Safety Program, 14th Annual Meeting of the Institute of Aeronautical Sciences*, New York City, January 1946, vol. 7.

Leveson, N. et al. 2004. Effectively addressing NASA's organizational and safety culture: Insights from systems safety and engineering systems. Paper presented at *the Engineering Systems Division Symposium*, MIT, Cambridge, MA, March 29–31, 2004.

Mahaffey, J. 2014. *Atomic Accidents: A History of Nuclear Meltdowns and Disasters: From the Ozark Mountains to Fukushima*. New York: Pegasus.

Nader, R. 1965. *Unsafe at Any Speed: The Designed-in Dangers of the American Automobile*. New York: Grossman Publishers.

2 Definitions and Concepts

Wouldst Thou—so the helmsman answered—
Learn the secret of the sea?
Only those who brave its dangers
Comprehend its mystery.

The Song of Hiawatha, 1855
Henry Wadsworth Longfellow

Divide each difficulty into as many parts as is feasible and necessary to resolve it.

Discourse on Method, 1637
Rene Descartes

Men willingly believe what they wish.

Julius Caesar, c 50 BCE
As quoted in The Adventurer No. 69, 1753
The Works of Samuel Johnson

The future comes like an unwelcome guest.

On Viol and Flute, 1873
Edmund Gosse

Before you can start applying system safety engineering tools in a practical way, you need to develop an understanding of system safety. Comprehending how accidents evolve from the initiating events, to their propagating effects, to the final consequences is paramount in designing safety into systems. Before you can apply methods to prevent them, you must understand how accidents occur. In developing accident avoidance schemes, or designing accidents out of the system, the practicing engineer must balance the cost and the benefit. A perfectly safe airliner will never leave the ground.

The central concept in system safety is the definition of a hazard. It is important to spend some time understanding what appears intuitive to all of us. For success, a design or production engineer has to be able to identify and correct or control these hazards. Once a hazard is defined, the system safety process can start to unfold and make sense. The hazard reduction precedence is the philosophical basis for most safety control systems across industry. And finally, engineering standards are part of the structure that ensures that all technological systems have some level of safety. System safety optimizes the safety process.

2.1 MAKEUP OF AN ACCIDENT

We may all say accidents happen. However, their occurrence may not only take human lives, destroy millions of dollars in property and lost business, they may also cost us our jobs and reputations. The Bhopal, India, accident in 1984 released methyl isocyanate and caused over 2500 fatalities. A petroleum refinery blew up in Houston, Texas, in 1989, killing 23 workers and damaging property totaling U.S. $750 million, spewing debris from the explosion over an area of 9 km. Many thought that after the Three Mile Island nuclear accident in the United States in 1979 and the Chernobyl nuclear power plant disaster in Ukraine in 1986, we would finally get a handle on how to prevent accidents. Unfortunately, the Fukushima nuclear accident in 2011 proved otherwise (see Picture 2.1).

Accidents don't just happen; they are a result of a long process, with many steps. Many times all of these steps have to be completed before an accident can occur. If you can prevent one or more of these accident steps from occurring, then you can either prevent the mishap or at least mitigate its effects. Part of system safety strategy is to intervene at various points along that accident time-line. The safety management system (SMS) is the management infrastructure that makes the system safety program sustainable and gives you the power to prevent accidents.

An accident is an unplanned process of events that leads to undesired injury, loss of life, and damage to the system or the environment. This means that death in war is no accident, but a jeep crashing on the way to battle is.

An *incident* or *near miss* is *an almost accident.* Three Mile Island was a radioactive near miss. No massive quantities of radioactivity were released to the environs, but they almost were. Figure 2.1 shows the events that lead to an accident.

PICTURE 2.1   (See color insert.) Fukushima nuclear accident. (Accessed from http://commons.wikimedia.org/wiki/File:Fukushima_I_by_Digital_Globe_crop.jpg.)

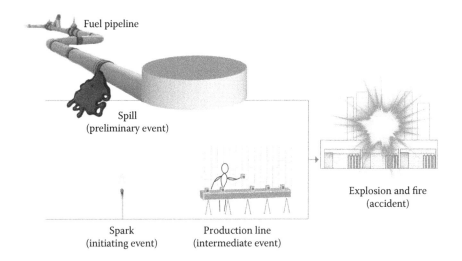

FIGURE 2.1   **(See color insert.)** Events that lead to an accident.

Preliminary events can be anything that influences the initiating event. Examples of preliminary events could be long working hours for chemical plant operators or poor or incomplete pump maintenance. Preliminary events set the stage for a hazardous condition. If we can eliminate the preliminary events or hazardous condition, then the accident cannot advance to the next step—initiating events. It is not unusual that there may be multiple preliminary events, and not just one.

The *initiating event*, sometimes called the *trigger event*, is the actual mechanism or condition that causes the accident to occur. It can be thought of as the spark that lights the fire. For example, a valve sticks open on a process feed line, an electrical short causes a spark at a fueling depot, a pressure regulator fails open in a cryogenic system, or a 220 V power feed is mated with a 110 V system.

Intermediate events can have two effects: They may propagate or ameliorate the accident. Functioning relief valves in a pressure system will ameliorate a system overpressurization. No pressure relief will propagate the hazardous condition and create an accident of system pressure rupture. Defensive driving on highways helps us protect ourselves from the *other* crazy driver or ameliorate the effects of his or her bad driving. Obviously, drunk driving does the opposite, propagating and intensifying an already dangerous situation.

Table 2.1 shows examples of how the elements of an accident fit together. Reading the table from left to right, you can see how an accident evolves. First, there is a hazardous condition—such as large quantities of flammable liquids. Then the initiating event occurs—for example, a valve sticks open. The effect of a valve-failed-open propagates a pressure rise in the system. Now, an in-line relief valve can mitigate the effects of the initial event. If not, an accident ensues—explosion. Chapter 11 discusses and details the events that lead to an accident. It also discusses James Reason's famous *Swiss cheese* accident model.

TABLE 2.1

Elements of an Accident

Hazards	Initiating Events	Propagating Events	Ameliorative Events	Accident Consequences
Significant inventories of	Machinery and equipment malfunctions	Process parameter deviations	Safety system responses	
Flammable materials	Pumps, valves	Pressure	Relief valves	Fires
Combustible materials	Instruments, sensors	Temperature	Backup utilities	Explosions
Unstable materials		Flow rate	Backup components	Impacts
Toxic materials		Concentration	Backup systems	
Very hot/cold materials		Phase/state change		
Inert gases				
Highly reactive	Containment failures	Containment failures	Mitigation system responses	Dispersion of toxic materials
Reagents	Pipes	Pipes	Vents	Highly reactive materials
Products	Vessels	Vessels	Dikes	
Intermediate products	Storage tanks	Storage tanks	Flares	
By-products	Gaskets	Baskets, bellows, etc.	Sprinklers	
		Input/output or venting	Relief valves	
Reaction rates especially sensitive to	Human errors	Material releases	Control responses, operator responses	
Impurities	Operations	Combustibles	Planned	

Process parameters	Explosive materials	Ad hoc
Maintenance	Toxic materials	
Testing	Reactive materials	
Loss of utilities	Operator errors	Contingency operations
Electricity	Omission	Alarms
Water	Commission	Emergency procedures
Air	Diagnosis	Personnel safety equipment
Steam	Decision making	Evacuations
		Security
External events	External events	External events
Floods	Delayed warning	Early detection
Earthquakes	Unwarned	Early warning
High winds		
High-velocity impacts		
Vandalism		
Terrorism		
Method/information errors	Method/information failure	Information flow
As designed	Amount	Routing
As communicated	Usefulness	Methods
	Timeliness	Timing

Source: Reproduced from Center for Chemical Process Safety, *Guidelines for Hazard Evaluation Procedures*, American Institute for Chemical Engineers, New York, 1985, pp. 1–3, Copyright 1985 by the American Institute for Chemical Engineers.

2.2 HOW SAFE IS SAFE ENOUGH?

The insurance industry functions by answering the question "How safe is safe enough?" Actuarial tables are based on the cost of an accident. One question you must answer is "How much am I willing to spend to protect myself from accidents (including lawsuits and lost business revenue)?" After the 2010 BP Deepwater Horizon offshore oil platform explosion and oil spill in the Gulf of Mexico, BP has budgeted around $40 billion for paying out of claims and other compensation. The U.S. National Safety Council publishes *Injury Facts* (formerly *Accident Facts*) annually with estimates of accident costs by industry. Their numbers include estimates for wage lost, medical expenses, insurance administration costs, and uninsured costs. The total cost (home, workplace, motor vehicle, etc.) to the U.S. economy of injuries and accidents in 2011 was $753 billion (U.S. National Safety Council, 2013). The average total cost of an accidental death across industries with total death compensation is about $1,420,000 per person (U.S. National Safety Council, 2012). This average cost means that companies have to have insurance policies that cover costs between $1 million and $5 million per fatal injury.

Of course, not all accidents or near misses result in personal injury or death. Part of the determination of the cost of safety is how much downtime the plant is willing to endure before replacing the broken machinery or cleaning up the mess. For example, a 1-week shutdown of an aluminum smelter can translate into up to 9 months of lost production, due to the operational nature of the plant. The real problem the engineer faces is how to make technology safe without it costing too much. We can make a car nearly totally safe, but we would never be able to use it. Part of the system safety engineering process and SMS is to help you identify the hazards, costs, and the associated risks. It is almost always much cheaper to *design out* the hazard while the product is still on the drafting table (or, nowadays, on the computer screen) than out in the field.

The UK Health and Safety at Work Act 1974 defined the concept of *as low as reasonable practicable* (ALARP). The ALARP principle is based on *reasonable practicability, which simply means that hazard controls are implemented to reduce residual risk to a reasonable level of practicality*. For a risk to be considered ALARP, it must be demonstrated that the cost in reducing the residual risk further would be grossly disproportionate to the benefit gained. Therefore, a risk assessment is conducted, and a cost–benefit analysis performed to determine how far to carry the hazard control. Of course, the challenge is deciding what is practical (e.g., cost, effort, time) balanced with how much benefit of lower residual risk the hazard control brings. Unfortunately, there is no standard method to demonstrate that the hazard control trade-off will meet ALARP. However, some of the following have been successfully used:

- Predefining hazard acceptance level criteria before analysis starts so that design and operations meet them
- Cost–benefit analysis that compares the costs of the hazard control to the perceived lower risk–benefit

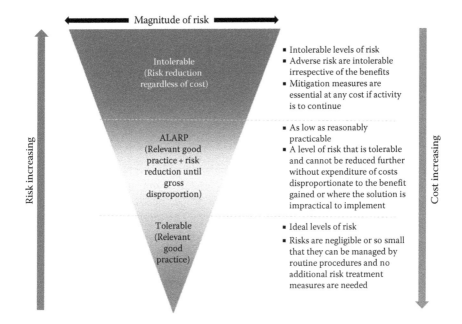

FIGURE 2.2 **(See color insert.)** ALARP principle.

- Comparing the design and auditing the implementation to accepted codes and standards
- Test and analysis by similarity
- Quantitative risk assessment

Many industries will justify the ALARP level with a quantitative risk assessment of how it will impact societal risk. In other words, how many additional lives are saved with this hazard control? This application is very common in the United Kingdom especially in rail safety but highly controversial in the United States. But this is changing to some extent in the United States. Trade-off benefits in pollution control in some industries are now using the ALARP method (Figure 2.2).

2.3   CASE STUDY: *BLACK SWAN* EXTREME EVENTS, FUKUSHIMA NUCLEAR DISASTER

But how far do you go in preventing an accident? When do you stop? You can't protect against everything, can you? What is a reasonable design limit? Do you need to worry about worst-case scenarios or just an average accident scenario? What about *black swan* events? These are critical questions that you have to answer before you start anything.

A *black swan event simply put is a surprise and totally unexpected very rare event that is highly improbable but with catastrophic consequences (very, very low probability of occurrence with very, very high impact).* These are very rare events that are impossible to predict. It is taken from the concept that almost all swans are white and that it is almost impossible to find a black one. Nassim Nicholas Taleb popularized the term in his 2001 book *Fooled by Randomness*. The Fukushima Daiichi nuclear accident on March 11, 2011, was a black swan event. It is worthwhile to look at what happened to better understand how to deal with black swan events.

2.3.1 THE ACCIDENT

On March 11, 2011, a 9.0 magnitude earthquake occurred, centered about 80 miles offshore from the Japanese coastal city of Sendal. It not only was much larger than other *typical* earthquakes in earthquake-prone Japan, but it was a rare and complex double quake that lasted around 3 min. The quake was so strong that it actually moved the main island of Japan 8 ft eastward and is estimated to have shifted the Earth on its axis between 4 and 10 in. The quake resulted in a 50 ft high tsunami that hit the power supply and cooling system of three Fukushima Daiichi reactors. All three nuclear reactor cores melted in the first 3 days of the event. At least three reactors exploded due to hydrogen gas overpressurization of the outer containment buildings after the cooling systems failed. The International Nuclear and Radiological Event Scale ranked the event a seven, major accident, the highest possible. The only other nuclear event with that ranking was Chernobyl in 1986. The quake and tsunami killed over 19,000 people, destroyed or partially collapsed over a million buildings, and forced over 100,000 residents to evacuate, with about 1,000 people killed during the evacuation itself. Unbelievably, to date, there have been no deaths due to radiation exposure (of course, time will tell if or how that changes). There were 11 nuclear reactors operating at 4 power plants in the region at the time of the event. All shutdown automatically, as designed, when the quake hit.

2.3.2 WHAT WENT WRONG?

In spite of the fact that the earthquake was the strongest recorded ever in Japan, and the fifth most powerful in the world since 1900, and another series of aftershocks measuring 7.0 hit almost a month later, the reactors were seismically robust. The earthquake created severe damage to the plant, as well as the rest of the region, but the real problem was the tsunami.

The plant's (World Nuclear Association, 2014) original design basis tsunami height was 10 ft, then in 2002 revised to 18.7 ft. The tsunami height that hit the plant was 50 ft high. When the tsunami hit, the power turbine halls were inundated with 16 ft of seawater. A tsunami created by the 1896 earthquake of magnitude 7.6 reached 124 ft in height. Should that have been considered in the design basis?

The earthquake forced the reactors into automatic shutdown, but it also destroyed all six external power supply sources. As designed, emergency diesel generators in the turbine halls kicked on to supply critical electrical power to the reactor cooling system, and under *normal circumstances*, the cooling system would have maintained temperatures at safe levels. But the first tsunami wave hit 41 min after the earthquake and a second 8 min later. This damaged and flooded the seawater pumps (located in the turbine halls) used to protect the main condenser and auxiliary cooling circuits. It also completely drowned the diesel generators used to provide emergency power to the cooling system. Battery supply and electrical switchgear, also located in the turbine halls, were also inundated, cutting off all electrical power to the control room leaving the plant dark and without any plant instrumentation (World Nuclear Association, 2014).

The reactor cores were producing their normal thermal power from fission decay. But without cooling and heat removal, excessive amounts of steam were produced that quickly mixed with a buildup of hydrogen gas, produced from steam reacting with the extremely high surface temperature of the zirconium cladding of the reactor fuel. Reactor designs have an emergency core cooling system to protect against scenarios like this, but unfortunately, they progressively failed from the tsunami effects. The exposed fuel temperature surpassed 5000°F (World Nuclear Association, 2014). The exothermic reaction quickly led to the hydrogen explosion, in spite of attempted venting of the containment building. Army units and volunteer fire fighters were later brought in to set up an emergency pumping scheme, taking seawater and pumping it on to the fuel to cool it. Nitrogen gas was injected into the containment building and pressure vessels to *inert* the explosive environment to prevent further explosions.

One of the long-term effects and continuous challenge will be managing the contaminated water created from the accident. Runoff from the emergency seawater pumping has flowed back to the sea, and there are measurable levels of radioactive contamination above acceptable levels. Also, contaminated water has seeped into the soil around the reactors. Approximately 1000 storage tanks (World Nuclear Association, 2014) have been set up to try to manage some of the contaminated water. But even some of these tanks have started leaking. Numerous schemes have been created to treat the contaminated water before it reaches the sea. Additional actions have been taken to pump groundwater out from under the reactors to avoid seepage into the reactor area and thus become contaminated. Monitoring and managing contaminated and noncontaminated groundwater will be necessary for many years to come. The Japanese government estimates that it will cost around $107 billion and 40 years to clean up the site (World Nuclear Association, 2014).

The Fukushima Nuclear Accident Independent Investigation Commission (NAIIC, 2012a) Chairman Kiyoshi Kurokawa stated,

> The earthquake and tsunami…were natural disasters of a magnitude that shocked the entire world. Although triggered by these cataclysmic events, the subsequent accident at the Fukushima Daiichi Nuclear Power Plant cannot be regarded as a natural disaster. It was a profoundly manmade disaster—that could and should have been foreseen and prevented. And its effects could have been mitigated by a more effective human response…how could such an accident occur in Japan, a nation that takes such great pride in its global reputation for excellence in engineering and technology?

2.3.3 Media Nightmare

On top of a devastating black swan event is the public information sharing with the media, which was another disaster. Many in Japan, and around the world, were very frustrated with Tokyo Electric Power Company (TEPCO) and government officials providing confusing, incomplete, and many times differing information. Many felt that TEPCO and the various government departments and regulators lacked the sense of responsibility both to adequately inform the public in a timely manner to save lives (especially during and immediately after the initial event) and to continue to remain opaque to the subsequent damage and potential short- and long-term health effects. In our age of instant media and microblogging, information opaqueness does not go away with time but can have a long half-life. Transparency and immediate information sharing is critical not only to save lives but also to manage the communities' perception of such an event.

2.3.4 Lessons Learned of What Could Have Been Done Differently

Of course, hindsight is 20/20. It is easy to look backward and identify what should have been done differently. We can't prevent such a large black swan event. But we can consider how to deal with the very rare and unexpected: prevent where possible, design appropriately, mitigate consequences that you can't control, and ensure a speedy and orderly long-term recovery. Though this was a black swan event, there are many themes that we can see from the accident (before and after) that you will see throughout this book. Unfortunately, black swan events and much smaller accidents seem to have a lot in common. The commission's report is devastating, but it is worth looking at some of their conclusions (NAIIC, 2012b) in their own words:

- After a 6-month investigation, the commission has concluded the following: In order to prevent future disasters, fundamental reforms must take place. These reforms must cover both the structure of the electric power industry and the structure of the related government and regulatory agencies as well as the operation processes. They must cover both normal and emergency situations.
- A *man-made* disaster: The TEPCO Fukushima Nuclear Power Plant accident was the result of collusion between the government, the regulators and TEPCO, and the lack of governance by said parties. …Therefore, we conclude that the accident was clearly *man-made*. We believe that the root causes were the organizational and regulatory systems that supported faulty rationales for decisions and actions, rather than issues relating to the competency of any specific individual.
- There were organizational problems within TEPCO. Had there been a higher level of knowledge, training, and equipment inspection related to severe accidents and had there been specific instructions given to the on-site workers concerning the state of emergency within the necessary time frame, a more effective accident response would have been possible.

- The situation continued to deteriorate because the crisis management system of the Kantei, the regulators, and other responsible agencies did not function correctly. The boundaries defining the roles and responsibilities of the parties involved were problematic due to their ambiguity.
- The safety of nuclear energy in Japan and the public cannot be assured unless the regulators go through an essential transformation process. The entire organization needs to be transformed, not as a formality but in a substantial way. Japan's regulators need to shed the insular attitude of ignoring international safety standards and transform themselves into a globally trusted entity.
- TEPCO did not fulfill its responsibilities as a private corporation instead obeying and relying upon the government bureaucracy.
- Replacing people or changing the names of institutions will not solve the problems. Unless these root causes are resolved, preventive measures against future similar accidents will never be complete.

The Institute of Nuclear Power Operations wrote an interesting addendum to their INPO 11-005, "Special Report on the Nuclear Accident at the Fukushima Daiichi Nuclear Power Station." They go on to discuss some of the lessons learned from the black swan event. Many of the lessons learned are heavily related to nuclear-specific design and operational issues; however, here are some that have more universal application and again are themes that we shall see throughout this book (INPO, 2012):

- *Prepare for the unexpected*: When periodic reviews or new information indicates the potential for conditions that could significantly reduce safety margins or exceed current design assumptions, a timely, formal, and comprehensive assessment of the potential for substantial consequences should be conducted. An independent, cross-functional safety review with a plant walk-down should also be conducted to fully understand the nuclear safety implications. If the consequences could include common-mode failures of important safety systems, compensatory actions or countermeasures must be established without delay.
- Plant design features and operating procedures alone cannot completely mitigate the risk posed by a beyond-design-basis event. Additional preparations must be made to respond if such an event were to occur.
- Corporate enterprise risk management processes should consider the risks associated with low-probability, high-consequence events that could lead to core damage and spread radioactive contamination outside the plant.
- Establish strategies for staffing operating crews, other key plant positions, and site and corporate emergency response organizations quickly in the initial stages of a multiunit event and over the long duration of the event response.
- Ensure primary and alternative methods for monitoring critical plant parameters and emergency response functions are available. Use drills and exercises to ensure emergency response personnel are able to use the available monitoring tools and methods.

- Clearly define and communicate the roles and responsibilities of emergency response personnel to help ensure effective postaccident communications and decision making.
- Communication methods and equipment should support accurate and timely information exchange, consistent and clear communications with the public, and information sharing between the utility and the government.
- Equipment required to respond to a long-term loss of all ac and dc power... should be conveniently staged, protected, and maintained such that it is always ready for use if needed.
- Plant modifications may be needed to ensure critical safety functions can be maintained during a multiunit event that involves extended loss of ac power and dc power.
- Conditions during and following a natural disaster or an internal plant event may significantly impede and delay the ability of plant operators and others to respond and take needed actions. The potential for such delays should be considered when procedures and plans for time-sensitive operator actions are being established.
- Behaviors prior to and during the Fukushima Daiichi event revealed the need to strengthen several aspects of nuclear safety culture. It would be beneficial for all nuclear operating organizations to examine their own practices and behaviors in light of this event and use case studies or other approaches to heighten awareness of safety culture principles and attributes.

2.4   WHAT IS A HAZARD AND OTHER IMPORTANT CONCEPTS

Over the years, there has been considerable confusion with the concepts of safety, risk, and hazard. The major problem is that many people tend to interchange the words as if they mean the same. On top of that, different industries often define the concepts differently.

The most important thing to remember is that *system safety engineering is a combination of management and systems engineering practices applied to the evaluation and reduction of risk in a system and its operation*. The objective of system safety is to identify hazards resulting from the use or operation of a system and to eliminate or reduce the hazards to an acceptable level of risk.

The *system is the combination or interrelation of hardware, software, people, and the operating environment*. In system safety engineering, you must look at the system from cradle to grave. In other words, the *system life cycle is the design, development, test, production, operation, maintenance, expansion, and retirement (or disposal) of the system*. A nuclear power plant is one large system with operators, pressure subsystems, electrical and mechanical subsystems, structural containment, safety systems, etc. A far simpler example is a boy riding his bike. The bike, the boy, the street (with all its traffic conditions), the weather, the time of day, and even other children make up the system of *boy on his bike*.

A succinct definition is that *a hazard is a condition that can cause injury or death, damage to or loss of equipment or property, or environmental harm* (Roland and Moriarty, 1990). Some typical hazards in various systems are electrical discharge or shock, fire or explosion, rapid pressure release, and extreme high or low temperature. Chapter 5 discusses the different types of hazards. Appendix B is a generic hazard checklist.

Of course, a hazard can be the result of a system or component failure, but it isn't always. Failure and hazard are frequently linked and oftentimes confused. There is an occupational hazard (no pun intended) to associating the two. A hazard can exist without anything failing. In other words, an engineer can actually (unintentionally of course) *design* in a hazard. Guns are very hazardous to life, especially when operated properly. To be successful in system safety engineering, we must look not only at failures and their associated hazards but also at the normal and emergency system operations and their hazards.

Hazard addresses only the severity or end result. Risk combines the concept of severity of the accident consequence and the likelihood of it occurring. In the simplest terms, *risk is the combination of the probability (or frequency of occurrence) and consequence (or severity) of a hazard.* There are always risks. There is a risk staying in bed and a different risk getting out of bed. As much as we would love to have zero risk, that is a practical impossibility. Because we cannot totally eliminate risk, we try to shrink it as much as possible. Lowering either the probability or the severity of the hazard or both can do this. So

Risk (consequence/time) = Frequency (events/time) × magnitude (consequence/event)

Another term that is becoming quite popular, especially after the September 11, 2001, terrorist attack in the United States is an all hazards approach. The *all hazards approach assesses the safety, security, and emergency management implications of an incident either intentional (security), accidental (safety), or natural disaster that can cause harm to people, property, equipment, or the environment.* Many embrace the term because they feel that the end result—a disaster—is the same, no matter the precursors.

However, many security experts feel that though responding to an all hazards event may be the same, the countermeasures or controls could be (and usually are) very different. The safety profession expounds on a very transparent approach to hazard management. Because of the nature of threat actors, those who have evil intent and capability to attack others, much security data must be kept confidential or classified, and information sharing is more controlled. This can make merging safety and security planning difficult. Most organizations have separate safety and security departments, though both leaders may report to the same superior.

Emergency response is important for any event. The Fukushima Daiichi disaster is an example of how a natural disaster created a massive safety accident. Chapter 4 details how emergency management is part of the SMS.

2.5   SYSTEM SAFETY VERSUS SAFETY MANAGEMENT SYSTEM

The term SMS is a relatively new term. As defined earlier, the concept of system safety is to merge the engineering with the management of the system development and operation into the concept of system safety. Over the last 20 years or so, the term SMS has been coined. Chapter 4 is dedicated to discussing SMSs. Briefly an *SMS is a sustainable, formal and structured, enterprise-wide safety program that appropriately manages safety risk comprehensively of products and the systems that produce them.*

Many feel that SMS is a new concept, but in reality, it is the same thing as system safety program management. However, the SMS term seems to be applied in a way that emphasizes the enterprise management of safety in all of its aspects and therefore is a much better term to describe the entire concept. You can think of it as follows:

* System safety is the process of identifying, evaluating, and controlling hazards and risks.
* SMS is the enterprise-level management structure that oversees that system safety process.
* System safety is a subset of SMSs.
* SMS is the superstructure or the infrastructure to appropriately manage your safety.

2.6   SYSTEM SAFETY PROCESS

The system safety process is really an easy concept to grasp. The overall purpose is to identify hazards, eliminate or control them, and mitigate the residual risks. The process should combine management oversight and engineering analyses to provide a comprehensive, systematic approach to managing the system risks. Figure 2.3 details this process.

Define objectives: As with any problem, the first step is to define the boundary conditions or analysis objectives. That is the scope or level of protection desired. You need to understand what level of safety is desired at what cost. You need to answer the question "How safe is safe enough?" Other questions to ask are as follows:

* What constitutes a catastrophic accident?
* What constitutes a critical accident?
* Is the cost of preventing the accident acceptable?

Most industries approach this step in the same way. However, how they differentiate among catastrophic, critical, minor, and negligible hazards may vary. You will need to modify the definitions to fit the particular problem. What is important is that these definitions are determined before work begins. A rule-of-thumb definition for each is the following:

Catastrophic—*any event that may cause death or serious personnel injury or loss of system* (e.g., anhydrous ammonia tanker truck overturns, resulting in a major spill)

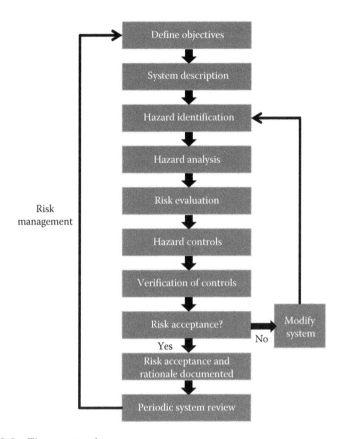

FIGURE 2.3   The system safety process.

Critical—any event that may cause severe injury or loss of mission-critical hardware or high-dollar-value equipment (e.g., regulator fails open and overpressurizes a remote hydraulic line, damaging equipment and bringing the system down for some days)

Minor—any event that may cause minor injury or minor system damage, but does not significantly impact the mission (e.g., pressure control valve fails open, causing pressure drops and increased caustic levels)

Negligible—any event that does not result in injury or system damage and does not affect the mission (e.g., lose commercial power, causing shutdown of plant cafeteria)

System description: The next step is system description. Some thought should be given to grasping how the system works and how the hardware, software, people, and environment all interact. If the system is not described accurately, then the safety analysis and control program will be flawed.

2.6.1 HAZARD IDENTIFICATION

Hazard identification is a crucial part of the system safety process. It really is impossible to safeguard a system or control risks adequately without first identifying the hazards. An all-too-frequent mistake in safety engineering is to skip over this step, or not to give it adequate attention. The hazard identification process is a kind of *safety brainstorming*. The purpose is to identify as many hazards as are possible and credible. Through this process, the engineer develops a *preliminary hazard list* (PHL) and later will assess the impact on the system.

To develop a PHL, you will want to use various methods to gather the most exhaustive list possible. This may include the following:

- Survey the site.
- Interview site personnel.
- Convene a technical expert panel.
- Analyze and compare similar systems.
- Review past accidents.
- Identify codes, standards, and regulations.
- Review relevant technical data (electrical and mechanical drawings, analyses, operator manuals and procedures, engineering reports, etc.).
- Analyze energy sources (voltage/current sources, high-/low-temperature sources, etc.).

PRACTICAL TIPS AND BEST PRACTICE

One of the most common mistakes at this juncture is not to verify that the system description matches what really exists. Most engineering projects go through various design changes *after* the design has been completed and reviewed. Frequently these changes are not well documented, so it becomes dangerous to base data gathering only on what is published in the plant library. It is very important for the engineer to make frequent site visits to verify what the system really looks like. This becomes especially important for old systems that have gone through numerous modifications over the years.

Hazard analysis: The next step is to analyze the hazards identified. A hazard analysis is a technique for studying the cause/consequence relation of the hazard potential in a system. The purpose is to take the PHL one level deeper and assess how each hazard affects the system. Is it catastrophic? Or is it critical? The hazard analysis will also assist you in further assessing which hazards are important and which are not and therefore do not need further study. Chapters 5 through 9 describe various hazard analysis techniques that are commonly used in different industries.

Risk evaluation: After hazards have been identified and analyzed, you need to control their occurrence or mitigate their effects. Evaluating the risks does this. Is the hazard likely to occur? If it does, how much damage will result from the incident? You need to understand the relationship between hazard cause and effect. With this information, the associated risks are then ranked, and engineering management is better able to determine which risks are worth controlling and which risks require less attention.

2.6.2 Hazard Control

After evaluating the risks and ranking their importance, you must control their effects. Controls fall into two broad categories: engineering controls and management controls. Engineering controls are changes in the hardware that either eliminate the hazards or mitigate their risks. Some example engineering controls include adding a relief valve to a 2000 psi oxygen system, building a berm around an oil storage tank, using only hermetically sealed switches in an explosive environment, or putting in hard stops in rotating machinery to prevent overtorquing.

Management controls are changes made to the organization itself. Developing and implementing a plant safety plan is a good method of applying management controls to hazards. Some examples are using production-line employees as safety representatives for their areas, requiring middle-management reviews and approvals of any plant or system modifications to consider safety implications, or assigning signature authority to safety engineers for all engineering change orders and drawings. Processes and procedures also are included in the management control area. Chapter 4 delves deeply into safety management and SMS. It should be remembered that management controls and engineering controls are just a small part of the overall SMS.

Verification of controls: Once controls are in place, a method needs to be used to verify that the controls actually control the hazards or mitigate the risks to an acceptable level. Verification of hazard controls is usually accomplished through the company or engineering management structure. The most frequent means is inspection. However, as we all know, inspection is also one of the most expensive ways to assure that controls are in place. An effective method of hazard control verification is the use of a closed-loop tracking and resolution process. Again, Chapter 4 addresses this issue in depth. It is important that verification of controls really entails: validating that the controls are adequate to control the hazard and then verifying that those controls are truly in place and operate effectively. The Fukushima Daiichi power supplies (in the turbine halls) did not have adequate controls because they weren't validated to be sure that they operate appropriately in all conditions.

2.6.3 Risk Acceptance

Safety is only as important as management wants to make it. At this point in the safety process, this becomes obvious. After the system has been studied and hazards have been identified, analyzed, and evaluated with controls in place, management must make the formal decisions of which risks they are willing to accept and which ones they will not. At this point, a good cost–benefit analysis will help management make that decision. Sometimes this is not easy. The Fukushima Daiichi accident is a great example of how difficult this can be. There were an earthquake and a tsunami in Japan's history that were very large. Should management design for that hazard or is that too much of an outlier?

Part of the risk acceptance process is a methodical decision-making approach. If the risks are not acceptable, then the system must be modified and the hazard identification process must be followed once again. If the risks are acceptable, then good documentation with written rationale is imperative to protect against liability claims. Chapters 13 and 14 address these issues in depth. Some people will use the term *risk appetite* to describe this process. The very first step—define objectives— should clearly define what is an acceptable risk (risk tolerance or risk appetite) that the company is willing to accept.

Modify the system: If the risk is unacceptable, then the system must be modified to reduce the risk, and the process of hazard identification starts again to make sure that the modifications don't obviate any safety controls and that they truly reduce the risk to an acceptable level.

Risk acceptance and rationale documented: Many times companies will not adequately document how and why a particular risk was accepted. The risk acceptance becomes too informal. This can be devastating if you are responding to a lawsuit or formal inquiry or accident investigation. Also, when you want to update a system and make changes, it becomes very important to understand how and why certain risks were accepted.

Periodic system review: Probably one of the key points of the system safety process is that it is a closed-loop system. This means that the engineering and management organizations periodically review the safety program, engineering processes, management organizations, and product field use. Actually, this should be part of the SMS. The American automobile industry has lost billions of dollars in automobile recalls due to safety problems, some of which possibly could have been avoided by periodic review of product use.

NOTES FROM NICK'S FILE

I was involved in looking at how an oil pipeline had internal corrosion and unexpectedly failed, leaking crude oil onto an environmentally sensitive area. The economics of oil drilling had changed, and the producer started injecting more and more seawater into old wells to get as much oil out of the well as possible. This is a common practice for mature wells. The problem was that no one had sufficiently studied the implications and understood that maintenance and inspection frequencies had to be increased to account for the much more corrosive operating environment from so much seawater added to the process.

2.6.4 RISK MANAGEMENT VERSUS SAFETY MANAGEMENT

Many people are confused by the terms risk management and safety management. Some use them interchangeably. They are very similar concepts, and you could say that safety management is a subset of risk management. Of course, safety management looks at how to manage operations safely. Risk can be interpreted to be broader and encompass how risk to the business is managed, with safety as a subset. Under this definition, risk management would also include managing nonsafety risks such as reputation, legal, business, financial, and market share. A forward-leaning company would have a robust SMS that is embedded in their enterprise risk management system.

2.7 HAZARD REDUCTION PRECEDENCE

NASA has developed an extremely useful hierarchy of hazard reduction (National Aeronautics and Space Administration, 1993). This method is applicable to all industries. The first step is to *design out* the hazard. If this is not possible, then the next best is to use safety devices. If safety devices are not sufficient control, then cautions and warnings can be applied. And finally, special procedures and training can be used to control the hazard or mitigate the consequences.

2.7.1 DESIGN OUT THE HAZARD

The first step, and always the best (from a safety-only viewpoint), is to *design out* the hazard—take the hazard out of the system completely. For example, four things are required for a fire: combustible material, oxygen, chemical reaction, and an ignition source. If one leg is taken out of the fire tetrahedron, then it is physically impossible to have a fire. Engineering talent really comes into play in trying to decide which of the four legs should be designed out. The system can be hermetically sealed and back-filled with dry nitrogen gas (taking away the oxygen). Or the system can be fabricated out of material that has a very high flash point (ensuring that the electrical cable can never reach the burning temperature of its insulation, under all electrical load conditions). Another solution is to disable the power so that electricity does not create an ignition source.

Figure 2.4 illustrates how easy designing out the hazard can sometimes be. The first part is a *typical* engineering design. Gaseous nitrogen is needed to support the operator's activities. However, as can be seen in the second block, the nitrogen feed bottles do not need to be located in the same area as the operator. The bottles have been moved outside the work area, thus alleviating the asphyxiation hazard.

Even though it may seem so, designing out the hazard is not always difficult. One common example is the hazard of mismating power cables. Many types of electrical instrumentation use various cable sizes both for power and for signal input and output. One hazard is mating a 208 V source to a 110 V system. If there is concern that cables could be mismated through human error, then the engineer needs to design out that hazard. Using cable connectors that are keyed differently and *scoop-proof* would ensure that it is physically impossible to mismate the cables.

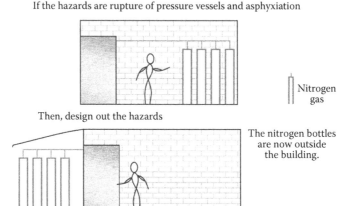

FIGURE 2.4   Designing out the hazard.

In the late 1980s, NASA experienced such a problem (National Aeronautics and Space Administration, 1989). After arrival at the Kennedy Space Center for Space Shuttle launch preparations, the *Magellan* spacecraft power control unit was reconnected. An extensive electrical power system check was necessary to ensure proper performance prior to applying power to other spacecraft subsystems. A technician was making a blind mate (mating cables without visual verification during the operation) of an electrical harness power cable to the *Magellan*. Because the connection was buried deep in the spacecraft, out of view of the technician, it was impossible to verify visually that the connection was correct. Sparks, flames, and smoke resulted from the cable connection operation.

2.7.2   SAFETY DEVICES

If the hazard cannot be designed out, because it is either impossible or cost-prohibitive, then the next best solution is to use fail–safe devices. An example is a valve that fails in a safe manner. In many chemical processes, it is critical that the temperature not exceed a certain level. If the control valve fails in an open position, then the system will still maintain cooling even if system power or control is lost for one reason or another.

The pressure-relief valve is probably one of the most common applications of a safety device. If the pressure within the line exceeds a predetermined level, then the pressure-relief valve opens, relieving the dangerous pressure buildup. Another example is a fuse in an electrical cable. If the cable is electrically overloaded, the fuse will blow and cut electrical flow before the cable overheats and burns or the system is overstressed.

Plastic injection molding machines are very common in many manufacturing plants around the world. Occasionally, the technician needs to reach inside the machine to clear out debris. Many hands and arms have been lost from trying to clean the debris while the machine was still powered on (though not running). An electrical interlock that automatically disconnects the power source when the inside envelope of the machine has been violated is another example of a safety device.

2.7.3 WARNING DEVICES

If the hazard cannot be designed out or sufficiently controlled, then the next best step is to warn operators of the imminent danger. A good example of a warning device is the smoke detector in a home. When a certain level of smoke is detected, an alarm is sounded, and we are alerted to immediately evacuate the area. Of course, there are pitfalls. Too many different warning devices or too many false warnings can be very confusing to people, and the warning itself could cause a hazard (too many burned hamburgers can give the false impression that the next smoke alarm is from dinner and not a serious fire). Other typical safety warning devices are stop signs, no smoking signs, oxygen alarms, fire alarms, and product labels (e.g., "WARNING: DO NOT USE NEAR OPEN FLAME").

Gas monitoring and warning detectors are very common examples of warning devices. Some typical gas detectors are ammonia, hydrocarbon fuels, hydrogen, methane, and even oxygen monitors. Care should be exercised in the use of warning devices. It is not uncommon that industrial facilities use too many warnings for different hazards. One manufacturing plant used a buzzer for ammonia leaks, a siren for nitrogen leaks, a horn for oxygen leaks, and a bell for fires. The problem is that the individual wastes precious time in deciding how to respond. A more desirable solution would be to connect all the different gas-monitoring sensors to one type of audio and visual alarm, such as a horn and flashing red light. The person then knows to immediately evacuate the area; it doesn't matter if it is because of a fire or an ammonia leak.

2.7.4 SPECIAL PROCEDURES AND TRAINING

If warning devices are not sufficient to control the hazard or mitigate its consequences, then special procedures and training are required. Because people tend to make mistakes much more often under stressful conditions, this obviously is the least desirable method of risk mitigation. Special procedures can be anything from emergency shutdown procedures when there is a fire in the facility to emergency response actions resulting from a hazardous material spill, such as an overturned tanker truck leaking anhydrous ammonia. Of course, like any new system, special procedures should be tested and periodically verified to assure their efficacy.

2.8 SAFETY MATURITY MODEL AND SAFETY MANAGEMENT SYSTEMS

It is useful to think of safety and SMSs that companies use as on a maturity curve, slope, or continuum. At the far left, or bottom, is a safety approach that is certainly pathological and highly illegal. Unfortunately, we have seen some companies do operate at this level, luckily not too many. At the top right, the most mature model is one that is constantly improving, evolving, and very adaptive to its changing environment. Figure 2.5 illustrates the safety maturity model. See Chapter 4 for an almost identical maturity model but that focuses on safety culture maturity.

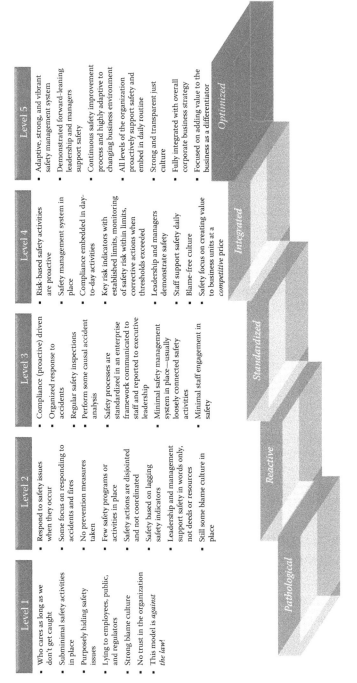

Level 1

- Who cares as long as we don't get caught
- Subminimal safety activities in place
- Purposely hiding safety issues
- Lying to employees, public, and regulators
- Strong blame culture
- No trust in the organization
- This model is *against the law!*

Level 2

- Respond to safety issues when they occur
- Some focus on responding to accidents and fires
- No prevention measures taken
- Few safety programs or activities in place
- Safety actions are disjointed and not coordinated
- Safety based on lagging safety indicators
- Leadership and management support safety in words only, not deeds or resources
- Still some blame culture in place

Level 3

- Compliance (proactive) driven
- Organized response to accidents
- Regular safety inspections
- Perform some causal accident analysis
- Safety processes are standardized in an enterprise framework communicated to staff and reported to executive leadership
- Minimal safety management system in place—usually loosely connected safety activities
- Minimal staff engagement in safety

Level 4

- Risk-based safety activities are proactive
- Safety management system in place
- Compliance embedded in day-to-day activities
- Key risk indicators with established limits, monitoring of safety risk within limits, corrective actions when thresholds exceeded
- Leadership and managers demonstrate safety
- Staff support safety daily
- Blame-free culture
- Safety focus on creating value to business units at a *competitive* price

Level 5

- Adaptive, strong, and vibrant safety management system
- Demonstrated forward-leaning leadership and managers support safety
- Continuous safety improvement process and highly adaptive to changing business environment
- All levels of the organization proactively support safety and embed in daily routine
- Strong and transparent just culture
- Fully integrated with overall corporate business strategy
- Focused on adding value to the business as a differentiator

Pathological *Reactive* *Standardized* *Integrated* *Optimized*

FIGURE 2.5   Safety maturity model.

Most companies seem to fall around Level 3. Safety is more than perfunctory. Management wants to do the right thing but may not know exactly how to do it. Regulators understand that they still need to push industry into a more forward-leaning position. Few companies operate at Levels 4 and 5, though that is where they should be targeting. Doesn't it make sense that safety is part of daily business decisions? This book will give many illustrations of how good safety systems will save companies' money and make them more competitive while improving their safety performance. The highest level also emphasizes an adaptive safety culture. That means that safety isn't static. You don't just design a safety process, manage it and you're done. The safety process or SMS must be adaptive to the changing business needs. This is important because business isn't static, and it too must be adaptive to the changing marketplace and operating and regulatory environments. Safety should too.

PRACTICAL TIPS AND BEST PRACTICE

- Use the safety maturity model as a *quick look* at your corporate safety posture. You can expand each of the bullet points and develop review protocols to gather more data. Then you can go to your management when you are looking for more money for your safety program.
- The safety maturity model can also serve as a template for more research on SMS's best practices and what your management goals should be.
- Combine the safety maturity model with the information in Chapter 4 to conduct a comprehensive diagnostic of your current safety health.

2.9   LEADING AND LAGGING SAFETY PERFORMANCE INDICATORS

Most companies, especially those at Level 3 or below in the safety maturity model, tend to focus their safety programs strictly on lagging safety performance indicators. They are based on historical data. They do, and should, track accidents, incidents, safety problems, and corrective actions and make sure that they are implemented appropriately or that the corrective actions or controls are still valid. This is all very important and must be done and not dropped. However, it is not enough.

Smart companies and organizations, those higher up on the safety maturity model, also look at where the next safety problems will be before they arise, or at least, before they get out of control. These companies use leading safety performance indicators to predict future safety problems. *Leading safety performance indicators are measures that help organizations identify future safety problems and proactively measure their future safety performance.* In other words, I want to know where my future problems are going to be and how bad they will impact my operations. Leading safety performance indicators are analogous to the business community using key performance indicators targeted to areas that will predict future problems. Businesses also want to predict future performance, and they too should be tracking leading key performance indicators.

Many, like the Organization for Economic Cooperation and Development (OECD, 2008), look at safety performance indicators from two perspectives:

1. *Outcome indicators*: Assess whether safety-related actions (policies, procedures, and practices) achieve their desired results and whether they lead to less likelihood of an accident. They are reactive or lagging (based on historical data). Typically, they measure change in safety performance over time or failures, such as number of slips, trips, and falls.
2. *Activities indicators*: Assess whether an organization is taking actions proactively to lower safety risks. OECD gives an example of measuring performance against a tolerance level that shows deviations from safety expectations. When tolerances are surpassed, actions should be taken.

The OECD activity indicators are very good, but we want to move them further toward predicting future problems. What we want to do is use our safety tools to identify where we will have safety problems or when safety controls are no longer effective. We want them to measure where our safety vulnerabilities lie but also when we will get new vulnerabilities. If the oil company in the example in *Notes from Nick's File* had designed leading safety performance indicators on safety and corrosion inspection, targeting or questioning periodically "Are my pipeline inspection and maintenance protocols still valid to detect corrosion?" then they would have a priori identified that increased use of seawater intensifies corrosion and therefore requires a change in maintenance and inspection protocols. The rest of this book will be very useful in helping you to develop protocols on how to develop leading safety performance indicators.

Here are some thoughts on how to design leading safety performance indicators. You will need to finish reading the book to fully comprehend and implement good leading safety indicators, but the following is an outline of a general process that you can follow:

Step 1: Identify key safety issues of concern—This can be done through different types of safety analyses. Chapters 5 through 9 discuss different safety analysis techniques that can identify where safety vulnerabilities currently exist and where you may have future problems if hazards aren't better controlled.

Step 2: Define outcome and activity indicators and determine leading indicators—Reviewing historical data and other outcome indicators will be very useful to map out a historical trend in safety performance. You can also add to this by reviewing information for activity indicators and understand how well your processes are meeting acceptable tolerance levels. But now you want to define the leading indicators. Those can again be found in your safety analyses. Avoid measuring what is easy and countable (quantitative). Try to look at what the root causes are for high-risk hazards. Chapter 5 details the interrelationship between hazard causes and their corresponding risks. You can use this information to design your leading indicators.

Step 3: Measure data—Once the leading indicators are determined, you will want to figure out how to monitor the data and any changes that occur in the data. Instrumentation, test, inspection, maintenance, and other activities will help you. But you will also want to understand the less quantitative information. Chapter 4 details how an SMS should be developed and managed. It will give you insight into how to measure its success. Also, Chapter 10 will give you ideas on where you can find additional historical data for your lagging indicators.

Step 4: Evaluate and repeat process—You will want to check on the success of your program, and Chapter 4 details how your SMS will do that. Of course, you will want to repeat this process on a periodic basis and ensure that your leading indicators are still valid. A very mature SMS will understand that you need to do this when material changes occur in your business but also on a periodic timeline to ensure that unforeseen conditions haven't changed that you did not pick up on.

2.10   USE OF STANDARDS IN SAFETY

The use of engineering standards must be considered in relation to company and industry standards; this also pertains to the application of standards to system safety engineering. In the recent past and even today, safety engineers have depended too much on safety checklists and have blindly followed established design standards. Also, remember standards weren't as a consensus of many and represent the lowest common denominator between all writers. Standards are *not* the highest level of safety, but rather an *agreed* level of safety. There really is no substitute for a safety analysis. However, this does not negate the need for and advantage of using engineering design and operation standards. Local, state, federal, and international laws also oblige the engineer to use certain standards. "A recent example of standards used in regulation, and the realization of interactions, is the area of vehicle passenger constraint and protection involving: seat belts regulation, child constraints, lap belts, air bag, crash zone responsiveness, side crash protection, secondary crash protection, egress restraint ..." (Schock, 1993). Some of the most well-known international safety standards are related to the International Civil Aviation Organization and International Maritime Organization, both United Nations bodies that have set air and maritime safety standards, respectively.

Standards are the common language that not only engineers but also industries, companies, and countries use to try to ensure a good, high-quality product. At the turn of the nineteenth century, water tanks would frequently rupture. Both commercial and home consumers could never be sure of what they were buying. As a result, the American Society of Mechanical Engineers (ASME) developed the Boiler and Pressure Vessel Code. These design and acceptance testing standards are now used the world over. Now, it is virtually impossible to legally buy pressure vessels that do not have the ASME "U" stamp or some equivalent.

Another design standard that affects our everyday lives, both at home and on the job, is the National Fire Protection Association (NFPA) National Electrical Code (NFPA-70). This design standard is used not only for home electrical wiring but also for operation of electrical systems in various explosive environments, from oil drilling platforms to grain elevators. Though these are American standards, they have been adopted the world over.

Engineers should look at standards not as a hindrance to design but as a tool to help assure that a product is safe for the user. Also, design standards are usually the minimum requirement. A design can still be unsafe even though it meets all required design standards. However, if engineers fail to follow certain *mandated* design standards, such as the ASME Boiler and Pressure Vessel Code, they can be held liable for designing an unsafe product. Four major standards used by engineers typically are government, industry, regional, and international standards. The following sections discuss these standards.

2.10.1 GOVERNMENT STANDARDS

Local, state, and federal agencies issue standards, specifications, and regulations that affect the safety of a system. As would be expected, the U.S. Department of Defense is the largest promulgator of standards. As an example, Table 2.2 lists some of the more commonly used U.S. government and military standards. Most of these documents can be obtained from the Internet website, U.S. Government Printing Office, or directly from the sponsoring agency. Of course, there are many others that cover other fields such as the Food and Drug Administration and the U.S. Department of Agriculture. Each country has its equivalent of this list.

The left column in Table 2.2 is the government acronym or alphanumeric designation for the standard. Typically a number follows the acronym. Each of these agencies has hundreds and in some cases thousands of standards. You should first ask for a detailed list of standards before ordering.

There is much discussion about the U.S. government getting out of the standards-writing business. This is occurring, especially with regard to American military standards. However, this does not mean that there will be no standards. In many cases, the same standard is published as a joint government–industry document. The fact that the government is phasing out standards means only that the periodic updating of standards will be performed by the joint committees that are embedded in standards-writing groups.

2.10.2 INDUSTRY STANDARDS

In the United States, industry follows a voluntary standardization program. Various industries have developed nonbinding standards and guidelines to help facilitate work and projects in their area. Industry realized that it was necessary to agree to some standard way of engineering certain systems. Otherwise, it would be difficult for a valve company to sell to a pipe fitting company. Accidents have also created the need to agree on how to build some systems safely.

Table 2.3 lists numerous technical or engineering organizations that write U.S. standards. Some of the organizations write joint standards. Most of the standards-writing organizations specialize in standards and guidelines for their particular industry. Again, contact the organization directly and ask for a list of standards before ordering.

TABLE 2.2
U.S. Government Standards

ACGIH series	American Conference of Government Industrial Hygienists.
Air Force AFOSH, AFR, AFISC, and AFSC series of documents	Numerous Air Force engineering design and system safety regulations.
Army EM and AMCP series	Numerous Army engineering design and system safety regulations.
CPSC	The U.S. Consumer Product Safety Commission writes various reports and studies and promulgates product safety standards.
DOD HDBK-764	System Safety Engineering Design Guide for Army Materials.
DOE order series	The Department of Energy has numerous engineering and safety-related documents.
DOT P 5800.5	Department of Transportation (DOT), Emergency Response Guidebook.
DOT	There are numerous DOT guidelines related to the transportation industry (including transportation of hazardous materials, i.e., compressed-gas cylinders, mass transit, intermodal transportation, and commercial space).
EPA	There are various Environmental Protection Agency (EPA) risk management handbooks and guidelines.
EPA/FEMA/DOT	Joint documents (EPA, Federal Emergency Management Agency, and DOT), Technical Guidance for Hazards Analysis, Emergency Planning for Extremely Hazardous Substances, and Handbook of Chemical Hazard Analysis Procedures.
FAA	Federal Aviation Administration (a part of DOT) promulgates numerous documents related to private and commercial aircraft.
Mil-Std 882	System Safety Program Requirements.
Mil-Std-1180	Safety Standards for Military Ground Vehicles.
Mil-Std-1472	Human Engineering Design Criteria for Military Systems, Equipment, and Facilities.
Mil-Std-454	Standard General Requirements for Electronic Components.
NASA	The NASA has various design engineering, system safety, and risk management guidelines for payloads, launch vehicles, and ground support equipment.
Navy	Numerous Navy engineering design and system safety regulations for Navy systems and facilities.
NRC	The Nuclear Regulatory Commission has various risk assessment regulations related to the nuclear power industry.
29 CFR 1910	Department of Labor, OSHA Regulations for General Industry.
29 CFR 1910.119	Department of Labor, Process Safety Management of Highly Hazardous Chemicals.
29 CFR 1926	Department of Labor, OSHA Regulations for Construction Industry.
29 CFR 3133	Department of Labor, OSHA 3133, Process Safety Management Guidelines for Compliance.

TABLE 2.3
U.S. Industry Standards-Writing Organizations

American Industrial Hygiene Association
American Institute of Chemical Engineers
American National Standards Institute
American Nuclear Society
American Petroleum Institute
American Society of Civil Engineers
American Society of Heating, Refrigeration and Air-Conditioning Engineers
American Society of Mechanical Engineers
American Society of Safety Engineers
American Society for Testing and Materials
Chemical Industries Association
Chemical Manufacturers Association
Compressed Gas Association
Electronics Industry Association (esp. G-48, System Safety Committee)
Factory Mutual
Industrial Indemnity Company
Institute of Electrical and Electronics Engineers
National Electrical Manufacturers' Association
National Fire Protection Association
National Institute of Occupational Safety and Health
National Safety Council
Safety Equipment Institute
Society of Automotive Engineers
System Safety Society
Underwriters Laboratories

After the Ladbroke Grove rail crash in the United Kingdom, the Rail Safety and Standards Board (RSSB) was established in 2003. It is an independent not-for-profit company that is owned by UK rail stakeholders such as infrastructure, rolling stock, and rail operating companies. Its purpose is to use the power of the rail industry to set common safety standards. Another example is Trakhees, Ports, Customs and Free Zone Corporation, owned by the government of Dubai in the United Arab Emirates. They have created a host of environment, health, and safety standards and construction safety regulations. Embedded in their other regulations such as industrial operations, port operations, green building regulations, and fire protection regulations are safety requirements. It makes sense that industry should join forces and take control of their safety standards. Clearly, when government sees that industry is not adequately policing itself, it tends to create more government regulations.

2.10.3 REGIONAL STANDARDS, EXAMPLE: INTEROPERABILITY IN THE EUROPEAN UNION

Though the International Organization for Standardization (ISO) (EN ISO 13849-1/2) and the International Electrotechnical Commission (IEC) (IEC 62061) are international bodies, they wrote two of the most rigorous machine safety standards in the world. The European Union (EU) standards bodies CEN and CENELEC have adopted these standards and mandated them across the EU. This is an important example of how international standards or other country standards (e.g., the ASME Boiler and Pressure Vessel Code) can be adopted by countries and regions. Regional safety standards, such as these machine safety standards, mean that any company that produces machines destined for the EU or in the EU will need to meet these standards. You can see how important this is to understand if you plan to sell your equipment to another country.

But regional standards also facilitate the economic growth and sustainability of an industry in that region. A great example is the rail interoperability standards in the EU. Europe is famous for its great railways. There is a reason for that—interoperability. The European Railway Agency (ERA) based in Lille/Valenciennes, France, was established in 2006 to develop harmonized and common technical specifications for railway design, operation, maintenance, and safety acceptance. They developed a common approach to evaluating and recognizing *safe* systems in railways. What this means is that to truly integrate Europe's rail network, it had to work so that you could essentially drive a train from one end of the EU to the other. And it also means that you can build a train system component in any country and sell it across the EU under one safety certification scheme, not dozens. This interoperability is defined essentially as the technical compatibility of rail infrastructure, rolling stock, signaling systems, and a host of other subsystems and operating requirements. Before, many European countries had different rail gauges, electrical standards, signaling systems, and other differences that made an integrated European rail network difficult.

At the center of this is the EN 50126, which deals with *reliability, availability, maintainability, and safety for railway systems.* EN 50129 applies to *safety-related electronic control and protection systems.* And EN 50128 applies to *safety-related software for railway control and protection systems.* These CENELEC standards cover all railway design, operations, and certification in the EU; though there are other related and applicable standards, these are the most important.

At the heart of EN 50126 is the concept of the safety case. The safety case documents the information that demonstrates that the particular product complies with the safety standards. Many will use a design safety case to demonstrate that the system was designed safely and the operational safety case to demonstrate that the system will operate safely under all conditions. Chapter 12 further discusses the use of the safety case. The EN 50126 standard is so powerful that it is not uncommon to see its approach used in many countries around the world including Asia, Australia, and the Middle East.

2.10.4 INTERNATIONAL STANDARDS

As engineering work crosses international borders and many companies have engineers working on the same project simultaneously from different parts of the world, a common understanding of which standards are to be used is paramount for success. The Uniform Building Codes (Uniform Building Code, 1994) of the International

Conference of Building Officials is an example of how the building construction industry has come to a consensus of *safe* building design. Obviously, to sell products in a foreign market, you must meet the local safety codes and standards. Most countries have a host of national standards, and some regions have even harmonized their national standards into a regional format. The European Community Product Safety Directives and Standards (EN-series) is one such example. Unfortunately, there is no single standards-writing body or source, so when doing business in a foreign country, it can take time to identify all of the required standards.

As regions start integrating their trade, standardization becomes more important. The ISO and the IEC (Strandberg, 1991) are the most recognized international standards-writing bodies. The ISO-9000 series documents in quality management and quality assurance are rapidly being adopted by governmental organizations and private companies the world over. ISO-9004 specifically addresses product liability and safety (International Standards Organization, 1987):

> The safety aspects of product or service quality should be identified with the aim of enhancing product safety and minimizing product liability. Steps should be taken to both limit the risk of product liability and to minimize the number of cases by
>
> 1. Identifying relevant safety standards in order to make the formulation of product or service specifications more effective
> 2. Carrying out design evaluation tests and prototype for model testing for safety and documenting the test results
> 3. Analyzing instructions and warnings to the user, maintenance manuals, and labeling and promotional material in order to minimize misinterpretation
> 4. Developing a means of traceability to facilitate product recall if features are discovered compromising safety and to allow a planned investigation of products or services suspected of having unsafe features

PRACTICAL TIPS AND BEST PRACTICE

- A practical way to circumnavigate the confusing world of codes and standards is to understand the difference between mandatory regulations and industry guidelines.
- Mandatory regulations are comprised of codes or standards that legally or contractually require the user to adhere to preestablished design, testing, and operations standards, and the document uses language such as *shall* or *must*. These standards are usually written by governmental organizations in the country where the product will be used or sold. OSHA regulations are examples of mandatory U.S. safety standards, and "the EC (European Community) directive on the safety and health requirements of the workplace" (Peters, 1992) is a European example.
- Industry guidelines are consensus standards and voluntary. Written by industry users, they tend to be of a detailed specification nature and offer a suggested method of compliance. The document uses language such as *should, recommended, may,* or *could.* Many times these guidelines are very

detailed and go into depth explaining how to calculate safety factors, testing conditions, acceptable construction materials, and even operational limits. Some examples of organizations with such guidelines are ASME, NFPA, and the Electronic Industries Association.

- It is critical that the engineer read the mandatory regulations closely. Many times the governmental regulations will cite industry guidelines and require the user to follow that guideline to the letter. The best-known example is the ASME Boiler and Pressure Vessel Code. An engineer cannot design, build, or sell pressure vessels in the United States that do not meet the ASME standard. NFPA 70 (the National Electric Code) is cited by OSHA as *the* industry standard to be followed for electrical systems.
- And remember standards and guidelines are updated periodically, so be sure to get the most recent version.

As stated earlier, each country has its own standards-writing body. Some common examples are shown in Table 2.4 (Fitzgerald, 1990). Note that these are government—not industry—standards. However, industry does have significant input into the standards-writing process in these countries.

Probably the most famous international safety standard is OHSAS 18001:2007, *Occupational Health and Safety Management Systems—Requirements*. Created in 2007, it is closely aligned with the well-known ISO 14001 *Environmental Management Systems*. The OHSAS 18002:2008, *Occupational Health and Safety Management Systems, Guidelines for the Implementation of OHSAS 18001:2007*, was also published. They were developed by the OHSAS Project Group, a consortium of 43 organizations from 28 countries, and include national standards bodies, certification bodies, and health and safety institutes. They are based on continual safety improvement through occupational safety and health policy, planning, and implementation; checking and corrective action; and management review through a closed-loop system. It should be remembered that these safety standards are voluntary and not mandatory, though some countries may cite these standards in their regulations and mandate their implementation. Many engineers will look at

TABLE 2.4

Some European Standards Organizations

France	AFnor—Association Francaise de Normalisation
Germany	DIN—Deutsches Institut fur Normung e.V.
Iceland	STRI—Technological Institute of Iceland
Spain	AENC—Asociacion Espanola de Normalizacion y Certificacion
United Kingdom	BSI—British Standards Institute
European	EN—European Community Standards
	CEN—European Committee for Standardization
	CENELEC—European Committee for Electrotechnical Standardization

ISO 14001, OHSAS 18001, and ISO 9000 (Quality) together to create an integrated management assurance system. Interestingly, the international safety community is doing just that.

In late 2013, an ISO project committee, ISO PC 283, comprised of 50 countries (but will surely increase over time), developed the first working draft of ISO 45000. It will replace the OHSAS 18000 series and closely align with ISO 9001 Quality Management and ISO 14001 Environmental Management. Because of globalization and too many different safety standards, it was recognized that there was a need for an internationally recognized global standard for health and safety worldwide. The standard is expected to discuss risk management much more than before and potentially standardize safety and risk terms. It will use the new ISO 9000 and ISO 14000 format and consider ISO 31000 *Risk Management* concepts and terms.

American companies have had significant reputational damage from using poor safety practices that don't meet U.S. workplace safety standards for their factories in other countries. Various European and other countries have suffered similarly. The intent of ISO 45000 will be to use a common approach across the globe. This will help better manage workplace safety and raise the bar internationally for countries to implement stronger SMSs. Of course, like all ISO standards, it will be voluntary, and individual countries can decide how they wish to implement the new standard.

This is a significant change. With the promulgation of this standard, workplace safety will be treated equally worldwide, the standard can easily be cited in international tenders, and it will serve as the international benchmark for health and safety. It should also facilitate international operations since companies can use a single, recognized standard. This should help ensure better SMS governance and the application of system safety concepts worldwide. The expected publication date is late 2016.

Other well-known international safety standards bodies are

- International Civil Aviation Organization
- International Maritime Organization
- International Labor Organization—They have numerous workplace safety standards
- UL (Underwriters Laboratories is actually a company) Standards
- International Atomic Energy Agency
- Many more

"'A more productive workforce and higher quality standards' are the primary benefits of global safety standards, say a majority of safety directors surveyed by Liberty International, Boston" (Upfront, 1995). A survey of Fortune 500 companies showed that 67% have global safety standards and that they continue to develop them because of the internationalization of the marketplace.

Safety standards are an important part of system safety engineering. However, you should not expect that if you build your system to an appropriate safety standard, it will automatically be safe. The only way to be sure that your system is safe is also to incorporate the other concepts discussed in this chapter and the rest of this book into your design and operations.

REFERENCES

Center for Chemical Process Safety. 1985. *Guidelines for Hazard Evaluation Procedures.* New York: American Institute for Chemical Engineers, pp. 1–3.

Fitzgerald, K. 1990, June. Global standards. *IEEE Spectrum.* Piscataway, NJ: IEEE Press, pp. 44–46.

INPO. 2012, August. Lessons learned from the nuclear accident at the Fukushima Daiichi Nuclear Power Station, INPO 11-005 Addendum, Revision 0. Atlanta, GA: Institute of Nuclear Power Operators. http://www.wano.info/wp-content/uploads/2012/08/11-005-Fukushima-Addendum2.pdf, downloaded March 28, 2014.

International Standards Organization. 1987. Quality management and quality system elements—Guidelines. ISO-9004. International Standard ISO 9004 was prepared by Technical Committee ISO/TC 176, Quality assurance. Geneva, Switzerland: International Standards Organization.

NAIIC. 2012a. The National Diet of Japan, Fukushima Nuclear Accident Independent Investigation Commission (Chairman: Kiyoshi Kurokawa), p. 9. http://www.nirs.org/fukushima/naiic_report.pdf, downloaded March 28, 2014.

NAIIC. 2012b. The National Diet of Japan, Fukushima Nuclear Accident Independent Investigation Commission (Chairman: Kiyoshi Kurokawa), pp. 16–21. http://www.nirs.org/fukushima/naiic_report.pdf, downloaded March 28, 2014.

National Aeronautics and Space Administration. 1989. Magellan Investigation Board Report on the October 17, 1988 spacecraft electrical mishap (March 24, 1989). Washington, DC: National Aeronautics and Space Administration.

National Aeronautics and Space Administration. 1993. Safety policy and requirements document. NHB 1700.1 (Vl-B). Washington, DC: National Aeronautics and Space Administration, pp. 1–3.

Organization for Economic Cooperation and Development (OECD). 2008. *Guidance on Developing Safety Performance Indicators Related to Chemical Accident Prevention, Preparedness and Response: For Industry,* 2nd edn., OECD Environment, Health and Safety Publications, Series on Chemical Accidents No. 19. Paris, France: OECD Environment Division, p. 5.

Peters, G. A. 1992. Hazard communication in perspective. *Hazard Prevention* (4th Quarter):9. The author cited Council Directive of November 30, 1989, concerning the minimum safety and health requirements for the workplace, 89/654/EEC. *Official Journal of the European Communities,* No. L393/1-12,30.12.89.

Roland, H. E. and Moriarty, B. 1990. *System Safety Engineering and Management,* rev. edn. New York: John Wiley, p. 6.

Schock, H. E., Jr. 1993. Regulation and system safety standards. *Hazard Prevention* (2nd Quarter):29.

Strandberg, K. 1991. IEC 300: The dependability counterpart of ISO 9000. In *Proceedings Annual Reliability and Maintainability Symposium.* Piscataway, NJ: IEEE Press, pp. 463–467.

Uniform Building Code. 1994. *International Conference of Building Officials.* Whittier, CA: International Conference of Building Officials.

Upfront. 1995, September. Global safety rules help. *Professional Safety,* p. 1.

U.S. National Safety Council. 2012. The total cost of injuries: How much do you really pay? http://www.nsc.org/news_resources/injury_and_death_statistics/Pages/Estimatingthe CostsofUnintentionalInjuries.aspx, downloaded, March 21, 2014.

U.S. National Safety Council. 2013. *Injury Facts.* Itasca, IL: National Safety Council. http://www.mhi.org/downloads/industrygroups/ease/technicalpapers/2013-National-Safety-Council-Injury-Facts.pdf, downloaded March 21, 2014.

World Nuclear Association. Fukushima accident, updated March 20, 2014. http://www.world-nuclear.org/info/safety-and-security/safety-of-plants/fukushima-accident/, downloaded March 21, 2014.

FURTHER READING

BS EN 50126-1:1999. Railway applications. The specification and demonstration of reliability, availability, maintainability and safety (RAMS). Basic requirements and generic process. Amended February 2007 and July 2010. London, U.K.: British Standards Institute.

BS OHSAS 18001:2007. Occupational health and safety management systems—Requirements. Geneva, Switzerland: International Standards Organization.

Ericson II, A. C. 2011a, August. *Concise Encyclopedia of System Safety: Definition of Terms and Concepts*. Hoboken, NJ: Wiley.

Ericson II, A. C. 2011b, September. *System Safety Primer*. Paperback. CreateSpace Independent Publishing Platform, Self-published.

International Association of Oil and Gas Producers. 2011, November. Process safety—Recommended practice of key performance indicators. Report No. 456. http://www.ogp.org.uk/publications/safety-committee/process-safety-recommended-practice-on-key-performance-indicators/, downloaded May 14, 2014.

ISO 31000:2009. Risk management—Principles and guidelines. Geneva, Switzerland: International Standards Organization.

Kltez, T. 2009, July 7. *What Went Wrong? Fifth Edition: Case Histories of Process Plant Disasters and How They Could Have Been Avoided*. Hardcover. Oxford, U.K.: Butterworth-Heinemann/IChemE.

National Aeronautics and Space Administration. 2001. Preparing hazard analyses for JSC ground operations. JSC 17773. Revision C. Safety and test operations division. Houston, TX: Lyndon B. Johnson Space Center.

Reason, J. 2008, December. *The Human Contribution*. Paperback. Surrey, U.K.: Arena Publisher.

Reason, J. 2013, November 28. *A Life in Error: From Little Slips to Big Disasters*. Paperback. Burlington, VT: Ashgate Publishing Company.

Stamatis, D. H. 2014, May. *Introduction to Risk and Failures. Tools and Methodologies*. Boca Raton, FL: CRC Press.

Taleb, N. N. 2005, August. *Fooled by Randomness: The Hidden Role of Chance in Life and in the Markets*, 2nd edn., Incerto. New York: Random House Trade Paperbacks.

3 Safety Analysis in Engineering
How Is It Used?

Everyman takes the limits of his own vision for the limits of the world.

Studies in Pessimism, 1851
Arthur Schopenhauer

All things are filled full of signs,
And it is a wise man who can learn about one thing from another.

Enneads, c 255
Plotinus

Out of this nettle, danger, we pluck this flower, safety.

King Henry the Fourth, Part One, 1623
William Shakespeare

In the United States, as in most countries, different industries tend to prevent accidents and protect the consumer in different ways. Occupational safety (popularly known as *workplace safety*) functions in the manufacturing industry differently from how the petrochemical industry prevents accidental spills or explosions. Though the aerospace and military industries are high tech, like the commercial nuclear power industry, safety controls are distinct. The mass transit industry protects riders from natural and man-made disasters in its own way. What is important for you to realize is that though the end result is to protect human life, the environment, property, and the bottom line, different sectors of the engineering community may approach that task in different ways. Also, what can be seen is how divergent industries solve the problem in much the same way. However, all industries can benefit from learning how others identify hazards and control risks. This cross-fertilization not only shares information but also helps you lower costs by identifying the most efficient and appropriate safety techniques available. We all benefit from sharing these safety best practices and determining which ones are portable to another industry, thus helping that industry to continue to learn and improve.

NOTES FROM NICK'S FILE

I've worked in many different industries over the years. One of the challenges that I regularly face is convincing people that safety and risk tools from other industries can benefit them too. But once I can show them actual results, they tend to be quickly convinced—engineers are a practical lot. For example, I've used petrochemical safety approaches with NASA and NASA approaches in mass transit. It has also been good for me as an engineer, forcing me to open my mind and look at various ways to view a problem and more efficient ways to mitigate risks.

It is important to also remember that safety is as equally important for small companies as large corporations and multinationals. In many countries, the economic and labor market growth engines are small- and medium-sized enterprises (SMEs). By some accounts, it can be almost up to 70% of all new jobs created in any given year. Safety tools and techniques are not just for big companies with a lot of money at stake but also for these SMEs. If the smaller companies don't take an equally methodical approach to safety, then it will become difficult for them to continue to grow in a sustainable way.

3.1 MANUFACTURING

Safety in the manufacturing industry is no longer just about slips, trips, and falls. Safety in manufacturing in the United States is based primarily on Occupational Safety and Health Administration (OSHA) compliance. The primary goal is workplace safety. In many manufacturing plants, an accident is usually more related to the worker than to the production line. If the line goes down, it rarely impacts the surrounding community unless large quantities of hazardous materials are part of the production process. In this case, the engineer is concerned with how the plant worker interfaces with the manufacturing environment. This means that the engineer focuses on occupational and health hazards to the worker. The handling of toxic chemicals in a microchip plant is one obvious example. The line engineer must control not only accidental spills but also the toxicity of respiratory poisons such as phosphine.

The four major areas of protection are facilities and workstations, material handling, workplace exposures and protection, and production operations. But it should be realized that hazards often cross over from one work area to another. Worker health issues, such as toxicity levels of certain chemicals, also can be related to explosive levels. Every country will have its own version of occupational safety and health standards. The U.S. Department of Labor OSHA standards are the most common compliance tools in the United States. The most important standards are 29 CFR 1910, *OSHA Regulations for General Industry* (and especially 29 CFR 1910.119, *Process Safety Management of Highly Hazardous Chemicals*, for companies working with dangerous chemicals); 29 CFR 1926, *OSHA Regulations for Construction Industry*; and 29 CFR 3133, *Process Safety Management Guidelines for Compliance.* Private industry also publishes numerous guidelines for voluntary compliance; see Chapter 2 for more on the U.S. industry standards.

The downside of depending only on compliance verification methods for safety is that you are not designing the optimal safety into the system. Also, compliance with standards and regulations is the minimal effort, many times not enough to prevent an accident or even best manage costs. Many engineers complain about using OSHA compliance standards such as 29 CFR 1910 because it is a huge compilation of seemingly unrelated requirements. Often an engineer is left wondering why something has to be 44 in. wide to be safe. What is much more useful is to introduce the concept of system safety analysis—analyzing the system to find the hazards—not just relying on compliance checklists. System safety in a manufacturing plant is a great way to optimize safety; the engineer is actually thinking about how the system could be made safer, not just blindly following regulations.

Even though the manufacturing industry is principally compliance based, it is changing. System safety engineering techniques and management styles are now being incorporated as well. For example, OSHA's Voluntary Protection Program (VPP) has demonstrated that VPP participants have far fewer lost workday injuries than nonparticipants. OSHA (Barab, 2012) states that "Data shows that site-based non-construction participants' Total Case Incident Rates (TCIR i.e., the total number of nonfatal recordable injuries and illnesses that occur per 100 full-time employees) of VPP members are 45 percent below the Bureau of Labor Statistics (BLS) rates. The Days Away from Work, Restricted Work Activity, or Job Transfer (DART, i.e., the rate of injuries and illnesses that result in workers having days away from work, restricted work activity, and/or a job transfer) rates are 56 percent below the Bureau of Labor Statistics (BLS) rates for their respective industries. For site-based construction and mobile workforce participants, TCIR are 60 percent below the BLS rates, and the DART rates are 56 below the BLS rates for their respective industries." These reduced rates don't just make government regulators and the public happy, but they also mean greater profits for employers because there are many fewer workers' compensation costs, lost-time costs, and other nonproductive costs.

The four major components of VPP (OSHA, 2013) are management commitment and worker participation, worksite analysis, hazard prevention and control, and safety and health training. The most frequently used safety analyses are safety checklists, compliance analysis, and process hazard analysis (in industries that handle hazardous chemicals). The great advantage of VPP Star Status (the highest level of accreditation) is not just a significantly lower accident rate and lower costs, but also OSHA will conduct on-site inspections once every 2½–5 years.

Also, with the advent of the 1992 OSHA Process Safety standard, companies are encouraged to use more sophisticated techniques for making systems safe. The process safety standard is most common in the chemical and process industries, but it is applicable to anyone who uses hazardous materials.

There is a global trend toward standardization. The intent is for the machinery manufacturer to design safety into the system. International manufacturing safety standards (or de facto international standards such as CENELEC in Europe) are being adopted by global companies that want to sell equipment worldwide. The European Union's (EU) mandate to follow international machine safety standards is one such example, which means that companies that sell equipment into or export from the EU must meet these standards. The EU mandates numerous rigorous machine safety standards such as ISO 13849-1/2 (*Safety of machinery, safety-related parts of control systems, and general principles of design*) and IEC 62061 (*Safety of machinery—functional safety of safety-related electrical, electronic, and programmable electronic control systems*). With these standards, companies are required to identify and document hazards and risks to the use of their machinery. An additional requirement is to include the mean time to probable failure and how it relates to the risk level of the equipment. Many companies have turned meeting the requirements into a business. For example, one company provides a functional safety service offering to manufacturing companies to help them appropriately size, buy, and install plant equipment.

PRACTICAL TIPS AND BEST PRACTICE

When possible, integrate safety with productivity, quality control, and reliability into the same program so that all the costs and benefits are understood and managed jointly. For example, some companies use an integrated risk management model with a productivity, quality, risk, and safety (PQRS) program for many of their industrial products, which helps improve profitability and sustainability in their manufacturing process. It is becoming much more common to see Integrated Management Systems that do just that.

It is also beneficial to use system safety tools from other industries. One such tool is hazard analysis, which is used to help identify and control hazards in a system. The technique, though invented in the military and aerospace industries and used in the mass transit industry, can easily be applied to the manufacturing world. In fact, facility hazard analysis is a specific use of hazard analysis in facility acquisition. The U.S. Navy has used it for many years in all of their facility constructions and renovations. The Navy has used it for such things as construction or modification of fuel depots, pier, and dry dock upgrades and for entire submarine bases.

Operations and support hazard analysis is a safety method that also has strong potential in manufacturing. The safety tool is applicable to any situation where human operators are an integral part of the process. For example, any plant that moves large amounts of material during the manufacturing process could benefit from using this tool. The analysis identifies the hazards at the critical points in the process where human error could have disastrous effects.

Though the food production and manufacturing industries are subsets of manufacturing, they are using very methodical system safety processes. In particular, the hazard analysis and critical control points (HACCPs)—like many other safety techniques—were borne out of the military and the National Aeronautics and Space Administration (NASA) (NASA used it to develop safe foods for the astronauts in space) and founded on key principles. First, a hazard analysis is conducted on the production process. Then critical control points (for safety monitoring) are determined, and their critical limits are established. Then a monitoring system is put into place to watch when the process is out of limits. Corrective actions are determined and verification procedures are followed to implement the corrective actions. Of course, the process is documented for record keeping.

PRACTICAL TIPS AND BEST PRACTICE

Governments and industry must work together to find the best solution to improving manufacturing safety. Government working in isolation creating new regulations without significant industry involvement historically has shown that it will most likely result in a high level of industry resistance. It is paramount for the correct balance of regulations with worker and public protection that solutions are developed jointly between government and industry. New Zealand's "Manufacturing Sector Action Plan to 2013" is an interesting example. The government brings key stakeholder groups together from across many industries to develop the manufacturing sector joint programs.

NOTES FROM NICK'S FILE

Once I facilitated a meeting between government regulators and industry on the transport of hazardous materials. The intent was to get industry to agree to voluntary standards and approaches on what to do with some aspects of the movement of hazardous goods. Of course, industry was very suspicious (and brought their lawyers to the workshops), but after much suspicion, debate, and then finally discussion, in the end, industry did make voluntary agreements and avoided mandatory regulations.

3.2 CONSUMER PRODUCTS

Consumer product safety picks up where production and manufacturing safety ends. Safety in the manufacturing plant protects factory workers during the production and manufacturing process. Consumer product safety focuses on protecting the end user—the consumer from the finished product. Probably one of the best-known entities is the U.S. Consumer Product Safety Commission (CPSC). According to the CPSC, over $900 billion is lost annually in the United States from consumer products that have resulted in deaths, injuries, and property damage. The CPSC evaluates consumer products and determines if they are safe for the public to buy and use. They evaluate not only products manufactured domestically but also those imported into the United States. The products cover the gamut of what is sold as a consumer product in the United States from nanoparticles in cosmetic products, power tools, ceramic lasagna pans, baby cribs, and cigarette lighters to toys and all terrain vehicles and many, many other products. Thousands of different products are regulated through the Consumer Product Safety Act and the Consumer Product Safety Improvement Act. The website http://www.saferproducts.gov was created in

2011 by the U.S. government and is a forum for the public to report unsafe products or unsafe handling procedures of consumer products.

The CPSC uses a mixture of regulations and standards (including banning of certain characteristics, for example, small piece parts are banned from products for children younger than 3 years old), testing, and recalls to ensure safety of consumer products. The way the CPSC certifies children's products is illustrative of the approval process. The Consumer Product Safety Improvement Act of 2008 requires that almost all children's products sold in the United States, whether imported or produced domestically, must

- Comply with all applicable children's product safety rules and regulations.
- Test for compliance by a CPSC-accepted laboratory.
- Obtain a Children's Product Certificate issued by the manufacturer that provides evidence of product safety compliance.
- Provide permanent tracking information affixed to the product and its packaging. (Manufacturers, importers, distributors, and retailers also are mandated to report defective products that could create a substantial risk of injury to consumers.)

Knock-off and counterfeit products are a new safety problem that is causing serious dangers around the world, especially with the significant increase in world trade. In some countries in the Middle East, for example, up to 40% of the pharmaceuticals sold to the public may be counterfeit. Knock-off products not only rob manufacturers from legitimate profits, but they many times pose a significant health hazard and safety danger. The crash of Partnair Flight 394 in 1989 was a result of counterfeit aircraft parts. The problem had become so big that the U.S. government passed the Aircraft Safety Act of 2000 to try to control the sale of knock-off aircraft parts. Counterfeit products, in general, are such a problem that international police organizations like Europol have set up elaborate tracking systems to interdict and prosecute violators.

Numerous bilateral and multilateral agreements are being forged to help maintain the integrity of the product development system. Rapid alert system for nonfood products posing a serious risk (rapid exchange of information [RAPEX]) is a multilateral agreement between EU Member States and others to facilitate rapid exchange of information to prevent or restrict products that could pose a serious risk to consumer health or safety (there is a separate system for food, pharmaceutical, and medical devices). RAPEX-China is a bilateral agreement with China that focuses on EU and Chinese products. As counterfeit and knock-off products become ubiquitous in some parts of the world, there is an ever-increasing reliance on these international agreements to ensure safe products are delivered to consumers.

Internationally recognized third-party testing labs are located in many countries around the world and in such diverse markets as the United States, Canada, China, Bangladesh, Pakistan, South Korea, Philippines, Indonesia, Spain, Brazil, India, Singapore, Peru, Mexico, Greece, Germany, and Turkey. The list goes on. Because there are so many of these recognized laboratories, there is no reason that exporters cannot access this accreditation process. An accredited branch of many of these third-party testing facilities is located especially in countries with heavy consumer product manufacturing. Consumer product safety works in the following way:

- Identify potential hazards early in the design and manufacturing process—gives manufacturers the ability to modify designs and manufacturing techniques before it becomes cost prohibitive
- Conduct a product hazard analysis—ensures that adequate design and manufacturing controls are in place to create safe products
- Perform a factory process control audit—verifies controls are adequate and still in place and controls process/manufacturing deviations that could create additional hazards
- Test the product at a accredited third-party lab—creates consumer confidence of impartial product verification
- Track the product through the supply chain and report (or recall) any significant deficiencies—ensures a closed-loop self-learning process for the product

In the end, the consumer benefits, as well as the product supplier, because recalls are minimized and brand protection is strengthened. We all benefit from this process. The automotive industry is probably one of the highest visibility manufacturers that use recalls.

3.3   CHEMICAL PROCESS AND OIL AND GAS INDUSTRY

The chemical process industry in the United States also falls under federal OSHA regulations for workplace safety. However, due to large quantities of hazardous chemicals, additional safety analyses and control are required. When 29 CFR 1910.119, *Process Safety Management*, was published in 1992, Acting Assistant Secretary of Labor Dorothy Strunk stated that once it is fully implemented, the regulation is estimated to prevent 264 deaths and 1534 injuries annually (Strunk, 1992)—a significant cost savings to the chemical industry. Until the regulation was promulgated, the chemical industry voluntarily followed the American Institute of Chemical Engineers' Center for Chemical Process Safety (CCPS), *Guidelines for Hazard Evaluation Procedures*, first published in 1985 (expanded in 1994 and recently updated in 2008). The CCPS has dozens of other very useful and practical publications for chemical process safety.

The 1992 OSHA regulation is a significant change in the chemical process industry. This regulation applies system safety engineering and management tools to the process industry. This change has required the chemical process industry to control not only workplace hazards but also risks that may propagate to the surrounding community and environment. It also requires a more formal safety analysis process and hazard management. The two most common safety analyses are hazard operability analysis (HAZOP) and safety checklists.

HAZOP is a team approach that uses brainstorming sessions to review the process drawings and identify any possible process deviations and their effects. The tool has been very useful to the industry, and though it is expensive, it is becoming more common throughout the industry.

Some in the industry use checklists, comparing the design against a set of questions or evaluation criteria. Checklists are really most useful for small projects. Because it is not a systematic approach to safety, it is easy to miss something in large or complex processes.

Risk assessments were first developed by the commercial nuclear power industry, but they are now being used extensively in the chemical process industry. The advantage of quantitative risk assessment is that it not only identifies hazards, it also gives you a way to decide how to manage those hazards. This becomes particularly important if a plant wants to better understand how a chlorine spill at a wastewater treatment plant would affect surrounding neighborhoods. The tool allows you to add on other models such as toxic cloud dispersion models.

Probably the most interesting aspect of risk assessment for the chemical process industry is that it is a very methodical tool for optimizing how much risk you are willing to accept. This is very important in designing chemical processes, because, if the process can be made more efficient and safer, significant cost savings can be achieved.

In spite of, or maybe because of, process industry accidents still occurring in various industries and the increasing complexity of industrial processes (especially with industrial control systems using cyber networks), the CCPS continues to push the envelope on process safety and safety in general with their Vision 2020. The purpose of Vision 2020 is to imagine what *perfect process safety* will look like. The vision is to have industry stakeholders to have demonstrable and actionable commitment to the competencies and processes to prevent, reduce, and mitigate process safety incidents (not just accidents). They do this through five tenets for industry and four societal themes (Center for Chemical Process Safety, 2013). Their five tenets for industry or core principles are

1. Committed culture
2. Vibrant management systems
3. Disciplined adherence to standards
4. Intentional competency development
5. Enhanced applications of lessons learned

Their four societal themes, which is a call to action for society in general, including our leaders and the public, are

1. Enhanced stakeholder knowledge
2. Responsible collaboration
3. Harmonization of standards
4. Meticulous verification

PRACTICAL TIPS AND BEST PRACTICE

Clearly, the American Institute of Chemical Engineers' CCPS Vision 2020 is a best practice. Other industries should look at this approach and use it as a template to help promote and inculcate safety into their industry. This can't and shouldn't be done with only industry leadership but also should include government regulators and oversight agencies to ensure that the *megacommunity of stakeholders* is involved in designing future safety processes (and oversight programs). It not only benefits the industry but also manages the appropriate balance of regulation and industry performance.

The oil and gas industry follows the same approach as the chemical process industry and uses the same safety tools discussed earlier. In fact, most training courses lump the two together since the hazards are very similar. HAZOPs are by far the most common safety tool used in the oil and gas industry. The OSHA *Process Safety Management* regulations are also prevalent. The BS Occupation Health and Safety Assessment Series (OHSAS) 18000 is an international series for health and safety management systems and is now becoming the de facto international safety, across multiple industries, for safety management systems. As mentioned in Chapter 2, BS OHSAS 18000 is expected to be replaced by ISO 45000 sometime in late 2016.

Oil and gas industries use many risk assessment tools beside the HAZOP. For example, failure modes and effects analysis (FMEA) is particularly common for evaluating how safety critical equipment can fail, such as a subsea blowout preventer used on drilling rigs. The bow tie model is another popular tool.

3.4   AVIATION

Because of the high visibility of airplane crashes and the fact that the commercial airline industry was borne out of the military air service, for many years, the airline industry has used system safety engineering and management tools for public safety. Many still manage their programs in a similar way that Douglas Aircraft Company did many years ago (Redgate et al., 1994). Douglas Aircraft Company is using the very effective four primary analyses: functional hazard analysis, failure mode and effects analysis, fault tree analysis, and zonal analysis.

Functional hazard analysis is the airline industry's name for hazard analysis. Failure mode and effects analysis and fault tree analysis are applied in the same way as in other industries. Zonal analysis is the verification of correct manufacture and installation. It starts by reviewing drawings and analysis and ends in the physical inspection of mockup, prototype, and production systems.

Though safety has been a concern since rickety aircraft first flew at the beginning of the twentieth century, it really wasn't until the Chicago Convention in 1944 that it was more seriously formalized (especially on an international scale). As World War II was winding down, many aircraft manufacturers and governments realized the potential for a strong commercial aviation market. To enable that industry, it was critical for an international agreement to not just support the industry but, more importantly, develop the international agreements and processes to safely manage aviation and airspace navigation. Fifty-two countries signed the original Convention on International Civil Aviation that quickly led to the United Nations body, the International Civil Aviation Organization, headquartered in Montreal, Canada, and comprised of 190 countries (there are still a few that have not signed on) in the United Nations.

What ICAO does is to set consensus international standards and recommended practices (SARPs) for safe air navigation, aviation infrastructure, flight inspections, and the protocols for civilian aviation accident investigations. The country signatory then will take these international SARPs and develop their own national standards and legally binding national civil aviation regulations. Currently, there are more than 10,000 SARPs included in 19 annexes to ICAO requirements. ICAO has served the world very well and has made commercial air travel one of the safest modes of transportation (much safer than driving a car). Individual national or regional aviation regulators, such as the Federal Aviation Administration (FAA) in the United States and the European Aviation Safety Agency (EASA), use a combination of design, maintenance, operations, training, and certification regulations as part of their safety oversight functions. Various type certifications are used to assure implementation and verify compliance. Civil Aviation Authorities in other countries follow a similar approach to meet ICAO requirements.

The aviation industry uses a combination of system safety engineering and regulatory compliance. Larson (1989), also from Douglas Aircraft Company, lists the industry standard as follows:

- Identify hazards and establish appropriate safety criteria for them [sic].
- Assess the design against these criteria.
- Establish and apply timely corrective action as required.
- Provide cognizant personnel with the information and data they need to evaluate trade studies and other design changes.
- Demonstrate compliance with the system safety criteria and design and operational requirements.

Because so much of aviation is controlled by people, human factor analysis tools are at the heart of the aviation industry. Different types of human factors' analyses are used in air navigation, such as air traffic control, crew resource management in the cockpit, and even appropriate design and maintenance of aircraft systems. Fault tree analysis, fault hazard analysis, FMEA, and different probabilistic risk tools are also used in the detailed design of safety critical subsystems.

3.5   MASS TRANSIT

The urban mass transit industry (both rail and bus) has followed a slightly different course in the United States. Most transit systems use regulatory compliance. With the popularization of system safety tools in the engineering community over the last 20 years, system safety engineering is now being incorporated into the fixed rail and bus industry. But large transit systems such as Bay Area Rapid Transport, Washington Area Mass Transit Administration, and the various New York transit systems have been using system safety analysis tools for quite some time. System safety techniques have also crept into places such as the U.S. Coast Guard and even into future technologies, such as magnetic levitation trains, both in the United States and abroad. With the U.S. Department of Transportation Federal Transit Administration taking on a much more vigorous and expansive regulatory role, system safety is quickly being promulgated to all transit systems across the United States.

PRACTICAL TIPS AND BEST PRACTICE

- When at conferences or meetings with engineers from other industries, ask them how they solve their particular safety problems. Don't associate only with people from your industry—learn from others.
- Have a company rep join one of the safety societies (e.g., the System Safety Society or the American Society of Safety Engineers), and see how safety is handled across different industries.

The primary system safety tools being used are hazard analysis and fault tree analysis. However, the transit industry could very much benefit from more human factors' safety analysis. Though the industry has used it before, it has never been applied to the same level of detail as it has in the commercial nuclear power industry or civil aviation. Even though quantitative human factors' safety analysis is still controversial, it could prove useful in the transit industry. Some countries, such as France, have already started to look more deeply into this.

Probabilistic risk assessment, again taken from the commercial nuclear industry, is another useful tool for the transit community. Because all transit systems and industries have to face public scrutiny, risk assessments are a very good way to communicate safety information to the public.

Europe, and especially the United Kingdom, more regularly uses quantitative risk assessments and other safety analysis tools. The European safety standard EN 50126, as described in Chapters 2 and 4, uses a safety case approach and quantitative safety risk targets for mass transit and other rail projects. There are a number of differences between how the United States and Europe (Asia and Australia tend to follow the European regulatory model). Though EN 50126 uses the safety case as the basis for safety accreditation, the United States does use a safety certification process that is similar to the safety case process. One difference is that the Americans tend to put design and operational safety analyses together in one hazard analysis, whereas the others separate them into two separate analyses. The other primary difference between the two is setting quantitative safety targets. Due to the litigious nature of the United States, they have shied away from using probabilistic safety targets due to their controversial nature.

3.6   MILITARY AND SPACE

The aerospace and military industries have been using system safety engineering since the 1960s. Mil-Std-882, *System Safety Program*, is the most famous system safety document in existence. Because accidental release of a nuclear warhead could have devastating consequences, it became imperative for the aerospace and military industries to develop and implement a comprehensive safety program. Identification of hazards early on in the program life cycle was paramount, because of the high costs of retrofitting mature systems. Typical safety analysis techniques are fault tree analysis, hazard analysis, operations and support hazard analysis, and FMEA.

Fault tree analysis was created in the military industry during the Minuteman missile program. It is a graphical tool used to identify the faults in a system and which events lead to that catastrophic event; they have proven so useful that other industries are starting to use them. The chemical industry is starting to use fault trees, especially in accident investigation.

Hazard analysis and operations and support hazard analysis are tools that go through the system methodically and identify all hazards to life and equipment and were created by the space and military community. They are now ubiquitous among almost all industries.

The aerospace and military communities could also benefit greatly from the chemical industry's use of HAZOP. Of course, HAZOP would have to be modified somewhat since the technique is really designed for process flows. But what is very interesting with HAZOPs is how it views processes. It could very easily be adapted to space and military manufacturing and operational processes.

FMEA is a reliability engineering tool created in the space and military community that has been appropriated by the safety community in all industries. It identifies what causes a component to fail and what the effects or consequences would be. FMEAs were developed to evaluate very complex and dangerous systems (e.g., nuclear warheads) and fully understand all the ways it can fail. The technique is expensive so it is important that engineers are judicious in its application.

Since the advent of the space era in the mid-1950s (though missile technology was developed earlier, it didn't really become spaceborne until this time frame), military safety tools were central. This was because national space programs were created out of military use and really were part of the space race. The NASA, European Space Agency, and other national space agencies took military know-how and applied it to civilian space safety. So, the same tools used by the military are also, and still are, used by civilian space safety.

But we are entering a very exciting era with the advent of truly commercial space travel. Shortly, you will be able to book a flight for a quick space trip. We have been launching commercial satellites and scientific spacecraft into space for decades. And of course, the most famous space trip was landing on the moon in 1969. Now technology is sufficiently inexpensive and space travel is well understood (at least short trips to low earth orbit) that we should see an increase in commercial space travel. Numerous private companies are vying to win customers to launch for relatively brief periods (at least for right now) in space. These companies are all using civilian and military space know-how and use the same safety tools that military systems use. Though organizations like the International Association for the Advancement of Space Safety (IAASS) have come into play, they essentially are using standard military and civilian space safety analysis tools.

What is interesting with the new commercial space race and soon-to-be space tourism is that the IAASS is trying to push the industry toward an ICAO-like set of international safety regulations. It is unclear if and when it will occur. To date, commercial space launches are regulated by the individual countries where the launches occur. But this leaves a host of legal issues that are still to be sorted. An ICAO-like international agreement could significantly move commercial space safety forward. As a minimum, it could help sort the international legal issues. Some of the legal gaps to be filled are as follows: who regulates launches from platforms in international waters, what happens when a commercial spacecraft explodes and falls on a third country, who regulates when space debris on orbit hit other spacecraft, and many others.

3.7 COMMERCIAL NUCLEAR POWER

You don't need to be reminded of the most recent nuclear accidents, principally Fukushima Daiichi in Japan in 2011. After the Three Mile Island accident in the late 1970s, the U.S. Atomic Energy Commission developed "WASH 1400, The Reactor Safety Study." The WASH 1400 report laid the foundation for the use of probabilistic risk assessments (called probabilistic safety assessments in Europe). According to Henley and Kumamoto (1991), probabilistic risk assessment involves studying accident scenarios and "numerically rank[ing] them in order of their probability of occurrence, and then assess[ing] their potential consequence to the public." Event trees, fault trees, and other risk–consequence tools are applied in developing and studying these scenarios. These techniques are extremely useful for the engineer but very expensive. The nuclear industry has been the leader in probabilistic safety analyses.

Reliability, availability, maintainability, and safety (RAMS) are at the heart of the commercial nuclear safety techniques. The industry has been very good at bringing these techniques, such as FMEAs, fault tree analyses, human factors analysis, and event trees together. They are either the primary components of the probabilistic risk assessment or significant feeder analyses. Other safety tools that they use are design basis accident analysis, master logic diagrams, human reliability analysis, and Markov analysis. What is very interesting of the commercial nuclear industry safety approach is the integration of reliability and safety analyses. Because safety control systems must be high-reliability systems, their reliability and availability are paramount to control safety hazards. This is also true for military and space systems. Other industries could benefit from the nuclear industry integration of RAMS, especially in high-hazard/high-risk industries such as mining and petrochemical industries. The advancement that the commercial nuclear industry has made in human reliability analysis could also be useful for other industries.

The nuclear industry could benefit from HAZOP studies. A nuclear power plant is conceptually very similar to a chemical plant. Because HAZOP looks carefully at process deviations, it would help identify hazards that might go unnoticed in a traditional probabilistic risk assessment.

Software safety analysis is a very new technique, but it also has promise for nuclear plants. Much of plant processes are computer controlled, yet until recently, little attention has been given to spurious commanding or sneak paths in the commanding. Software safety tools could help identify these problems and offer solutions.

As this chapter demonstrates, all industries take safety seriously. There is some variety as to which safety tools they use or whether they are more compliance focused or they use system safety engineering and management. However, what is obvious is that all industries are already using many system safety engineering and management tools and many are slowly but surely incorporating more of them every day.

It also is obvious that many system safety tools are already creeping from one industry into another. The process is very slow, however, and it is hoped that this book can accelerate it somewhat and demonstrate that there is something to be learned from other industries and applied to your own. Nevertheless, it is very advantageous to use system safety methods proven in one industry in another application. This cross-fertilization will help make systems safer and at less cost.

REFERENCES

Barab, J. 2012, June 28. Deputy Assistant Secretary for Occupational Safety and Health, U.S. Department of Labor testimony before the Subcommittee on Workforce Protections Committee on Education and the Workforce, U.S. House of Representatives. https://www.osha.gov/pls/oshaweb/owadisp.show_document?p_table=TESTIMONIES&p_id=1542, downloaded May 11, 2014.

Center for Chemical Process Safety. 2013. Vision 2020 process safety: The journey continues. New York: Center for Chemical Process Safety, American Institute of Chemical Engineering. https://www.aiche.org/sites/default/files/docs/pages/vision2020.pdf, downloaded May 11, 2014.

Henley, E. J. and Kumamoto, H. 1991. *Probabilistic Risk Assessment.* New York: IEEE Press, The Institute of Electrical and Electronics Engineers, p. 4.

Larson, M. N. and Hann, S. 1989. *Safety and Reliability in System Design.* Needham Heights, MA: Ginn Press, p. 254.

Occupational Safety and Health. 2013. VPP corporate application instructions. www.osha.gov, application_coroprate.doc, downloaded May 11, 2014.

Redgate, M. L., McKelvey, M. H., and Jolly, C. L. 1994. Implementation of an integrated safety program: The MD-90 antiskid system. In *Proceedings Annual Reliability and Maintainability Symposium.* Piscataway, NJ: IEEE Press, pp. 52–58.

Strunk, D. L. 1992, February 14. U.S. Department of Labor news release 92-84 (as quoted in *System Safety Analysis Handbook*, July 1993, System Safety Society, pp. 5–1).

FURTHER READING

Center for Chemical Process Safety. 2008. *Guidelines for Hazard Evaluation Procedures*, 3rd edn. New York: Wiley.

Foley, M., Fan, Z. J., Rauser, E., and Silverstein, B. 2012. The impact of regulatory enforcement and consultation visits on workers' compensation claims incidence rates and costs, 1999–2008. *American Journal of Industrial Medicine*, 55(11):976–990. doi:10.1002/ajim.22084. 2012.

Government of New Zealand, Department of Labour. 2013. Manufacturing sector plan to 2013. Auckland, New Zealand: Ministry of Business, Innovation, and Employment, Government of New Zealand. http://www.dol.govt.nz/whss/sector-plans/manufacturing/01.asp, downloaded May 16, 2014.

Government of Western Australia. 2014. Work health and safety model regulations and codes of practice consultation regulation impact statement information and issues paper. http://www.safeworkaustralia.gov.au/sites/swa/model-whs-laws/public-comment/pages/draft-model-whs-regulations-and-cop-for-mines-public-comment, downloaded May 16, 2014.

Haviland, A. M., Burns, R. M., Gray, W. B., Ruder, T., and Mendeloff, J. 2012. A new estimate of the impact of OSHA inspections on manufacturing injury rates, 1998–2005. *American Journal of Industrial Medicine*, 55(11):964–975. doi:10.1002/ajim.22062.

International Civil Aviation Organization. 2007, July. ICAO global safety plan. http://www.icao.int/WACAF/AFIRAN08_Doc/gasp_en.pdf, downloaded May 16, 2014.

International Electrotechnical Commission. 2005. IEC 62061:2005, Safety of machinery—Functional safety of safety-related electrical, electronic and programmable electronic control systems. Geneva, Switzerland: International Electrotechnical Commission.

International Standards Organization. 2006. ISO 13849-1/2:2006, Safety of machinery; Safety-related parts of control systems; General principles of design. Geneva, Switzerland: International Standards Organization.

Levine, D., Toffel, M., and Johnson, M. 2012. Randomized government safety inspections reduce worker injuries with no detectable job loss. *Science*, 336(6083):907–911.

Mansdorf, Z. S. 1993. *Complete Guide to Industrial Safety in Manufacturing.* Upper Saddle River, NJ: Prentice Hall.

PharmaTutor. 2014. Product safety methods. http://www.pharmatutor.org/articles/industrial-hazards-and-safety-measures, downloaded May 16, 2014.

Reactor Safety Study. 1975. Reactor safety study—An assessment of accident risks in U.S. Commercial Nuclear Power Plants. WASH-1400. Washington, DC: U.S. Nuclear Regulatory Commission.

Shehane, R., Huan, X., and Ali, A. 2011, August. A framework analysis of the action plan for import safety. *Journal of International Business and Cultural Studies*, 5:1–19.

U.S. Consumer Product Safety Commission. 2013. Consumer Product Safety Improvement Act. https://www.cpsc.gov/en/Regulations-Laws--Standards/Statutes/The-Consumer-Product-Safety-Improvement-Act/, downloaded May 16, 2014.

U.S. Department of Defense. 2012. Military standard, system safety program requirements. MIL-STD-882E. Washington, DC: U.S. Department of Defense.

U.S. Food and Drug Administration. 2014. Hazard analysis and critical control points. http://www.fda.gov/food/guidanceregulation/haccp/default.htm, downloaded May 16, 2014.

U.S. Nuclear Regulatory Commission. 2014. Nuclear reactor regulation. http://www.nrc.gov/about-nrc/organization/nrrfuncdesc.html, downloaded May 16, 2014.

4 Safety Management Systems

Gonna change my way of thinking, make myself a different set of rules.
Gonna put my good foot forward and stop being influenced by fools.

Gonna Change My Way of Thinking, 1979
Bob Dylan

Management of many is the same as management of few. It is a matter of organization.

The Art of War, 6th Century BCE
Sun Tzu

A change of fortune hurts a wise man no more than a change of the moon.

Poor Richard's Almanack, 1756
Benjamin Franklin

"This is a safe place," Hammond said, "no matter what that damn mathematician is saying—" "It's not—" "And I'll demonstrate its safety—"

Jurassic Park, 1990
Michael Crichton

Almost 20 million passengers in 2012 used the Eurotunnel (all services) and 10 million on the Eurostar alone between Britain and France (Groupe Eurotunnel, 2012). The French government built La Cite de l'Europe, a small city next to the tunnel terminal for shoppers detraining in Calais, France. Three parallel tunnels each 31 miles long connect the French and British coasts. Because safety is critical to the Eurotunnel's 24/7 operational success, a binational Channel Tunnel Safety Authority was established. This Anglo-French entity is "responsible for ensuring that all appropriate national and international safety requirements are implemented...The agreement establishes a framework of organizational, procedural, and technical requirements for constructing and operating the fixed link" (Cunliffe, 1994). The Eurotunnel is an excellent example of the importance of creating and maintaining a viable safety management structure. Not only does the Safety Authority have to confront the technical safety issues involved in building and operating one of the most important civil engineering projects of the twentieth century, it must also successfully manage the complex interactive social and organizational differences between two countries— and two languages.

To achieve this goal, the engineering organization must incorporate system safety into the system life cycle. Many times, this is precisely where engineers have failed.

Not only does safety have to be designed into the product life cycle, it must also have management commitment to thrive and persevere. A good system safety management program does significantly lower business-operating costs. According to the U.S. Occupational Safety and Health Administration (U.S. OSHA, 2013), there are 60%–80% fewer lost workday injuries than the national average by using system safety as part of the Volunteer Protection Program (VPP). According to OSHA (Gibbs and Lahale, 2014), it is a quite significant reduction:

- 54% below the Bureau of Labor Statistics (BLS) Total Case Incident Rate (TCIR) for their industries.
- 53% below the BLS days away, restricted and transfer (DART) rate for their industries.
- 354 VPP sites experienced zero recordables.

In the same report, Gibbs and Lahale give examples of how safety saves money:

- Within 1 year of achieving star status, Lockheed Martin's Moorestown site's worker compensation costs decreased by 75% from over $740,000 per year to $188,869 per year. VPP participation continued to positively impact the site's bottom line, and in 2006, their workers' compensation costs were about $94,000 annually.
- In 2002, MYR Group, one of the largest electrical transmission and distribution companies in the United States had a TCIR of 7.3 and a DART rate of 3.9; they were also facing significant enforcement actions. Through concerted efforts throughout the organization, safety and health has been transformed into a key corporate value. By 2007, MYR's TCIR was reduced to 2.7 (a 63% decrease) and their DART to 1.2 (a 69% decrease). Their 2007 TCIR and DART rates are also 53% and 63% below the 2006 BLS national average for the power and communication line construction industry, respectively.

The company's system safety program, as executed through a safety management system (SMS), is the umbrella organization that will apply the safety concepts discussed in Chapter 2 to real-life situations. How the safety program is implemented and managed is critical to its success. A closed-loop process is really the only way to track and resolve identified hazards adequately. Safety reviews and audits are also part of this process. A good system safety management program (SMS) fits neatly into the new voluntary protection programs (VPPs) that various governments are implementing.

4.1 SAFETY IN THE SYSTEM LIFE CYCLE

Even today in the United States there are about 13 deaths daily in the workplace and 4 million injuries per year (Barab, 2012), which means we have to better integrate system safety into all engineering aspects. "Studies conducted at Stanford University estimate the cost of accidents for users of commercial and industrial construction at $1.6 billion annually. Hidden costs were found to be two to 18 times higher. Researchers also found that construction safety research over a 10-year period showed irrefutable evidence that accidents are controllable, to some extent, by all levels of construction management" (Gallagher, 1993). The UK Health and Safety Executive (Institution of Engineering and Technology, 2012) studies show that even minor accidents have ongoing costs, which can be quite high. A few examples they cite are as follows:

- In one organization, accident costs (when all factors were included) amounted to 37% of profit
- In another, it was 8.5% of tender prices.
- In the third, it was 5% of running costs.

Of course, the statistics vary from industry to industry; what is important to note is that accidents do happen, even minor ones, but they can be prevented.

The primary method of preventing accidents is through a comprehensive and systematic approach to safety management. The most cost-effective way to control risks is to implement and maintain a comprehensive SMS throughout the product or system life cycle—from cradle to grave. This will lower failures or mistakes (by both people and machines), prevent gaps in analysis (avoid overlooking valuable data and warning signs), and provide demonstrable safety coverage (meeting federal and local laws and creating a legally defensible audit trail). If you have a comprehensive SMS, you can protect yourself from liability claims in the most cost-effective manner possible.

4.1.1 System Life Cycle

Every industry has its own way to define the system or product life cycle. For example, the facility construction industry life-cycle phases are requirements, planning, design, construction, activation, operation, and disposal. For purposes of discussion, the system life cycle is defined as (cradle to grave) follows:

Concept: The idea of the project is *hatched*. During this phase, engineers develop an idea to accomplish the goal. More than one concept can be presented at this phase.

Definition: At this phase, the selected concept is amplified to exactly how the product or system is to be built. Preliminary design is performed.

Detailed design: The critical design is conducted. Detailed drawings and calculations are performed.

Development: The design is now mature and the system is constructed. As every engineer knows, there can be various iterations back and forth between the detailed design and development phases.

Test and evaluation: Once the system is built, it must be tested and evaluated. If serious problems still exist, the product can go back to the design and development phases.

Production: The system or product now enters the manufacturing phase.

Deployment: At this phase, the system is placed into its service location. Preoperation (or prerevenue) tests and adjustments are made at this time.

Operation: The system performs its intended function.

Modification (optional): Sometimes, the product or system must be changed
due to design or operation deficiencies that were not identified earlier.
Disposal: The system is retired from operation and decommissioned.

For each life-cycle phase, the hazards should be assessed and controlled. The earlier in
the life cycle this is done, the cheaper it is to accomplish. An SMS is a sustainable management system that is built around this process. Section 4.2 will discuss this in detail.

4.1.2   Safety and the System Life Cycle

Of course, system safety tools can be applied during any one of the aforementioned phases; however, you need to be judicious in deciding when, where, and how
much intervention is required. Probably all of the activities shown in the following
should be performed, but not necessarily to the same level of detail. You will need
to tailor the safety engineering to the appropriate level of operation. Obviously, the
earlier in the project design phase this is done, the better. If the design is still on
the computer-aided design system (what used to be the drawing board), it is 10- to
100-fold less expensive to modify the design than after deployment in the field.

What is also important to realize is that different activities are done during the
various life-cycle phase. Some of the activities are repeated and some are not. Also,
some of the work is performed as a management function and some as an engineering
function. The rest of this book details how each of these safety tools can be applied,
their pitfalls, and suggestions for maximum utilization at the most reasonable cost.

Table 4.1 shows where each of the major safety milestones fits into the system life
cycle. Safety activities are listed on the left-hand side. The remainder of the table
describes the kinds of activities to be performed during each phase of the life cycle.
As you can see, many of the activities are repeated throughout the life of the system,
such as implementing and maintaining the system safety program plan (SSPP).

4.1.3   Case Studies of Poor Application of Safety in the System Life Cycle

We all remember numerous technological *disasters* where management or engineering intervention might have averted a catastrophic situation. To refresh your memory
with just a few choice examples, the column on the left of the table in Table 4.2 was
taken from the first edition of this book, and the column on the right represents
more recent accidents. One quickly realizes that in spite of the many year intervals
between the two editions, we are still having horrendous, preventable, disasters.

As society becomes more and more complex and our engineered systems are more
difficult for one individual or group to master and control, it becomes important for us
to understand how some of these accidents came about. Many of today's disasters not
only affect the plant but also many times cross international borders. If you are following appropriate engineering and safety processes and standards, the likelihood of such
a catastrophic disaster is lowered. It is hoped that you will find that your own plant
or system is not following along the same path of the infamous accidents mentioned.

The Flixborough explosion and the *Challenger* accident are two case studies of poor
safety management. What can be seen is that the accidents were *destined* to occur.

TABLE 4.1
System Safety in the Life Cycle

Safety Activities	Life-Cycle Phases									
	Concept	Definition	Detailed Design	Development	Test and Evaluation	Production	Deployment	Operation	Modification (optional)	Disposal
Develop system safety program	Write preliminary program				Review program	Review program		Periodically review program	Review program; change if necessary	
Review (sub) contractor safety program		Review contractor's program	Modify if necessary							
Research and develop safety design criteria	Review various data sources									
Develop hazard tracking and risk resolution system		Develop tracking system	Maintain up-to-date tracking system	Continue	Continue	Continue	Continue	Review field data; maintain tracking system up to date	Continue	
System Safety Program Plan (SSPP)		Finalize program plan	Implement plan	Continue	Continue	Continue	Continue	Continue	Write new plan if necessary	Continue

(Continued)

TABLE 4.1 (*Continued*)
System Safety in the Life Cycle

Safety Activities	Life-Cycle Phases									
	Concept	Definition	Detailed Design	Development	Test and Evaluation	Production	Deployment	Operation	Modification (optional)	Disposal
Conduct safety analysis	Develop preliminary hazard list	Update preliminary hazard list	Conduct preliminary hazard analysis	Conduct subsystem/system hazard analyses	Update hazard analyses (perform other safety analyses if necessary)	Update hazard analyses	Update hazard analyses	Conduct operations hazard analysis	Update hazard analysis if necessary	Perform hazard analysis
Define safety design requirements	Initial definition	Final definition								
Conduct design reviews	Concept design review	Preliminary design review	Critical design review	Continue	Test safety review	Review production process	Review deployment plans	Review operational procedures	Review modifications	Review disposal method and site
Evaluate risks	Develop risk evaluation criteria	Evaluate risks	Continue	Continue	Continue	Continue	Continue	Continue	Continue	Continue
Identify safety equipment			Identify safety equipment							

Activity									
Verify controls in place	Develop hazard control methods	Verify controls included in design	Continue	Inspect system to verify controls	Periodic audit of controls	Verify controls in place	Continue	Verify controls still valid	Verify controls in place
Safety review of engineering change orders	Review change orders	Continue	Continue	Continue	Continue	Continue	Continue	Continue	Continue
Perform safety tests on system			Perform safety tests	Perform safety tests					
Conduct safety training and awareness	Identify training and awareness goals		Develop training and awareness programs	Test the training and awareness effectiveness	Put into place	Conduct programs	Continue	Review and update if necessary	Conduct programs
Audit safety program	Conduct audit	Periodic audits if necessary		Periodic audits if necessary	Periodic audits		Periodic audits	Conduct audit if necessary	Audit process
Participate in mishap investigations	Participate as required	Continue	Continue	Continue	Continue	Continue	Continue	Continue	Continue

TABLE 4.2
Selected Industrial Disasters

1974–1989	1996–2013
Flixborough, United Kingdom (1974)—vapor cloud explosion—killing 28 people, $232 million in damage	Charkhi Dadri, India (1996)—mid-air collision between Kazakhstan and Saudi airliners, world's deadliest mid-air collision, killing all 349 passengers on both flights
Mexico City LPG (1984)—liquefied petroleum gas explosion—300 fatalities, $20 million in damage	Space Shuttle Columbia (2003)—disintegrated while re-entering Earth's atmosphere, killing all seven crew members, leaving only three operating Space Shuttles and costing approximately $13 billion
Bhopal, India (1984)—toxic material release—2500 fatalities	BP Texas City (2005)—explosion and fire killed 15 and injured 170, third largest refinery in the United States
Chernobyl, Ukraine (1986)—fire and radiation release—immediately killing 31 people, permanently evacuating 140,000 people from their homes, radioactive particles riding the wind currents to over 20 nations, affecting 3 billion people living in the northern hemisphere	Deepwater Horizon, Gulf of Mexico (2010)—explosion and rig sinking killed 11 workers, sea floor oil gusher continued for 87 days, largest marine oil spill in history
Challenger, United States (1986)—Space Shuttle Challenger disintegrates, killing all seven astronauts onboard, grounding the manned space program for 2½ years, and costing NASA over $5 billion	Pike River Mine Disaster, New Zealand (2010)—killed 29, country's worst mine disaster since 1914 and largest single accidental loss of life since 1979
Phillips 66 Chemical Plant, Houston, Texas, United States (1989)—Explosion and fire, killing 23 people, $750M in damage, and an additional $700M in lost revenue	Fukashima Daiichi, Japan (2011)—earthquake-induced tsunami created accident and caused nuclear meltdown and radioactive release, worst nuclear disaster since Chernobyl, no deaths from radiation but over 100,000 people permanently evacuated
	Fertilizer Plant Explosion, Texas (2013)—15 killed, 160 injured, 150 buildings damaged or destroyed
	High-Speed Train Derailment, Spain (2013)—79 killed, worst Spanish rail disaster since 1972
	Garment Factory Building Collapse, Bangladesh (2013)—1129 killed, 2515 injured, 1 year after 2012 Dhaka garment factory fire killed 117

PICTURE 4.1 The Flixborough explosion. (From American Institute of Chemical Engineers, *Process Safety Management with Case Studies: Flixborough, Pasadena and Other Incidents,* American Institute of Chemical Engineers, New York, 1994. Copyright 1994 by the American Institute of Chemical Engineers; reproduced by permission of Center for Chemical Process Safety of AIChE.)

Going through a brief synopsis of the mishaps, think about your own individual plants and systems. Are there any similarities? What is your organization doing differently? How would you have averted these disasters? Something to also think about, unfortunately, many of the recent accidents in the figure earlier had similar safety management mistakes (Pictures 4.1 and 4.2).

The Flixborough Explosion

In our day and age, 24 h chemical plants are the norm. With a process plant operating a continuous flow with significant financial investments at risk, the management pressure to keep production going can be very intense at times. This was precisely the psychological atmosphere at Nypro's Flixborough Works (Parker et al., 1975; Sadee et al., 1976), situated between the villages of Flixborough and Amcotts, United Kingdom, on June 1, 1974.

In 1967, the ammonium sulfate plant had been converted to production of caprolactam, a product used to make nylon. A second plant on the site was completed in 1972, increasing the original 20,000 tons/year capacity to 70,000 tons/year, crowding an already congested plant. The plant control room was located near the cyclohexanone production facilities. Caprolactam is based on the oxidation of cyclohexane to cyclohexanone, a highly exothermic reaction at 8.8 kg/cm^2 (126 psig) and 155°C (311°F).

Because of the low conversion efficiency (6% per pass), six reactors were necessary. Temperature control was normally achieved by the evaporation and condensation of cyclohexane in the off-gas stream. The off-gas stream also contained nitrogen and unreacted oxygen from the air feed to the reactors. Two oxygen analyzers were

PICTURE 4.2   Aftermath at Flixborough. (From American Institute of Chemical Engineers, *Process Safety Management with Case Studies: Flixborough, Pasadena and Other Incidents*, American Institute of Chemical Engineers, New York, 1994. Copyright 1994 by the American Institute of Chemical Engineers; reproduced by permission of Center for Chemical Process Safety of AIChE.)

in the off-gas line, set at 4% (slightly below the lower explosive limit). An automatic nitrogen purge would be pumped into the system if the oxygen level rose above 4%.

A few days before March 27, cyclohexane was observed leaking from Reactor 5. The plant was shut down. As was common practice, river water was flowed over the outside of the reactor to cool it off for inspection. A half-inch-wide crack was found. On March 28, it was noted that the crack was actually 6 ft long. At this time, the management decided to take the vessel out of service and connect Reactor 4 directly to Reactor 6 to minimize downtime. The chief engineer of the plant had resigned some time prior to the incidence and thus was not part of the plant management team. A replacement had still not been found prior to the accident. The engineer coordinating the effort was an electrical engineer with little maintenance experience and no mechanical design background.

Reactors 4 and 6 were at different elevations, so a *dogleg* connection was needed between the two vessels. Even though the reactor inlet and outlet diameters were 28 in., the only spare piping available was a 20 in. diameter. Plant personnel considered this a routine plumbing job. No calculations were conducted, and the only drawing was a chalk schematic on the workshop floor. The bellows design guide was not consulted during the construction of the expansion joint, and thus, it was not designed to withstand the 38 tons of hydrostatic thrust. Temporary scaffolding supported the hastily completed bypass leg. Because of the excessive plant downtime (now 2 days), no pressure test was performed on the assembly before installation.

The plant was brought back up and the 4 kg/cm^2 leak test found a small leak. After its repair, a 9 kg/cm^2 leak test indicated no leaks. The plant was back on line on April 1. The plant had been shut down twice during the month of May to repair minor leaks in the system. On May 29, a leak was discovered in one of the vessels and the circulation and heat were stopped. However, the leaks seemed to *cure* themselves, and the plant restarted, in spite of the fact that there was an unexplained abnormal pressure rise in the system.

New leaks were found in the system at 4 a.m. on June 1. The leaks were left unrepaired because explosion-proof tools were locked up for the weekend. The system pressure increased rapidly and consumed an inordinate amount of nitrogen. By now, the plant had depleted most of its high-pressure nitrogen supply. However, the unusual pressure rise did not increase the temperature appreciably. Actually, Reactor 1 had reached only 110°C and Reactor 6, 50°C. More nitrogen would not arrive for at least 14 h.

It appears that there was too little nitrogen in the system to allow operating conditions to be reached. Because of this shortage, cyclohexane was circulated for several hours before oxidation could be initiated.

At 4:53 p.m. on June 1, the dogleg assembly ruptured, releasing approximately 30 tons of cyclohexane at 300°F. The resulting vapor cloud exploded with an equivalent force of 15–45 tons of TNT. Twenty-eight plant employees were immediately killed, thirty-six hospitalized, and hundreds more treated for minor injuries. The plant was completely destroyed, and 1821 houses and 167 shops were damaged. The fire burned for 10 days.

Lessons Learned

In reviewing what happened, it is important not to attach blame to individuals or groups, but to understand how you can avoid the same kind of mistakes in your plant or system. The first step in studying your SMS is to find weaknesses—not blame. Some interesting conclusions from the accident are addressed as follows. How many of them sound familiar?

The original plant had been modified for new and increased production. The converted plant was not designed for the original intent. The plant expansion forced the crowding of buildings too close together. Even the plant control room was next to the cyclohexanone production process. In the event of a disaster, as was the case, the control room is destroyed with everything else—not only are people killed, but also critical process control is immediately lost. If the control room had survived, it might have helped mitigate the effects of the original explosion. Vital records, necessary for accident investigation and reconstruction, would not have been lost.

Because the reaction process was very inefficient, six reactors were required. One possible design control would be to replace all six reactors with a single flow reactor, minimizing the inventory of hazardous materials, but maintaining the same production rate.

There was a single source of nitrogen for both production and emergency purging. When the nitrogen source was depleted, not only production was degraded, but also, more importantly, there was no safety purge if the mixture got above the 4% lower explosive limit. It is critical that safety systems (such as the nitrogen purge system) be totally and completely independent. That means separate supply and pumping stations for the nominal process and emergency situations.

River water was used routinely to cool the reactors down before visual inspection. The river water running next to the plant contained dissolved nitrates, which ultimately cause stress corrosion cracking. How materials are used in the entire system is very important. Obviously, plant engineering looked at how the process affected the reactor materials. However, they overlooked the use of untreated river water on the steel.

Reactor 5 was taken out of service because of the crack, but no one thought to inspect the other reactors to see if they suffered from the same *common cause* failure. In fact, there was no one in the management team with enough clout to stop the activities because the critical position of chief engineer had not been refilled. The electrical engineer in charge of coordinating the work had limited experience in maintenance and none in mechanical engineering. Actually, there was no qualified mechanical engineer on the team who could recognize the dangerous design fix.

Because the repair was considered a routine plumbing job, no torque, shear, or thrust calculations were conducted. The chalk diagram on the shop floor was not an adequate drawing. Of course, there was no design review or sign-off at a higher level of management. This is a classic example of the engineering team and management team not communicating.

The original bellows manufacturer's design guide was not consulted, nor did the team follow the applicable British Standards for piping design and testing. As was stated in Chapter 2, engineering standards are only a minimal requirement; however, in this example, if the standard had been followed, the accident may have been averted.

The dogleg assembly was not pressure tested before being attached to the *temporary* scaffolding. When leak and pressure tests were conducted on the system, they were done below the system maximum allowable working pressure of 11 kg/cm^2. It is important to remember that pressure tests are performed to check system integrity, but they must be conducted at high enough levels to actually test the system.

The plant was shut down twice in May, but safety tests failed to identify any problems. Likewise, when one leak seemed to *cure* itself, this did not raise any concern. Necessary tools for repairing any leaks, such as the *spark-proof* tools, were inaccessible to plant personnel.

And right before the explosion, the unusually low reactor temperatures and the large temperature variance between reactors did not stop the process for investigation.

The financial incentives and stress on plant operators to keep a plant on line without interruption is very strong. For this reason, it is extremely important that a good safety management structure be in place to identify hazards, control them, and mitigate their effects. A sound management organization will have the safety organization in the chain of command. As could be seen earlier, numerous errors were committed. But by far the worst was that there was no management structure that would assure that safety would be in the system at all times. In our next example, we will see how a silent safety organization can lead to catastrophic events.

Unfortunately, today, we are still making very similar mistakes that we saw at Flixborough decades ago and other disasters. For example, the accident investigation into the BP Texas City Refinery explosion and fire in 2005 found eerily similar accident causes and precursors:

- Worker trailers were located too close to hazardous operations.
- Supervisor was absent.
- Process safety not a priority—more safety focus on occupational safety.

- Strong financial pressures impacted operational safety.
- Abnormal start-up.
- Operator inattention.
- Poor communication during shift handover.
- Did not take advantage of opportunities to replace inefficient and more hazardous equipment.
- Some instrumentation did not work.
- Lack of follow-up or follow-through from previous incidents.

The Space Shuttle Challenger Accident

"The accident of Space Shuttle Challenger, Mission 51-L, interrupted for a time one of the most productive engineering, scientific and exploratory programs in history, evoked a wide range of deeply felt public responses" (Report, 1986a). "Flight of the Space Shuttle Challenger on Mission 51-L began at 11:38 a.m. Eastern Standard Time on January 28, 1986. It ended 73 s later in an explosive burn of hydrogen and oxygen propellants that destroyed the External Tank and exposed the Orbiter to severe aerodynamic loads that caused complete structural breakup. All seven crew members perished" (Report, 1986b).

After extensive investigation, analysis, testing, and hearings, the Rogers Commission (named after Chairman William P. Rogers), in their Report of the Presidential Commission on the Space Shuttle Challenger Accident, found that the *Challenger* disaster was caused by a failure in the aft field joint between the two lower segments of the right solid rocket motor. Hot gases *blew by* the two 0.280 in. thick O-rings impinging on the external tank, causing it to disintegrate due to severe aerodynamic load.

During the firing of the solid rocket motor, the joint tang and clevis assembly expanded, creating a gap. Normally, the two O-rings follow or track that expansion and prevent gases from escaping. Typically, primary and secondary gap increases are 0.029 and 0.017 in., respectively. On that day, however, the joint temperature was about 28°F (ambient air temperature at ground level was 36°F, 15° colder than any previous launch). The day of launch the joint was exposed to rain. Ice probably formed in the joint and prevented the O-ring from tracking properly, allowing a hot gas plume to impinge against the attachment ring, destroying its structural integrity and eventually weakening the external tank, causing it to break apart. "The failure was due to a faulty design unacceptably sensitive to a number of factors. These factors were the effects of temperature, physical dimensions, the character of materials, the effects of reusability, processing, and the reaction of the joint to dynamic loading" (Report, 1986c). However, what is of interest to us here is the management failure.

Silent Safety Program

There is no doubt that the decision to launch the *Challenger* was flawed. If the decision makers had known all of the facts, they most likely would not have launched. The crux of the management failure appears to have been communication. Engineering data and management judgments often conflicted, information was often incomplete or misleading, and the routes of communication in the NASA management structure did not allow this crucial information to be passed to the appropriate launch managers. If the decision process had been structured to emphasize safety concerns, then the waiving of launch constraints might have been avoided. Engineers were saddled with the requirement to prove that the Space Shuttle was unsafe—not that it was safe. Unlike American jurisprudence, designs should be guilty until proven safe. As with Flixborough, a process designed for success added a lot of pressure to the management system.

The *Challenger* launch had been postponed three times and scrubbed once from its initial launch date. The entire country was training its eyes on the "Teacher in Space" program, personified by Christa McAuliffe. A critical NASA communications satellite was to be deployed, and a short launch window existed for a Halley's Comet experiment. The weather had been uncooperative at emergency landing sites around the world. And of the most intangible, NASA was under severe budgetary and political pressures to prove that the Space Shuttle was *operational* and no longer an experimental vehicle.

The Rogers Commission found that the *modus operandi* of NASA engineers and managers was to try to solve technical problems in house rather than communicating problems to upper layers of management or other NASA centers or managers. They often failed to notice design flaws, at times failed to fix them, and then finally treated them as acceptable flight rules.

The most well-known problem was the solid rocket booster (SRB) joint design. The failure modes and effects analysis and critical items list (FMEA/CIL) treated the second O-ring as Crit 1R, a criticality of 1 and redundant—meaning that it was a redundant feature to the first O-ring and both their failures would result in loss of the shuttle. However, a 1980 NASA Space Shuttle verification and certification committee, with the purpose of assessing shuttle flightworthiness, found that the leak tests of the O-ring were in the wrong direction from actual motor pressurization. According to the Rogers Commission, NASA tests in 1977 found that joint rotation during launch could cause loss of the seating of the secondary seal. It was unknown if the second seal would hold if the first was lost and the motor case pressure exceeded 40% of its maximum expected operating pressure. But memos to that effect got lost in the bureaucracy.

In 1982, NASA did change the classification of the secondary O-ring from Crit 1R to Crit 1—meaning that it was not a redundant feature and its loss could cause loss of the shuttle. The successful use of the design on the *Titan* missile program was the rationale for retaining the design. However, psychologically, engineers still treated the second seal as redundant to the first. In fact, the discussion over its retention was so fierce (while the shuttle continued to fly) that the calculations predicting joint opening went to a *referee* for testing but were not concluded until after the *Challenger* accident.

Further flights found some erosion in the primary O-ring, but this just reinforced the confidence that the second seal would hold. The problem was tracked but not considered serious enough to warrant launch constraints. The Rogers Commission found (Report, 1986d) that, of the prior 24 shuttle flights to 51-L, 13 had seen some erosion or blow-by in at least one field joint. Of those 13, eight had erosion or blow-by in more than one field joint.

The April 29, 1985, launch found that the primary O-ring was compromised and the secondary had seen some erosion. The launch was at an ambient temperature of 53°F. This then became a launch constraint (the problem and disposition must be assessed before launch) for subsequent shuttle flights. However, it was waived for future flights, including 51-L. The rationale was, "it is safe to continue flying existing design as long as all joints are leak checked with a 200 psig stabilization pressure, are free of contamination in the seal areas and meet O-ring squeeze requirements" (Report, 1985). The commission reported that "tracking and continuing [sic] only anomalies that are 'outside the data base' of prior flight allowed major problems to be removed from, and lost by, the reporting system" (Report, 1986e).

The Rogers Commission was surprised that in the many hours of testimony, NASA's safety staff was never mentioned. They also discovered that there were no safety representatives on the mission management team that made key launch decisions on January 20. The investigators found (Bunn, 1986) four significant failures in the NASA safety program: *lack of problem reporting requirements, inadequate trend analysis, misrepresentation of criticality, and lack of involvement in critical discussions*. The commission also concluded that the safety team was not *independent* for creating a set of checks and balances in the system.

Analysis of the safety program found that personnel reductions in the safety office also affected flight safety. There was a lack of people adequately versed in the engineering designs involved. To further save money, the problem-reporting system, which originally tracked all problems, was streamlined to include only those problems that dealt with common hardware items or physical interface elements. This eventually led to eliminating reporting of flight safety problems, flight schedule problems, and some problem trends.

Cuts in the safety, reliability, and quality assurance staffs were justified because the shuttle was now operational and no longer an experimental vehicle. Human flight into space was considered routine.

The laws of physics destroyed the *Challenger*, killing all aboard, but a good SMS could have avoided this type of catastrophic accident. After studying the *Challenger* accident, we can see some important issues, salient points for a good safety program.

The first and most important is the *law of safety*, which is the following: *the design and its operation must be proven to be safe—it is not the engineer's task to prove that it is unsafe*. The hazard must follow the Napoleonic Code: guilty until proven innocent. There is a significant difference. It is psychologically more difficult to prove that something is safe than unsafe; that is why many engineering organizations tell the engineer to *prove that it is unsafe*. The reason is that the organization is building toward success, but as we have seen, that rush to completion leaves many holes. All safety programs, to be effective and to insulate themselves from liability claims, must have this as their foundation.

We all talk about communication and how important it is, but we don't always practice it. The safety program must have a vehicle for engineers, or any staff for that matter, to have a way to report and track their concerns. It is typical for humans to want to solve problems in house, not to air dirty laundry. Fear of ridicule and punishment is very strong. Engineers (nonsafety) should also have an appropriate problem-reporting system that allows them to enter data without reprimand. The tracking of problems (or, better yet, anomalies) in the system is critical to any safety program—really, to any program that wishes to consistently produce a superior product. The tracking system must follow not only the most obvious problems but also others that may not have obvious consequences. Trending data are crucial for success.

Another important factor is an independent safety staff. If the safety staff must report to the engineering director, then there is much room for unspoken pressure to cause them to accept the unacceptable. However, if the safety engineers are overly cautious, they will lose credibility. The safety engineer must be as well versed about the system as the systems engineering staff is. It is a very fine razor's edge that must be walked. All engineering organizations should have a safety engineer, or someone designated to follow through on any safety issues. That individual must also have the power to stop work if necessary and the clout to report to the appropriate management authority to do it.

NOTES FROM NICK'S FILE

I worked for many years as a system safety engineer on various NASA payloads that flew on the Space Shuttle and on expendable rockets. One battle I came up against constantly was proving that I understood the system design as well as the other NASA engineers. The only way that I could be credible to them was to spend the time and effort necessary to understand the design and operation of the payloads. It was the only way to convince them that my suggestions for hazard controls were viable. On one international payload, the project manager asked me to sit down and help them redesign the safety control system to adequately control the hazard. I had to be part of the team; otherwise, I would be an outsider and safety would suffer.

Challenger did prove the danger of continuing to operate as normal when uncertainty or ambiguous data are involved. Again, we should treat the issue as the following: prove that it is safe. If safety rules are established, then they should be followed and not routinely waived. If constraints are always waived, then they are inappropriate, or there is a blatant disregard for their necessity.

Again, it is very eerie to compare the similarities of the *Challenger* (1986) and *Columba* (2003) Space Shuttle accidents, especially with a focus on safety management. Like the previous example and the list of accidents in the Table 4.2, we must all be humble in recognizing that though we may understand the causes of accidents, we don't always know how to not repeat them. Some of the causes of the *Columbia* accident, which were similar to *Challenger*, are the following:

- Inadequate safety culture across NASA organizations and undervalue of system safety engineering and system safety engineers.
- Working-level engineers identified safety issues but did not get an adequate hearing in front of more senior management—barriers to dissent.
- Performance pressure on operations.
- Significant budget pressures and dwindling resources.

- Lack of independence of the safety organization.
- Misunderstanding and lack of adequate analysis of previous anomalies and incidents.
- Engineers did not push their safety concerns due to fear of impact to their careers.

And of course, one of the biggest issues for any organization is cost: reducing budgets affects the process. The trick is to be sure that shrinking costs do not impact what is important. All of these factors and others are important for an appropriate SMS; the remainder of this chapter will discuss how to design a good SMS that is appropriate for you.

4.2 DEVELOPING A ROBUST SAFETY MANAGEMENT SYSTEM

To create a robust SMS, we need to first understand what SMS is. The term has become quite common throughout the world, especially in the last 20 years. Most industries including transportation, petrochemical, mining, and manufacturing use the term and concept as the center of their safety program. Even in occupational health and safety SMS is ubiquitous. Government regulators also use the concept to ensure that their regulated industries are adequately managing safety and protecting the public. See Chapter 12 for its discussion on safety regulations and how SMS fit into it.

There are many ways to define SMS and most of them are fairly good. Unfortunately, many organizations (especially government regulators) sometimes are vague on what constitutes an SMS (especially a good or robust one). Chapter 2 states that *system safety engineering is a combination of management and systems engineering practices applied to the evaluation and reduction of risk in a system and its operation.* It is important to apply it during the entire program (or product) life cycle (not just at the end). On top of that, a sustainable management structure must be in place to successfully run system safety during that entire life cycle. With a management structure also comes management and leadership commitment and safety promotion, which is evidenced by a vigorous safety culture across the organization. But to be sure that safety risk is well managed, it means that the SMS must also incorporate a strong safety assurance system, which includes a process that assures that safety risks are identified, evaluated, controlled, verified, and tracked to closure in a closed-loop fashion. Performance must be measured by leading safety metrics with results feed back into the SMS. Bringing all this together, here is a succinct definition:

> Safety Management System (SMS) is a sustainable, formal and structured, enterprise-wide safety program that appropriately manages safety risk comprehensively of products and the systems that produce them.

Figure 4.1 illustrates the four primary parts of an SMS. All parts must work in tandem and be designed and operated together to have an effective SMS. If one or more parts are weak, the entire SMS will be dysfunctional, and the system will collapse. In reality, the four parts overlap quite a bit and therefore impact each other significantly.

FIGURE 4.1   (See color insert.) SMS.

The four primary SMS parts are the following:

1. *Safety governance*: The set of management practices and decision rights that administer and control how safety risk is evaluated, reduced, and managed. Safety governance is the set of levers that a company can operate to efficiently manage safety risk.
2. *Safety organization*: The group that owns and manages the safety process. However, they are *not* responsible for safety: *employees, frontline managers, and leaders are always responsible for safety.* The safety organization is responsible to manage the processes and tools to ensure that the company is identifying, evaluating, and managing safety risk appropriately.
3. *System safety program*: An adaptive, rigorous, and comprehensive set of engineering and management processes and infrastructure that ensures that safety risk is appropriately identified, evaluated, reduced, and managed. The system safety program involves people, processes, technology, and infrastructure to appropriately manage safety (of people, environment, and equipment) throughout the entire product or system life cycle. This is the *keystone* of any SMS and is its most critical element.
4. *Safety culture*: Is the complete suite of enterprise employee, management, and leadership attitudes toward safety risk. Without a strong safety culture, the system safety program cannot be successful and is just a piece of paper.

The rest of Section 4.2 will further elaborate.

4.2.1   ELEMENTS OF A SAFETY MANAGEMENT SYSTEM

Definitions are important to encapsulate a concept. But of course, it is important to decompose the concept into discrete elements so that an SMS can truly be understood and successfully applied. The author was very involved with the investigation of the 2003 Waterfall rail accident in New South Wales, Australia, and led the team that conducted a comprehensive audit of both the rail operator SMS and safety regulator's oversight program. A total of 29 elements were identified as critical to a robust SMS of an industry system or operator and 11 elements for evaluating the competency of government oversight of industry using SMS (Bahr, 2005a). Chapter 12 lists and discusses a slightly generalized version of 10 elements for evaluating government oversight programs.

Both element sets were based on an international best practices review across various high-hazard industries. Though most of the sources are from the transport industry, the elements are equally applicable to any industry. Some of the sources used for input included

- Discussion with SMS experts from multiple industries
- Qantas Airways system safety audit process (which is based on ICAO requirements)
- American Public Transit Association (APTA) Manual for the Development of Rail Transit System Safety Program Plans, 1999
- AS 4292 (all parts)—Australian Standard: *Railway Safety Management*, 1995
- New South Wales Rail Safety Act 1993, 2002, and Transport Administration (Safety and Reliability) Act 2003
- Civil Aviation Safety Authority (Australia) regulations—notice of proposed rulemaking part 119
- BlueScope Steel Occupational Health and Safety Management System
- U.S. Department of Transportation. Federal Transit Administration. *Handbook for Transit Safety and Security Certification.* DOT-FTA-MA-90-5006-02-01, 2002
- ISO 9001:2000 Quality Management Systems
- U.S. Department of Defense. Military Standard System Safety Program Requirements. Mil-Std 882C, 1993
- U.S. Occupational Safety and Health Administration, *Voluntary Protection Program.* 1996
- Center for Chemical Process Safety. *Guidelines for Hazard Evaluation Procedures.* New York: American Institute of Chemical Engineers
- U.S. National Aeronautics and Space Administration (NASA). NSTS 1700.7B—Safety Policy and Requirements for Payloads Using the Space Transportation System (STS), 1999
- McCormick, N.J., Reliability and Risk Analysis: Methods and Nuclear Power Applications. London, UK: Academic Press, 1981
- Bahr, N.J. *System Safety Engineering and Risk Assessment: A Practical Approach.* London, UK and New York City, United States: Taylor & Francis, 1997, First edition

TABLE 4.3
SMS Key Elements

Safety Governance

  1.0 Governance, Policy, and Objectives
  2.0 Mgt. Review, Responsibilities, Accountabilities, and Authorities
  3.0 Incident/Accident Reporting System
  4.0 Change Management
  5.0 Safety in the System Lifecycle

Safety Organization

  6.0 Safety Organization
  7.0 Safety Representative and Personnel
  8.0 Safety Committees

System Safety Program

  9.0 Hazard Identification and Risk Management
  10.0 System for Managing Requirements and Changes
  11.0 Document Control
  12.0 Record Control and Information Management
  13.0 Procurement of Goods and Services
  14.0 Management of Contracted Goods and Services
  15.0 Supply Chain Traceability of Goods and Services
  16.0 Internal Audit
  17.0 Incident/Accident Investigation
  18.0 Analysis and Monitoring
  19.0 Emergency Management and Response Procedures
  20.0 Medical Issues
  21.0 Human Factors
  22.0 Measuring Equipment and Calibration
  23.0 Equipment Maintenance

Safety Culture

  24.0 Management Commitment
  25.0 Safety Culture and Awareness
  26.0 Training, Education, and Competence
  27.0 Management and Staff Recruitment and Retention
  28.0 Customer Feedback
  29.0 *System Safety Program Plan* (*SSPP*)

James Reason, an internationally recognized SMS and human factors expert, noted that the SMS review methodology "…constitutes one of the most exhaustive, detailed and sophisticated examinations of an organisation's safety practices and thinking I have yet seen" (Donaldson and Edkins, 2004). Table 4.3 illustrates the 29 elements that are key to a robust SMS for industry. The list has been modified from the original Waterfall elements.

Note, Element 29, "System Safety Program Plan," is the compendium of engineering and management processes that manage all four primary SMS parts.

In designing a robust SMS, it is imperative to use the 29 elements described previously. Each of the elements has associated protocols that can be used to help conduct a review of one's SMS and its effectiveness. Here is a brief description of each of the 29 SMS elements, grouped into the four SMS primary parts.

4.2.1.1   Safety Governance

Safety governance, policy, and objectives: Defines the management, decision rights, and communications infrastructure to effectively manage safety risk; it provides a clear definition and governance, which all can understand, of how system safety is applied in the company. This also includes regular review updating (as appropriate) of the SMS itself.

Management review, responsibilities, accountabilities, and authorities: Describes how management reviews and approves all phases of the system life cycle for system safety compliance (internal requirements and external mandates) and appropriateness of meeting the SMS. It also defines management responsibilities, who are held accountable, how they are held accountable, and their authorities and decision rights.

Incident/accident reporting system: Is the process in which incidents or notifiable occurrences are reported and corrective action tracked to closure.

Change management: Describes how the company manages changes (market, operational, cultural, financial, etc.) in company operations to ensure that system safety is appropriately incorporated.

Safety in the system life cycle: Describes the gated review, approval, and iterative process that define how system safety is incorporated into the system life cycle, especially during design, development, implementation, and operations.

4.2.1.2   Safety Organization

Safety organization: Is the formal group within the company that manages the SMS and its tools. It is an enabling organization to the rest of the company activities. It is *not* responsible for safety but is the owner of the processes and tools.

Safety representative and personnel: Identified individuals in the company responsible for ensuring that the company follows a robust SMS (they are *not* responsible for safety, which is the responsibility of each and every employee). They also help facilitate efficient safety interface coordination and consultation.

Safety committees: Is the scope and purpose of a group of individuals in the company that review and support safety procedures and ensure that safety is incorporated into daily activities and integrated across the enterprise. They support the safety representatives. Safety committees can be ad hoc, temporary, or permanent, depending on the need and purpose.

4.2.1.3   System Safety Program

Hazard identification and risk management: Is the closed-loop process that identifies and evaluates hazards and prioritizes the corresponding risks so that that can be adequately managed.

System for managing requirements and changes: Is the closed-loop process that documents how requirements and changes in system design, operation, and other aspects of the system life cycle are documented. It also includes process controls to production. It works hand in hand with document and record control.

Document control: Is the company's system for configuration control of system design and operation.

Record control and information management: Is the closed-loop process that records how decisions are made and carried out and information is managed. It records how safety decisions and safety corrective actions are validated to be appropriate and verified to be in place. It also documents all safety compliance for internal and external compliance.

Procurement of goods and services: Is the formal procurement program and process for goods and services and managing their procurement risks. It is critical that system safety is incorporated early into the procurement process (especially in the tendering process and especially as part of tender requirements), along the entire value chain.

Management of contracted goods and services: Is the system that performs day-to-day management of outside vendors and members of their supply chain (contractor and subcontractor management) and ensures that they employ appropriate SMS techniques (within their contractor and subcontractor activities) to ensure adequate system safety protocols are in place. It is also important to have close safety collaboration with outsider vendors. Remember, even if contractor and subcontractors are at fault for an accident, the procuring company will still be held liable.

Supply chain traceability of goods and services: Is the program that actively documents how contracted goods and services are managed through the entire supply chain and demonstrates appropriate system safety protocols. For example, the safety and quality control of key raw materials used in manufacturing have important safety implications, especially once those materials are on-site within the factory gates. It is the reporting and tracking function of the management of contracted goods and services.

Internal audit: Is the process of independently reviewing programs to ensure that system safety protocols are in place and followed.

Incident/accident investigation: Defines how incidents, notifiable occurrences, and accidents are investigated and evaluated for corrective action.

Analysis and monitoring: Is the approach to evaluating and trending safety and risk profiles over time. This includes regulatory compliance, safety performance indicators, and appropriate process controls and asset management.

Emergency management and response procedures: Describes how the company will respond to and manage an emergency event. It also covers business continuity operations in the event of a significant disruption (which may or may not be a safety event).

Medical issues: Is the system that ensures that employee medical safeguards are in place and that employees are working in a healthy environment. Environmental protection and occupational health are described here. It also includes fatigue, drugs and alcohol controls, and general health and fitness programs. Note many companies have a separate and elaborate environmental protection program.

Human factors: Is the tool to ensure that the human–machine interface is accounted for and ensuring that the SMS appropriately accounts for how people actually work.

Measuring equipment and calibration: Discusses how control equipment of the system life cycle is defined and ensured to be within specification, especially as it relates to system safety.

Equipment maintenance: Is the system that ensures that equipment is maintained according to the appropriate periodicity, including scheduled, unscheduled, and emergency maintenance. It should also include design for ease of maintainability.

4.2.1.4 Safety Culture

Management commitment: Describes how executive and senior management demonstrate their commitment in time, money, and available resources to safety.

Safety culture and awareness: Is the process by which all levels of staff, management, and leadership understand the SMS and how safety should be managed.

Training, education, and competence: Is the company program that gives safety awareness and understanding of how to use system safety as part of each employee's daily activities and ensuring each employee's safety competence is appropriate for their job function.

Management and staff recruitment and retention: Is the process for hiring and retaining staff and ensuring they follow the company SMS.

Customer feedback: Is a formal process to invite input from customers and understand how system safety can be impacted. Customers are a company's eyes and ears and can be very helpful ensuring a robust SMS.

And the last but certainly not least of the 29 elements is the following:

System safety Program Plan (*SSPP*): Is the formal document that describes the SMS and how all the pieces fit together and operate within the company. It also includes general engineering and operational system safety requirements.

4.2.2 CONDUCTING A DIAGNOSTIC OF YOUR SAFETY MANAGEMENT SYSTEM

There are many terms that various industries use to evaluate their SMS program: audit, evaluation, program review, etc. But probably the most descriptive is to conduct a *diagnostic of your SMS*. As the term *diagnostic* implies in medicine, the purpose is to determine the health and well-being of an organization's SMS.

A diagnostic of a safety management system is a holistic approach that reviews an organization's management infrastructure and determines if the appropriate processes are adequate and in place to ensure that safety activities are carried out correctly and risks appropriately managed.

FIGURE 4.2   SMS diagnostic approach.

An SMS diagnostic is very useful to determine and measure an organization's SMS health, but its true value is to ensure that the appropriate management infrastructure is in place that will help ensure that safety is appropriately managed, which will result in the reduction of the impact (on people, operations, and direct and indirect costs) of safety incidents and accidents. Another benefit is that a documented and viable SMS can be used as evidence of an organization's *best faith* effort in effectively managing safety. Figure 4.2 illustrates the SMS diagnostic approach.

Phase 1: Develop diagnostic goals and schedule—It is critical to first determine what the SMS diagnostic is trying to measure and set specific goals and objectives to the diagnostic. This will ensure that the diagnostic is not over- or underscoped and is to actually measure what is required. It should also determine any risks to the schedule or diagnostic approach.

Phase 2: Compile background information—Before any SMS diagnostic starts, it is important to gather as much data as possible and to understand what constitutes the organization's SMS and how it fits into the organization.

Phase 3: Prepare diagnostic plan and template—Like an audit, an SMS diagnostic should follow a formal and methodical review plan. It should be well defined before the diagnostic begins. The remit of the diagnostic should be clearly defined. Documents that will be reviewed; staff, managers, and executives that will be interviewed; and operations that will be observed should be all defined before the diagnostic launches. Of course, the review template should be created.

Phase 4: Prereview meeting—The purpose is to bring the diagnostic team together and ensure that they understand the diagnostic plan, goals, and objectives, the template use, data sources, procedure for conducting the diagnostic, and how results will be collated and communicated to senior management. The team should be briefed on team member roles and, where appropriate, practice key (or controversial) interview questions. This prereview meeting should discuss prediagnostic, diagnostic, and postdiagnostic activities, preparations, and expectations.

Phase 5: Conduct on-site diagnostic—This is the actual SMS diagnostic. Of course, all applicable organizational parties should be informed before the diagnostic begins.

NOTES FROM NICK'S FILE

Many people think that only unannounced inspections and audits will truly turn up safety management failures. I've done lots and lots of SMS reviews and in my experience, it is very difficult to hide SMS problems. A good or bad SMS is so overt and part of daily operations (for better or worse) that it is actually pretty easy to determine an SMS well-being.

The diagnostic should focus on causes rather than symptoms. The on-site diagnostic steps are the following:

1. Review related documents (not just SMS-related but other documents that should mention how SMS may be applied, such as quality assurance plans).
2. Interview cross section of representative staff, managers, and executives that can give insight into the *true* SMS operations, not just what may be documented. Experience has shown that most SMS programs do not follow 100% what they have written into their plans. Only through interviews and observations can you determine if the written SMS follows the actual SMS.
3. Observe a cross section of key SMS activities or general activities where the SMS can be observed in action. It should include, where appropriate, selected working practices, project schedules, working conditions, and areas where problems or incidents/accidents have occurred in the past (this can help illuminate if corrective actions are effective).
4. Data gleaned from the prior three steps should be included in the diagnostic template.

Phase 6: Prepare final report and briefing—This final phase will summarize the diagnostic template worksheets, collate the data into understandable summaries, and recommend corrective actions and options for improving the SMS. The report and briefing to senior executives should identify improvement opportunities and prioritize them into actions and reasonable timelines (along with expected costs). Of course, if activities observed during the diagnostic are life-threatening, they should be immediately reported to senior managers for action. Key sections of the final report should include

- Executive summary of findings and recommendations
- Diagnostic scope
- Diagnostic findings, including description, impact, and evidence gathered
- Discussion of findings, description of what the findings mean, trends, implications to the organization, and implications to safety in general
- Conclusions and recommendations (these should include actions, timelines, and expected budget impact)

NOTES FROM NICK'S FILE

When we were conducting the Waterfall rail accident Safety Management System audit on the railway and regulator, we found too much data. The team was inundated. We met at the end of every day and held a *hot wash* reviewing each other's notes and impressions of the audit. This was very helpful to sort and prioritize what was important and share information.

Typically, an experienced team of around five people (with diverse experience and with backgrounds, such as engineering, operations, management) conducts the SMS diagnostic (of course, depending on organization size, complexity, and nature of the industry risk profile) in about a month or so, less if it is a small organization. If the diagnostic template has not been created, the team is new, and an SMS diagnostic has never been conducted before it could take longer. The actual on-site diagnostic activities should be around 1 week. Of course, if operations are expansive and large and span around the globe, it would be necessary to visit various sites to understand how they implement corporate SMS policies locally, and that will increase on-site timelines.

PRACTICAL TIPS AND BEST PRACTICE

- Many people jump into the SMS diagnostic too fast without the upfront work done first. Don't do that. Spend the time defining the diagnostic goals and objectives and carefully plan out all aspects of the diagnostic. People are busy and the company has work to do; don't waste people's time by being overly intrusive.
- Definitely don't believe everything that you read. It is *very* rare that an organization exactly follows their SMS procedures. Observing how SMS activities are completed and interviewing staff are critical to determining ground truth and not just *written truth*.
- Spend the time training the team. Too many audits and diagnostic teams are poorly trained and seem to run around without really knowing what to do.
- Be careful with data overload. A lot of data do not always equal useful information. Be selective and be practical in what you can gather.
- Don't worry about people *hiding problems*. It is very hard to hide structural SMS problems. The diagnostic approach looks at the SMS from so many angles that it would be very difficult to hide all evidence of poor SMS performance.
- Try to select a diverse diagnostic review team. If everyone has the exact same background, you will tend toward a groupthink of what is going on. Diversity in background, education, and knowledge of the organization will help ensure that the SMS is evaluated from many angles and not just one.
- Don't use the diagnostic like a checklist. This is not a *tick the box* exercise. The template is a tool to gather data, but the data are only data in the tool. It still must be sorted and time spent making sense of what it all means.
- Make sure to follow the data to the root cause. It is very easy to document symptoms and not the cause of the symptoms. You wouldn't want your doctor to do his diagnostic of your health and look only at what your symptoms and complaints are; we don't want the SMS diagnostic to do that either.

4.3   ORGANIZATIONAL MANAGEMENT AND SAFETY

Numerous organizations around the world do have good safety programs. Some of them, like NASA, have suffered tragic losses on the road to an effective program and continue to learn and evolve. However, we don't have to wait for a disaster to employ effective programs. One of the 1995 winners of the Innovations in American Government, sponsored by the Ford Foundation and the John F. Kennedy School of

Government at Harvard University, is the *Maine Top 200 Experimental Targeting Program, the Occupational Safety and Health Administration*. This is a program that "encourages employers to identify in-house workplace hazards and take corrective action before they lead to injury and illness" (Barr, 1995).

As the award implies, industry and government do not always have to be at loggerheads to support safety programs that are both effective and cost conscious. OSHA's Maine 200 program allows interested employers to choose between working with OSHA to reduce injuries and illnesses or face increased enforcement. The companies received support in developing good safety management programs; in return, they were given the lowest priority for inspection. Inspections occurred only if there were complaints about serious accidents. Since the program started in 1993, "employers self-identified more than fourteen times as many hazards as could have been cited by OSHA inspectors. Nearly six out of ten employers in the program have already reduced their injury and illness rates, even as inspections and fines are significantly diminished" (U.S. OSHA, 1996a). OSHA is expanding the program nationwide. The key is companies and OSHA working together, and the first step is company management commitment (U.S. OSHA, 2013).

4.3.1 MANAGEMENT COMMITMENT

Lack of management commitment directly affects the bottom line and market share. A good example is to look at how U.S. Federal Aviation Administration (FAA) actions can impact air carriers. The FAA bars all planes that do not meet FAA safety regulations from flying into the United States. The FAA has publicly listed countries that are not allowed into the lucrative U.S. market because of lax safety controls and management. Numerous countries are listed as conditional, and various have been listed as following U.S. safety rules (the most strict in the world) but being lax in other countries. Not only are certain air carriers disallowed to fly in the United States, but also non-U.S. flights drop in number because of the adverse international publicity. The lack of FAA safety approval not only denies air carriers lucrative U.S. routes, it also may bankrupt an airline.

For any organization to have an effective safety program, a safety culture must be created. See Section 4.9 for more details on safety culture. That means that from the CEO to the production-line operators, safety must be part of their consciousness. Safety has to be an everyday item. For this to truly take effect, the leaders must lead. If the top leader honestly believes that safety is important, and practices what he believes, then the safety program will be successful.

The upper management team at NASA's Johnson Space Center (JSC) (with over 12,000 private and public workers) in Houston, Texas, is one good example of how safety-conscious management can become. Of course, they already have an active civil service and contractor safety organization. What is of note is some of their more unorthodox methods of showing the community that they are truly involved. Obviously, great publicity doesn't hurt. The center's weekly newspaper has a full-page safety column, where safety experts answer workers' everyday questions. The column showcases the good and bad of their safety program and how things can be improved. JSC also holds an annual safety awareness day, when top-level civil service and contractor managers attend safety information programs; fire department and emergency response team demonstrations are conducted; safety motivational talks are given; various NASA contractors set up safety information booths; and guest speakers and music are brought in. The safety outreach program is now ubiquitous across all NASA centers, as well as their contractor base.

4.3.2 SUGGESTED IDEAS TO ENHANCE MANAGEMENT INVOLVEMENT

There is much said about getting employees to make safety part of their everyday culture. But getting senior executives and leadership involved is at the heart of it. It can be a daunting task to get one's management involved and committed to a sound safety program. Of course, if upper management only pays lip service, the program will get only lip service. Listed as follows are a few ideas on how to *motivate* senior executives and leadership to take safety seriously and put some money into it. For more information on how to motivate the workforce, see Section 4.9 (see Table 4.4).

And last, but certainly not the least, it's the ethical and right thing to do.

4.3.3 SAFETY MANAGEMENT SYSTEM ORGANIZATION

Even though this book is dedicated to how the practicing engineer can integrate safety into the engineering and management process, it is worthwhile to take a little time to look at the SMS organization and how safety engineers fit into a company's operation.

In some countries, the president of the national airline is the chief of the national civil air safety agency, a situation that prevents the necessary autonomy to be an effective oversight body. In fact, in the developing world, it is still common to see the transport regulator and operator under the same roof. The *Challenger* accident forced NASA to rethink its safety organization and its role in the NASA mission and realign the safety manager's reporting from the head of engineering to the head of NASA. The *Columbia* accident compelled NASA to revisit its bureaucracy to understand how better to integrate safety into all processes. The SMS organization must be independent enough from the engineering and operations departments to be able to ensure that the design of the product and the production process are safe. However, the system safety engineer must understand the system well enough to be able to influence the design and operations in a positive way.

Again, the *post-Challenger* NASA JSC safety program is a good example of how to merge the seemingly irreconcilable independent safety organization and the hands-on, systems-knowledgeable safety engineer. As indicated by the author (Bahr, 1988),

> The JSC ground test safety program functions as a two-tier system. The JSC Safety Division must be an integral team member from program concept through test (and ultimately through flight hardware deployment) while maintaining independent safety oversight. Test safety engineers must operate at the "nuts and bolts" level and fully understand all systems and subsystems that will be tested. They also work with members of various divisions to help reach the common goal of achieving a successful test. The Safety Division is completely autonomous of any test organization and reports to the JSC Director. This maintains the necessary independence that is required for appropriate oversight. Even though there can never be a perfect safety program, reconciling these mutually exclusive relationships is key to providing a meaningful safety function.

TABLE 4.4

Suggested Ways to Motivate Senior Leadership

Motivating Senior Leaders to Take Safety Seriously

Run the calculations of how much accidents cost in lost time, workers' compensation, insurance costs, lost product, schedule slip, lawsuits, inefficient use of resources, downtime, etc. See Section 14.3 for more details.	Being unsafe is terrible for the company image and bottom line. Loss of market share is one of the biggest factors.
Use your safety calculations to lower insurance costs. Show them to your insurance carrier, along with all the documentation that indicates how safety has been institutionalized.	Fewer injuries mean lower workers' compensation costs and lower medical costs.
Bring the company lawyer in to explain how a better safety record and a systematic approach to safety will protect the company from employee, community, and governmental lawsuits.	Accidents and near misses slow down the production schedule. And time is money.
A good safety program (including appropriate documentation) will make government oversight a breeze. And that means less money to spend on government audits.	Being known as a safety company also helps keep the regulators at bay.
Identify what the competition is doing in the way of safety and how that affects their competitiveness.	Demonstrate how using systematic and methodical analysis and a management approach to safety will also help identify other system deficiencies and inefficiencies.
Safety sells. Volvo is probably the best example. When people think of the car, they think of safety.	Show that minor perturbations in the system won't shut you down the way they used to.
Being safe is very good publicity and gives a positive corporate image, as much as being a part of the community does.	Tell top management, and show them, how employees are more productive and positive in their job attitude when they feel that they work in a safe environment.
Have someone from top management oversee a regular safety column in the company paper.	Hold a safety awareness day, and use it for the CEO and other top managers to get out and see the troops. This also helps to show that they will take time out of their busy schedule to address safety.
	Suggest that positive personnel evaluations be tied to good safety records.

The key operative words are the following: work at the *nuts and bolts* level; fully understand all systems and subsystems; and work with members of various divisions to achieve a successful test. These concepts are critical; if the system safety organization is obstructionist, then nothing will be accomplished except adversarial interactions within the company itself. Unfortunately, this all too often is the case. How many engineers are frustrated with their safety staff for shutting down their operations without even taking the time to understand what the implications are? Many readers can cite numerous horror stories of the safety engineer being totally ignorant of how the manufacturing process works but nonetheless still stops everything.

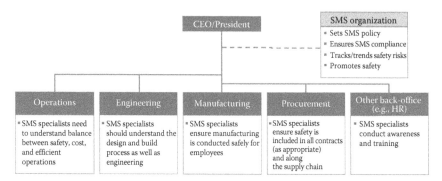

FIGURE 4.3   Simplified organizational structure.

Likewise, an over-accommodating safety organization does not perform its function and is only a facade to achieving real safety goals. Safety engineers must understand the system design as well as the design engineer, must know when to stand their ground, and must know when to yield to reach an appropriate consensus.

There are numerous ways to set up an SMS organization. Shown in the following is just one idea (though the most frequently used). Figure 4.3 is a simplified organizational chart. Obviously, most organizations are more complicated, but there really is no need to show large, complex organizations to illustrate where safety fits in. It is important to note is that the SMS organization does not work under any other company division. If system safety is not independent, then it will never be able to conduct honest internal critiques of how safe the product or system is.

In particular, the SMS organization must be independent from operations and engineering/manufacturing. The reason is that the SMS organization needs to be able to inform top leadership of the safety health of the company and its operations. If the SMS organization reports to a division head, then it is likely that intended or unintended pressures will influence the system safety process. The SMS organization must also cover the entire system life cycle.

The biggest challenge to an independent SMS organization is understanding the design, build, and maintenance of a system or product sufficiently to be able to challenge current practices. It is critical that the SMS organization report directly to the CEO/president of the organization. This is the only way that you can ensure that system safety is truly independent. An analogous situation is the financial services industry. The chief risk officer most times reports to the CEO. He or she is responsible for managing the financial institution's risk profile. The same should be done with the safety organization. The 29 SMS elements must also be incorporated into the SMS organization and the rest of the company processes. Section 4.4 details some of the key SMS organizational activities.

4.4   SYSTEM SAFETY PROGRAM: KEYSTONE TO A ROBUST SAFETY MANAGEMENT SYSTEM

As much as we all (especially engineers, who are a pragmatic and practical lot by nature) hate institutions and bureaucracy, they are necessary if we want to *design* safety in. We all seem to dislike bureaucracies because they appear to run amok and not help us complete our mission—which is building and running technological systems. Institutionalized oversight, such as a safety program or SMS, at times can be wrought with overindulgence and a lack of understanding of how engineering organizations and businesses function. Engineers many times feel that they are Joseph K., Franz Kafka's antihero in *The Trial*, fighting unseen forces, staving off unarticulated critiques and criticisms from secret bureaucrats. Like Joseph K., engineers feel that their safety organizations (including federal and local government oversight) are secretly accusing them of being callous, but they won't tell them exactly what they are doing wrong or how to fix it. And on top of that, they appear to hinder a pragmatic solution to the problem. This is the single most important negative that must be overcome in developing an effective SMS. In reality, it is not that difficult.

There is absolutely no reason why a good SMS, which is both effective and understanding, cannot be put in place. It really is quite easy. There are many sources for information in designing and running SMSs; numerous existing successful safety programs you can copy, tailor, and implement as your SMS; and various organizations willing to help. The rest of this chapter is dedicated to demonstrating how to set up a successful SMS, debug it, and keep it running. A sample SSPP (which is the cornerstone of the SMS) is included, distilled from various safety organizations in different industries. It is included here for you to use and tailor to your needs.

SMSs, whether developed for the aerospace, marine, food, or any other industry for that matter, all have certain concepts and elements in common. As explained in detail in Chapter 2, especially Figure 2.2, the safety organization within the SMS must follow a certain process. To refresh your memory, first, understand what you want to do (or protect, i.e., lives, hardware); then identify the hazards in the process; analyze the hazards; evaluate the risks (including the costs and benefits); control or mitigate the unacceptable risks; verify that it is in place; and finally, document and periodically review the entire system. The SMS institutionalizes this system safety process. Simply put, *if the safety organization does not apply the system safety process, then it is not a viable SMS.*

4.4.1 ELEMENTS OF A SYSTEM SAFETY PROGRAM

There are almost as many safety programs as there are technological programs. The vast majority, in the United States, however, is rooted in two primary sources: the U.S. OSHA and the U.S. military. As stated in prior chapters, the need for safety regulations arose in different ways in different countries. However, because so many of them rest on the foundation and promulgation of OSHA and the military standards, this chapter focuses only on those two. The 29 CFR 1910, U.S. Department of Labor, *OSHA Regulations for General Industry*, and Mil-Std 882, *Military Standard System Safety Program Requirements*, are the universally recognized and most often-cited safety standards.

The promulgation of the OSHA Act of 1970 has resulted in positive results for safety: the overall workplace death rate has been cut in half, brown lung disease has been virtually eliminated in the textile industry, and deaths from trench cave-ins have declined by 35%.

In spite of much debate in Washington, DC, about government safety oversight and how it hinders business, OSHA will not disappear. What is more likely is that it will be slowly reconfigured into a more modern bureaucracy. OSHA is already working toward a more flexible goal of increased safety with far fewer dollars. It also is seriously trying to forge new partnerships with states and businesses. The adversarial role is starting to change. The new OSHA is attempting fundamental organizational and operating change throughout the entire system. The cornerstone of that structure is the new OSHA partnership with states and businesses.

To enact this, OSHA is encouraging companies to put in place a strong and effective health and safety program. OSHA now is attempting an incentive-based approach. As the OSHA web page (U.S. OSHA, 1996b) states,

> To encourage employers to reduce death and injury by implementing safety and health programs on a continuing basis, OSHA will grant an array of penalty adjustments based on the vigor and effectiveness of the program.

> If, for example, OSHA finds, during the course of a workplace inspection that an employer has implemented a *superior* safety and health program, it will grant large reductions—up to 100%—in the penalties that would otherwise be assessed for violations found. For employers who have less effective programs in place but are making good-faith efforts, OSHA will grant a sliding scale of incentives. To qualify, the employer's program must include each of the recognized elements of a good safety and health program, which must be effective in practice and not just on paper. As evidence of a program's effectiveness, OSHA will expect to find that the workplace has a low injury and illness rate, that the employer has in fact found and fixed most hazards, that the workplace has not been cited in the past three years for the gravest type of violations, that the inspection was not prompted by an employee fatality or catastrophic accident, that any violations found in the current inspection are comparatively minor, and that the employer is prepared to correct any violations found.

An example is in the construction industry. Again, from the OSHA web page, if OSHA finds an effective safety program, OSHA will only inspect for the top four hazards: falls, electrocution, crushing injuries, and being struck by material or equipment. "If these hazards are well controlled, the inspector closes the inspection promptly and leaves the site. Conversely, where a safety and health program has not been established or is ineffective, OSHA conducts a complete site inspection, with full citations" (U.S. OSHA, 1996b).

For many years, the OSHA standard never really suggested very clearly what should be included in a safety management program. Safety management was based more on regulatory compliance of individual regulatory provisions, such as walking–working surfaces; means of egress; powered platforms, man-lifts, and vehicle-mounted work platforms; occupational health and environmental control; fire protection; electrical; and hazardous materials. Each of these work areas had its own safety program compliance system. There was no overall safety program management standard other than the OSHA Act of 1970, which states in part: "To assure safe and healthful working conditions" (Senate and House of Representatives of the United States of America in Congress, 1970). Emphasis was injury record keeping and regulatory notification, in other words reactive, not preventive, management. This is where a lot of the Kafkaesque feelings originated in industry.

In 1995, OSHA launched its New OSHA Initiative (U.S. OSHA, 1996b):

The new OSHA: OSHA will change its fundamental operating paradigm from one of command and control to one that provides employers a real choice between a partnership and a traditional enforcement relationship.

Common sense regulation: OSHA will change its approach to regulations by identifying clear and sensible priorities, focusing on key building block rules, eliminating or fixing out of date and confusing standards, and emphasizing interaction with business and labor in the development of rules.

Results, not red tape: OSHA will change the way it works on a day-to-day basis by focusing on the most serious hazards and the most dangerous workplaces and by insisting on results instead of red tape.

It goes on to further state that it will nationalize the *Maine Top 200* program. OSHA has further evolved its government–industry into a strategic partnership (U.S. OSHA, 2013):

- In a partnership, OSHA enters into an extended, voluntary, cooperative relationship with groups of employers, employees, and employee representatives (sometimes including other stakeholders and sometimes involving only one employer) in order to encourage, assist, and recognize their efforts to eliminate serious hazards and achieve a high level of worker safety and health.
- Partnering with OSHA is appropriate for the many employers who want to do the right thing but need help in strengthening worker safety and health at their worksites. Within the OSHA Strategic Partnership Program (OSPP), management, labor, and OSHA are proving that old adversaries can become new allies committed to cooperative solutions to the problems of worker safety and health.
- OSHA and its partners can identify a common goal, develop plans for achieving that goal, and cooperate in implementation.
- OSHA's interest in cooperative partnerships in no way reduces its ongoing commitment to enforcing the requirements of the Occupational Safety and Health Act. While employers in partnership remain subject to OSHA enforcement, the OSPP provides them an opportunity to work cooperatively with OSHA and workers to identify the most serious workplace hazards, develop workplace-appropriate safety and health management systems, share resources, and find effective ways to reduce worker injuries, illnesses, and deaths.
- Most of the worksites that have chosen to partner with OSHA are small businesses.

Many businesses felt that OSHA can come in and close them down for no apparent reason. It was difficult for businesses to be proactive in safety management and therefore avoid OSHA audits. The 1992 publication of 1910.119, *Process Safety*

Management of Highly Hazardous Chemicals, changed all that. It is a safety management program based on OSHA's hazardous waste regulations. Although it pertains to management of hazardous chemicals, it is extremely useful for other industries. OSHA says the major parts of a safety program should include (U.S. OSHA, 1992)

- Employee involvement in process safety management
- Process safety information (right to know)
- Process hazard analysis
- Operating procedures and practices (the need for written procedures that incorporate safety controls)
- Employee training
- Contractors
- Pre-start-up safety
- Mechanical integrity (re: maintenance programs)
- Nonroutine work authorizations
- Managing change
- Investigation of incidents
- Emergency preparedness
- Compliance audits

Look at Figure 2.2 again. You can see that the OSHA standard approaches the SMS much better than it did before, but it still doesn't really give you a way to manage the risk in the system or process. The process hazard analysis clause comes closest to that concept, but it still does not take into account the risks of the system—it only looks at the hazards. As we remember from Chapter 2, the hazards may be enormous, but the risk may be infinitesimal (e.g., another planet hitting the earth). Risk management is a necessary ingredient in the appropriate use of system safety. That is how you can still be safe *and* keep costs at a reasonable level.

State OSHA programs have further developed the concept. For example, the Maryland Occupational Safety and Health office states that the steps to developing an effective program are (Report, n.d.) the following:

Step 1: Develop a plan of action that includes management and employee involvement.
Step 2: Designate a person to be responsible for safety and health.
Step 3: Determine the safety and health requirements for your particular workplace and operation.
Step 4: Conduct a hazard assessment of the workplace.
Step 5: Correct identified hazards.
Step 6: Keep your workplace hazard-free; develop emergency procedures.
Step 7: Train employees in safety and health.
Step 8: Keep your program up to date and effective.

The U.S. military uses Mil-Std-882, which is a much more complete definition and application of SMS (see Table 4.5). Don't be put off by the size of this matrix; the key is to tailor your particular operation to what is appropriate and drop the rest.

TABLE 4.5

Application Matrix for System Program Development

Task	Title	Task type	MSA	TD	EMD	P&D	Q&S
				Program Phase			
101	Hazard Identification and Mitigation Effort Using the System Safety Methodology	MGT	G	G	G	G	G
102	System Safety Program Plan	MGT	G	G	G	G	G
103	Hazard Management Plan	MGT	G	G	G	G	G
104	Support of Government Reviews/Audits	MGT	G	G	G	G	G
105	Integrated Product Team/Working Group Support	MGT	G	G	G	G	G
106	Hazard Tracking System	MGT	S	G	G	G	G
107	Hazard Management Progress Report	MGT	G	G	G	G	G
108	Hazardous Material Management Plan	MGT	S	G	G	G	G
201	Preliminary Hazard List	ENG	G	S	S	GC	GC
202	Preliminary Hazard Analysis	ENG	S	G	S	GC	GC
203	System Requirements Hazard Analysis	ENG	G	G	G	GC	GC
204	Subsystem Hazard Analysis	ENG	n/a	G	G	GC	GC
205	System Hazard Analysis	ENG	n/a	G	G	GC	GC
206	Operating and Support Hazard Analysis	ENG	S	G	G	G	S
207	Health Hazard Assessment	ENG	S	G	G	GC	GC
208	Functional Hazard Analysis	ENG	S	G	G	GC	GC
209	System-of-Systems Hazard Analysis	ENG	n/a	G	G	GC	GC
210	Environmental Hazard Analysis	ENG	S	G	G	G	GC
301	Safety Assessment Report	ENG	S	G	G	G	S
302	Hazard Management Assessment Report	ENG	S	G	G	G	S
303	Test and Evaluation Participation	ENG	G	G	G	G	S
304	Review of Engineering Change Proposals, Change Notices, Deficiency Reports, Mishaps, and Requests for Deviation/Waiver	ENG	n/a	S	G	G	G
401	Safety Verification	ENG	n/a	S	G	G	S
402	Explosive Hazard Classification Data	ENG	n/a	S	G	G	GC
403	Explosive Ordnance Disposal Data	ENG	n/a	S	G	G	S

Source: U.S. Department of Defense, Military standard, system safety program requirements, Mil-Std-882E, U.S. Department of Defense, Washington, DC, 2012, A-90.

Notes:

Task type—ENG, engineering; MGT, management.

Program phase—MSA, material solution analysis; TD, technology development; EMD, engineering and manufacturing development; P&D, production and deployment; O&S, operations and support.

Applicability codes—G, generally applicable; S, selectively applicable; GC, generally applicable to design change; n/a, not applicable.

The purpose of the 100-series tasks is to set up an effective system safety program in the corporate or organizational structure. These tasks establish the detailed program elements but also set up the safety organization, lines of organizational communication (as they pertain to system safety), and program milestones and establish the authority for resolution of identified hazards.

The SSPP is written by the contractor (or organization) to

- Describe program scope and objectives
- Describe the system safety organization
- Explain system safety program milestones
- Address general system safety requirements and criteria
- Describe the hazard analysis techniques and methodologies to be used
- Describe the approach for collecting system safety data
- Describe the safety verification process
- Explain the audit program
- List the type of safety training conducted for each category of employee
- Explain the incident reporting system (accident reporting)
- Identify the system safety interfaces between all the other engineering and corporate disciplines

The other 100-series tasks address in more detail the major elements of the SSPP. The 200-series tasks explain engineering safety analyses to be performed. These tasks are actual system safety engineering analytical tools used to identify hazards and their controls in any technological system. Chapters 5 through 9 detail the numerous kinds of system safety engineering analytical tools available and give application examples.

The 300-series tasks focus on evaluating the risks in a program and the safety review of the engineering design process. The 400-series tasks (excluding the self-explanatory explosive safety tasks) concentrate on system requirements compliance and verification of safety controls.

In October 1972, the U.S. government promulgated the Consumer Product Safety Act. Where OSHA is concerned primarily with workplace safety, the Consumer Product Safety Commission is concerned with product safety. Specifically, the purpose is to "protect the public against unreasonable injury risks, assist consumers in evaluating the product safety, develop uniform safety standards and promote research into the causes and prevention of product-related deaths, illnesses and injuries" (Kitzes, 1991). With so many different requirements for worker and product safety, how do you set up a good in-house program? The following section answers that question.

4.4.2 Setting Up a System Safety Program

A system safety program documents the SMS. Table 4.6 is a sample SMS SSPP and is a best practice and practical compilation of the 29 SMS Elements; OSHA; Mil-Std-882C; Consumer Product Safety Commission; OSHA Voluntary Protection Programs; numerous UK, European, and Australian national SMS guidelines; and various industry SMS SSPPs. Use all sections listed in the following.

TABLE 4.6
Sample SMS SSPP

Part I: Safety Management System Program and Policy Administration

CEO Statement on Management Commitment to SMS

Introduction—Policy and Purpose
Purpose
Scope
Policy
Objectives
References

Safety Governance

Safety Governance
Safety Policy
Safety Objectives
Description of Safety Governance Structure
Safety Decision Rights
Review and Updating the SMS

Management Review
Management Functions and System Safety
Management Responsibilities, Accountabilities, and Authorities
Management Review and the System Life Cycle
SMS Review and Compliance
Safety Waiver Review and Approval

Incident/Accident and Near-Miss Reporting System
Notifiable Incident/Accident Reporting Process
Other Incident/Accident Reporting
Near-Miss Occurrence Reporting and Tracking
Incident/Accident and Near-Miss Record Keeping
Corrective Action Verification Process, Tracking, and Record Keeping

Change Management
Change and System Safety
Managing External Changes to Company Operations (e.g., Market Forces, Regulatory)
Managing Internal Changes to Company Operations (e.g., Operational, Financial, Cultural)

Safety in the System Life Cycle
System Safety in the System Life Cycle
Gated Review and Approval Process and Authorities

Safety Organization

Safety Organization
Safety Organization Description and Organization Chart
Safety Organization Functions, Responsibilities, and Authorities
Safety Engineering Staff
Safety and Engineering Standards and Best Practices

Safety Representative and Personnel
Safety Representative Functions, Responsibilities, and Authorities
Safety Representatives and the Company SMS

(Continued)

TABLE 4.6 (*Continued*)
Sample SMS SSPP

Part I: Safety Management System Program and Policy Administration

Safety Representative Interface with Company Divisions

Safety Committees

Safety, Health, and Environment Committees

Safety Committee Functions, Responsibilities, and Authorities

Permanent Safety Committees

Ad Hoc and Temporary Safety Committees

System Safety Program

Hazard Identification and Risk Management
    System Safety Methodology
        Hazard Reduction Precedence
        Hazard Inspection and Abatement
        Hazard Resolution Process
        Closed-loop Hazard Tracking
    Safety Assessment
        System Safety Life Cycle Safety Activities
        System Safety Analyses
        Safety Trend Analysis
    Safety Verification Tasks
        Design Verification
        Inputs to Specifications
        Acquisition Tests
        Operational Tests
        Safety Tests
        Inspections
    Risk Management
        Risk Management and the SMS
        Risk Assessment
        Risk Assessment Criteria

System for Managing Requirements and Changes

Change Order Review and Signature Authority

Change Management Board

Document Control

Company Documentation Control System

Design and Operations Configuration Management Documentation

Record Control and Information Management

Safety Decision Documentation and Control System

Information Management, Communications, and Safety

Documenting Safety Corrective Actions and Safety Verification Tracking

Safety Compliance Documentation Control

Procurement of Goods and Services

Company Procurement Process

System Safety in the Procurement Process

(*Continued*)

TABLE 4.6 (*Continued*)
Sample SMS SSPP

Part I: Safety Management System Program and Policy Administration

Management of Contracted Goods and Services
Contractor and Subcontractor Safety Programs Review and Evaluation
Oversight and Methods
Self-Evaluation
Evaluation of Transient Contractors

Supply Chain Traceability of Goods and Services
System Safety and the Supply Chain

Internal Audit
Safety Audit Process
Safety Audit Schedule
Other Company Audits that Impact Safety
External Safety Audits
Safety Audit Results Tracking and Incorporation into SMS

Incident/Accident Investigation
Reporting the Accident (Internal)
Forming the Investigation Board
Documenting the Accident
Corrective Action
Verification of Corrective Action
Updating the SMS after an Accident
Informing the Public

Analysis and Monitoring
Safety Performance Indicators
Safety and Asset Management
Safety and Production Process Controls
Evaluating and Trending Safety and Risk Data

Emergency Management
Emergency Preparedness, Response, and Business Continuity Analysis
Emergency Response and Community Services
Emergency Management and Business Continuity Training
System Safety and Business Continuity
Emergency Management and Business Continuity Plans

Medical Issues
Environmental Protection and Occupational Health
    Hazardous Materials and Chemicals Control
    Environmental Compliance Control
    Occupational Health
    Interfaces with Company Environmental Protection Program
Fatigue
Drugs and Alcohol Policy
General Employee Health and Fitness Program

(*Continued*)

TABLE 4.6 (*Continued*)
Sample SMS SSPP

Part I: Safety Management System Program and Policy Administration

Human Factors
People, Process, and Technology Interfaces
Human Factors and SMS
Human Factors Considerations for Designing, Operating, and Maintaining Equipment

Measuring Equipment and Calibration
Process Control Equipment and Impact on System Safety
Maintaining Equipment Calibration

Equipment Maintenance
Safety Implications to Equipment and System Maintenance
Scheduled, Unscheduled, and Emergency Maintenance

Safety Culture

Management Commitment
    CEO System Safety Commitment Letter (Signed)
    Executive Leadership and Management Safety Commitment Letter (Signed)
    Management Commitment Program to Safety and the SMS
    Management Commitment in Time, Money, and Resources to Safety

Safety Culture and Awareness
    Safety Culture and the SMS
    Safety Culture and All Employees
    Safety Awareness in the Workplace
    Disseminating Safety Information

Training, Education, and Competence
    Safety Promotion Programs
    Safety Competencies in Each Job Function
    Employee's Right-to-Know
    Employee Safety Awareness and Training
    Personnel Certification for Hazardous Operations
    Internal and External Safety Training and Education

Management and Staff Recruitment and Retention
    Recruitment, Retention, and Safety

Customer Feedback
    Customer Feedback and Safety Improvements
    Soliciting, Analyzing, Trending, and Tracking Customer Feedback
    Customer Feedback into the SMS

Part II: Operational Safety Requirements

Operational System Safety Plans and Procedures
Hazardous Operations Procedures
Construction Safety Procedures
Safety Design Requirements
Human Factors Design Requirements

Emergency Response Procedures
Emergency Response Procedures
Business Continuity Procedures (with Safety Impact)

However, the actual customizing to your own operations is what will drive the level of detail needed in the plan. A large chemical processing plant will produce a fairly large document(s), whereas a small textile mill may only have one small manual. Sections can be just a paragraph long if the operations are not overly complex or risky. But both will be complete and will cover safety for workers and consumers.

Your company should use this SMS SSPP to describe how system safety is maintained for all processes in the company and document the complete SMS. The SSPP is divided into two sections: Safety Management System Program and Policy Administration and Operational Safety Requirements. The first part describes the safety organization, safety philosophy, analysis, control methodology, and program control. The second part details the actual safety design and operational requirements and procedures.

NOTES FROM NICK'S FILE

I was evaluating a transport company's SMS and found it was great on paper. But when we interviewed employees, we found that nobody followed it, not even the CEO. Remember: write down in the SMS System Safety Program Plan only what you really do.

4.4.2.1 CEO Statement on Management Commitment to SMS

This is a formal commitment from the CEO that he or she will ensure that SMS is given the adequate resources and attention that it deserves. This is critical because it sets the standard by which the company will operate in respect to safety. If the CEO and his or her team are not onboard, the SMS will not work nor be sustainable.

4.4.2.2 Introduction: Policy and Purpose

This is a general section, which describes the purpose, scope, policy, and objectives of the SSPP document. It describes why it was created, the scope of the SSPP, how it is updated, control authorities, and the document objectives. Many companies will include a list of relevant references such as codes, standards, and other company documents that impact the SSPP. A list of other company and government references is very useful for the reader to locate more information and to verify that the SSPP complies with appropriate regulations.

4.4.2.3 Safety Governance

The aim of this section is to describe the company's safety governance structure and processes and specifically its stated and well-articulated safety policy and philosophy. This section should also clearly state the objectives of the system safety program. Who and how safety decisions are approved needs to be overtly declared; this so that all can understand that safety decisions are made consciously and not by default or inaction. Safety decision making needs to be part of the fabric of company operations and not just an afterthought or only *check the box* compliance. This section also describes the process and periodicity of updating the SMS and republishing the SSPP.

4.4.2.4 Management Review

How senior executives and line managers interact in the system safety process is critical for a sustainable SMS. Their responsibilities need to be clearly defined and how they are held accountable. For example, it is imperative that safety performance is part of each employee's job description, in particular, the managers. The company must explain the manager's authorities to ensure a safe workplace and how safe products for consumers are produced. This is incorporated through a formal management review and approval process that documents management review in the system life cycle. Managers are responsible for complying with the SMS and in extraordinary circumstances when safety waivers that deviate from the SMS but be implemented that they have had formal management review and approval.

NOTES FROM NICK'S FILE

A huge challenge I faced with one aerospace company was getting them to review all the safety waivers that had been given over the years and redesign the SMS so that these lessons learned could be incorporated. Waivers or deviations from safety rules should not be common. They should be very rare and given only during extreme circumstances.

4.4.2.5 Incident/Accident and Near-Miss Reporting System

The SSPP has two incident/accident sections. The first one is dedicated to explaining how accidents, especially those that are notifiable (that are required by law to report to regulators) are reported. This section delineates the formal reporting and tracking process (internal company processes as well as external reporting processes). Not only should the program plan address government notification to OSHA, EPA, or other agencies, it should also explain how the company would investigate itself. But not only should accidents be investigated and reported, much can be learned from near misses—those occurrences that almost resulted in an accident but did not. This section also describes how corrective actions are documented and validated to be adequate and verified to be in place. Chapter 11 discusses this further.

4.4.2.6 Change Management

Companies are constantly under change pressures. Some are external such as market forces (change in the price of oil) or changes in regulatory law, or even market competition. But companies also face internal or cultural changes over time. Company operations periodically change: company reorganizations, consolidations, acquisitions, or even downsizing can create havoc and impact safety. This section describes how the company recognizes that external and internal change is inevitable and addresses how changes can impact system safety and what the company will do when these types of changes are taking place. Note that the SSPP section, System for Managing Requirements and Changes, is different and focuses on internal operational requirements changes.

4.4.2.7 Safety in the System Life Cycle

This section explains how the company understands system safety and how it fits into the system life cycle and should be written with very practical language regarding company views toward how system safety is managed during the entire life cycle. Here, discussion of the gated review and approval process and authorities is addressed.

4.4.2.8 System Safety Organization

This section describes the safety organization functions, its responsibilities, and its authorities. It elucidates where the system safety organization is in the company hierarchy, demonstrating both safety's independence and integration into the company. This is a hard balancing act. The safety organization should be independent and report to the most senior executive, yet it has to be technically competent to be able to interact with the more technical parts of the company and support its operations. The safety engineering staff is a group of safety engineers dedicated to overall *system safety*. In other words, they are *at-large* professionals whose responsibility is to assure that the company is traveling down the right path to effective system safety management. The safety engineering staff has a very difficult job. They must be invested with enough authority to stop production if an imminent catastrophic hazard exists, but they must also be judicious enough to know when to use their power and when not to. The perennial problem of safety engineers is when to acquiesce and when to dig in their heels. If company safety policies and objectives are sufficiently clear, the safety engineer will almost always know where he or she stands.

The use of engineering and safety standards should be clearly explained. Company, local, national, and international standards for the various stages of product development and process management should be cited. We all know that there are times that it is undesirable to follow certain standards or even company policy. This is not always bad. In fact, at times that flexibility is what allows a company to compete. What is important is to have a policy and regulation waiver process, which is followed, as described in the Management Review section of the SSPP. Safety organization representatives should sit on the waiver approval committees. The process should be well documented and reviewed periodically. As stated earlier, if the same requirement is waived regularly, should it still be a requirement, or are you just becoming too lax? This is precisely what led to the *Challenger* accident and again to the *Columbia* accident.

4.4.2.9 Safety Representative and Personnel

Many firms have found that having a safety officer (a nonsafety professional suffices) assigned to each major work area and/or work group on the property can be very effective. The safety officer is responsible for assuring that safety rules are enforced in his or her area. More importantly, the safety officer, being a line worker, understands the production process and thus can best implement the appropriate safety controls and not useless controls that may meet regulations but do not protect people and property. These individuals then attend regularly scheduled safety working groups (once a month or once a quarter), where they can study safety issues and how they affect the company, review current policies and procedures and determine if they still make sense, and fine-tune the system they already have in place.

However, the corporate system safety engineer should chair the group. This section describes the safety representative functions, responsibilities, authorities, how they are part of the SMS, and how they interface and interact across company divisions. It should always be emphasized that safety is everyone's business and responsibility. Many times, company managers or employees will want the safety organization or safety representative to be the responsible party. That should never be the case. The people directly involved with the process and responsible for managing a particular process are always responsible for safety. They cannot hide from this responsibility.

An important point to note is that companies must allow any worker to stop the production line, without suffering corporate retribution, if someone's life (or bodily injury) is at risk. Workers should not have to wait until the safety engineer notices them and their plight. Many accidents are avoided by the immediate intervention of a coworker during a dangerous situation. This is precisely why we want to analyze to the best of our ability the safety implications of all our processes. Companies that do not do this will surely suffer serious reprimands and fines from the regulator, not to mention possible employee lawsuits.

4.4.2.10   Safety Committees

Many times, safety (or environment, health, and safety) committees are formed to manage or oversee a particularly hazardous condition or operation. This section describes how safety committees are formed, their functions, responsibilities, and authorities. Many companies will have permanent safety committees (e.g., safety audit or inspection teams). But they will also have ad hoc or temporary committees to solve a discrete problem or issue. What is important to remember is that safety committees are multidisciplined and not just the safety engineers from the safety organization. The best safety committees are comprised of many different divisions that impact a particular issue. For example, if there are safety issues related to mixing hazardous materials together, the safety committee involved in reviewing or auditing the process should include procurement (to make sure that contracts specify the correct safety requirements), engineering (to be sure that processes are designed safely), receiving and warehousing (to make sure that hazardous and nonhazmat goods are stored appropriately), and production and quality control (to ensure that processes are followed safely and safety steps aren't skipped).

4.4.2.11   Hazard Identification and Risk Management

This is the heart of the SSPP and what really makes the SMS tick. The System Safety Methodology is very important. This is where the company lays out the specifics of what constitutes a hazard, how it is identified and controlled, and how residual risks are mitigated. As shown in Chapter 2, the Hazard Reduction Precedence should be described in detail, explaining how the company applies that philosophy. Hazard inspections and control are also described.

The hazard resolution process is a critical component of the SMS. If employees and OSHA (or other regulatory) inspectors are to feel secure that the firm is following a sane safety program, it is important to explain how hazards are resolved. How are hazards controlled? Who has signature authority verifying that the hazard is controlled adequately? Many corporations use the system safety committee as a controlling body.

It is important to maintain an institutionalized closed-loop hazard tracking system. This not only allows the firm to perform trend analyses to study accident patterns (or near-miss patterns), it also provides documentation of not only accident data but also hazard data for regulatory compliance. If a company can show an effective hazard tracking and resolution system to regulatory inspectors, such as OSHA, most of the battle is already won.

The Safety Assessment section summarizes what safety activities are performed and when. Also, the types of safety analysis techniques used are listed. Chapters 5 through 9 detail numerous system safety analysis tools. The reader may wish to pick a few of the techniques addressed in Chapters 5 through 9 and briefly describe them in this section.

The key to safety is to design it in. To fine-tune the SMS, you need to periodically review how work is performed. This part of the SSPP explains how that trend analysis will be performed and what will be done with the results.

The Safety Verification Tasks assure that the hazard control is validated to be appropriate and verified to be in place. Verification of the implementation of system safety processes into company operations is very important, not only for regulatory audit and inspection survival but also to be sure that money is spent wisely. Verification is done through review and approval of the design process, input to specifications that call for verification schemes, and various tests (i.e., acquisition, operational, safety) to physically test hazard control adequacy. Of course, physical inspections (destructive and nondestructive) are part of the verification process. This is extremely important in product safety, but also very important in any plant safety. Explaining the testing of safety-critical systems and testing systems to assure they operate safely is very important.

The Risk Management process is important not only for government regulators; it is even more important to the firm itself. Effectively managing risks is what makes a system safety program cost-effective and worthwhile. It is also what gives additional benefit to the company by saving money. This section delineates how risks are assessed and what the corporate acceptance criteria are. As stated earlier, risk is the likelihood of an event occurring coupled with the severity of the consequences. Specific details should be given here stating how the elements of risk are defined, how they fit into the SMS, the risk assessment process, and the criteria for assessing risks. Chapters 13 and 14 discuss risk assessments in detail.

4.4.2.12 System for Managing Requirements and Changes

This section is different from the Change Management section of the SSPP. This section focuses on how engineering and operations requirements are controlled and changes to the requirements are reviewed, tracked, and approved in a closed-loop process. In particular, it explains configuration control in the system life cycle, especially during the design process. Who reviews and authorizes engineering and management change orders? This is an important question; failure here was a significant contributing factor to the Flixborough accident and many other accidents. When material changes are made to the engineering design or operations (or other critical) processes, a change management board should be created comprised of senior managers, and they should review and approve the change order.

4.4.2.13   Document Control

It is important for a company to have a document control system. This system is a formal system that controls how documents are updated so that there is always a record of how designs or operations evolve over time. The document control system is the company's primary document control system and is not just for safety information management. That is covered in the next section.

4.4.2.14   Record Control and Information Management

This process works in tandem with the document control system described earlier but focuses on safety-relevant information. It is critical not just for compliance but good safety management to clearly document safety decision making, safety communications, safety corrective action and any other safety compliance material that must be shared with outside auditors. Though this seems redundant to the document control system or should be subsumed into it, it should not. Safety compliance documentation and control is so important not just for demonstrating safety competence to regulators but also avoiding lawsuits. It is worthwhile to have a very clear safety record control and information management process. Not all companies have a separate section in their SSPP for safety compliance. However, if the fear of OSHA audits is great or there is a history of safety problems, it may be useful to have a separate section describing the documentation process. Many companies also use this section to state clearly how documentation is controlled and audits are performed.

4.4.2.15   Procurement of Goods and Services

System safety must be instituted into the facility and product acquisition process (including retrofitting and minor modifications). This section delineates the system safety milestones and safety products in the procurement process. This section should describe how the general procurement process works and how system safety should be involved in this process. For example, the safety organization should make sure that safety requirements are part of the procurement tendering process and well defined for bidders and vendors. Poor safety requirements definition has led not just to contract disputes but more importantly has many times led to poor safety management and ultimately to contributing to accidents.

4.4.2.16   Management of Contracted Goods and Services

Many companies outsource using contractors and subcontractors. Their safety programs have a direct impact on yours. This section details how the company reviews and maintains oversight of its contractor safety program. Because many subcontractors are small operations, a large plant may require on-site contractors to also participate in self-evaluations. How those programs are administered should be described. Companies must also evaluate short-term transient contractors and their safety programs. Family members of injured transient contractors have sued many firms. Remember, the only company that people remember related to the 2010 Deepwater Horizon oil spill in the Gulf of Mexico is BP and not any of the other contractors or companies that were on-site that day. The site owner is always responsible for any activities conducted on his or her property. But he or she is also responsible for any goods or services that he or she contracts for his or her operations.

4.4.2.17 Supply Chain Traceability of Goods and Services

All companies worry about their supply chain and how they impact their operations. This section describes how system safety should be involved in the supply chain. It should describe how vendors along the entire supply chain are selected, vetted, and managed and safety requirements are assured.

NOTES FROM NICK'S FILE

Working with anhydrous ammonia, I learned the hard way how important understanding where my valve soft goods came from and exactly what material they were made from. The soft goods supplier claimed that they were anhydrous ammonia compatible; after testing and a big mess in our piping system, I found out they were not. Lessons learned—validate any uncertainties in the supply chain that could impact safety ahead of time.

4.4.2.18 Internal Audit

Internal safety audits are critical to ensure that safety systems are in place and operating as designed. This section describes the audit process, periodicity, objectives, and audit team. Safety audits may be conducted in tandem with other company audits (e.g., quality assurance), but you should be careful not to go overboard with audits and over fatigue the company or employees by being in constant audit mode. Regular safety audits can also be part of the hazard control verification process described earlier. Of course, it is imperative that this section address how safety audit results are plugged back into the SMS. Interestingly enough, the author's experience has found that many accidents could have been prevented if safety audit results were actually implemented.

4.4.2.19 Incident/Accident Investigation

Clearly stating how accidents are reported, investigated, documented, and corrected is a cornerstone to any system safety management program. An earlier section in the SSPP describes how incidents and accidents are reported and how records are kept. This section focuses on conducting the investigation. Many companies define different levels of accidents and their corresponding investigation board mandates. An institutionalized corrective action process is also critical to ensure that the accident is not repeated. This section describes who should be part of the investigation board and the detailed process for investigating the accident. If the accident is serious enough, it may be important to have an independent division investigate separately from those who were involved in the event. The accident should investigate to the root cause and not be satisfied with just symptomatic evidence. Of course, the investigation will give recommendations for avoiding repeat accidents and corrective actions to current operations. But if these results are not incorporated into the SMS, they are for naught. How the public is informed is very important and should be discussed here. The public has a right to know of safety-critical events that could impact their lives, but the company must think judiciously about how to release the information and who can speak for the company. Chapter 11 further discusses investigating incidents and accidents.

4.4.2.20   Analysis and Monitoring

As Socrates said, "an unexamined life is not worth living," so too an unexamined SMS is not worth having. This section describes how safety data are analyzed and safety performance indicators are set and monitored. It is imperative that safety performance indicators are forward leaning, which means that they are predictive safety indicators not just historical or lagging indicators. Both are needed to truly understand how well the SMS is actually working. Safety in asset management is part of this process and is described here. Likewise, understanding the deviation of production process controls and their impacts on safety is important. This means that evaluating and trending these data and other safety and risk data are important.

4.4.2.21   Emergency Management

Local and federal laws require companies to have an emergency preparedness plan explaining how hazardous chemical spills will be contained. In fact, every company should have some sort of emergency preparedness plan and business continuity plan that addresses how the company will respond to both natural and man-made disasters or major operational disruptions. This section discusses what constitutes an emergency, how company processes are analyzed to identify potential emergencies, what the contingency plans are, what kind of community services may be required to respond adequately, and how the firm practices its emergency response. Many times, this is a fairly large plan and is published under a separate title. Also, usually, companies publish pamphlets and posters telling employees what to do in the event of certain emergencies.

Though business continuity plans are not necessarily safety focused, many times, they have huge safety implications and for that reason should be described here (though they may be published separately). Emergency management plans are more tactical than business continuity plans and usually focus on immediate safety implications to employees or the public. Actual emergency response procedures are also usually published separately and are collected in Part II: Operational Safety Requirements.

4.4.2.22   Medical Issues

Environmental protection and occupational health corporate policies and procedures are documented here. Because environmental protection laws are very involved, many companies will have a separate environmental management systems volume. In that case, this section would describe how that system would interface with the SMS. Also, the SSPP should clearly outline the hazardous materials and chemicals control program and processes used in the plant. Remember to discuss hazardous material handling from procurement to disposal.

Occupational health programs such as noise control, lighting, and indoor air quality are also addressed in this section. Occupational health should not be separate from the system safety process. Protecting workers in all facets is important. Most companies cite or excerpt the OSHA requirement that pertains to the particular health hazards in the plant and may include it in this section. Fatigue, drugs, and alcohol are very important factors that can impact safety. This section describes the policy and processes related to these items and how they are controlled and monitored. General employee health and fitness also is important and is explained here if the company has any special programs.

4.4.2.23 Human Factors

This can be viewed as the combination of people, processes (the procedures they follow, both management and technical), and the technology that they operate to perform their job. If the balance between the three is not well thought out, it can create serious safety problems. This section describes how the company views human factors within the SMS and addresses how it is considered during designing, operating, and maintaining equipment and systems. Chapter 8 discusses human factors techniques in more detail.

4.4.2.24 Measuring Equipment and Calibration

This section describes the impact that process control equipment can have on system safety. It details activities that are performed to keep equipment and instrumentation within calibration and how that is verified.

4.4.2.25 Equipment Maintenance

Keeping equipment properly maintained is critical to safety. This section addresses the safety implications, how systems are maintained, and the periodicity of that maintenance, especially for safety-critical equipment.

4.4.2.26 Management Commitment

Though the first page of the SSPP is the CEO Statement on Management Commitment to SMS, this section will include not just the CEO's statement but also a signed letter by him or her delineating his or her commitment. It will also contain his or her executive leadership and management team commitment letters and describes the management commitment program to safety in general and the SMS in particular. Of course, none of this matters, if time, money, and resources are not adequately allocated to the SMS. This section will also address how that is demonstrated.

4.4.2.27 Safety Culture and Awareness

Much of this fits into the employee right-to-know law. Part of designing safety into a system is ensuring that employees know which operations are inherently safe and which are not. This section should not only address that information but also detail what is considered good safety culture and how safety culture is central to the SMS (and *everyone's* business). This section also discusses the importance of formal safety awareness programs for the workplace, as well as how safety information is disseminated to all employees.

4.4.2.28 Training, Education, and Competence

People want to be safe. They don't want to hurt themselves or others. But what happens is that many times people don't know how to do that. This section addresses the importance of training, educating, and making all staff competent in safety. Formal safety promotion programs are described here. Also, this section describes the process for determining and ensuring that each job function and each individual has been properly evaluated to determine their safety competency. Where there are gaps, the education and training program is used to bring employees up to the correct level of safety competence.

A very useful component of safety awareness is instituting a personnel certification process for hazardous operations. OSHA requires it. Typical certification programs are hazardous material handling, confined space entry, and crane operations. Chapter 10 discusses both safety training and awareness.

4.4.2.29 Management and Staff Recruitment and Retention

Many may question why this is important to safety and part of the SMS. Many accident precursors have been directly tied to severe staff turnover. This section describes how company staff retention impacts safety and any remedial actions that are taken when turnover exceeds the norm.

4.4.2.30 Customer Feedback

Again, many may wonder how this impacts safety. But customer feedback has proven to significantly impact how well companies operate, not just outwardly to them, but also internally. Customers can help serve as eyes and ears toward safety, and their feedback improves safety. This section details how customer data are solicited, analyzed, trended, tracked, and then fed back into the SMS.

Part II of the SSPP is the operational safety requirements, procedures, and plans. Large, complex organizations and operations may wish to publish this as a separate, stand-alone document(s). A small operation can incorporate this into the single program management plan.

4.4.2.31 Operational System Safety Plans and Procedures

This section is a compendium of individual, tactical, short system safety plans or procedures that detail specific hazardous operations and processes and how they are protected. Some typical operations addressed are

- Cryogenic agents and refrigerants
- Pressure vessels and pressurized systems
- Mechanical equipment
- Electrical equipment
- Electrical lockout/tagout
- Batteries
- Confined space entry
- Work in high places
- Toxic substances
- Flammables and combustibles
- Radiation
- Material-handling equipment
- Hand tools

- Welding and cutting operations
- Hazardous material handling
- Explosives handling
- Automated lines, systems, or processes
- Fire protection
- Facility safety
- Construction safety

Detailed design and test safety requirements and procedures are compiled in this section. For example, certain parts of the National Electric Code that are especially important to electrical safety are cited here. Any ASME requirements used in pressure system design and operation are also addressed here. The types of available personnel protection clothing and equipment supplied should also be described.

This section also includes construction and operations safety procedures, safety design requirements, and human factors design requirements and considerations.

4.4.2.32   Emergency Response Procedures

Detailed emergency response procedures, call plans and call trees, and other emergency response actions are included here. Though business continuity is related to emergency management, those business continuity actions and activities that have safety implications are also included here.

4.4.3   EVALUATING CONTRACTORS AND SUBCONTRACTORS

A June 28, 1993, explosion and flash fire occurred during a resin reactor building maintenance project…Two of those hospitalized were plant employees, two were contract workers…A major oil company is currently implementing settlement terms reached regarding OSHA citations alleging that the company did not periodically evaluate its contractors' performance…A major petroleum company is contesting an OSHA citation resulting from an explosion that killed a contract worker…According to OSHA, the oil company had not developed safe work practices to control hazards during hot work operations (Roughton, 1995).

No matter which way you look at it, employers can be (and are being) held liable for injuries and accidents on their premises. OSHA may cite the employer even if his or her own employees are not exposed to the hazard, but contract employees are.

The three major OSHA standards most frequently cited to employers regarding contractor safety are *Process Safety Management of Highly Hazardous Chemical* (29 CFR 1910.110), *Control of Hazardous Energy (Lockout/Tagout)* (29 CFR 1910.147), and *Hazard Communication Standard* (29 CFR 1910.1200). These regulations pertain to the hiring of outside contractors and subcontractors to perform repair work, plant modification, equipment maintenance, etc. Many countries around the world have their comparable standards.

Contracts written to outside firms should specify that the contractor meet certain minimal safety and health requirements. Contractor system safety should be included in the construction process from prebid to project acceptance. The contract should also require that the contractor enforce an effective safety and health program. Bid solicitation documents should require that safety costs be included in the actual bid. Sometimes, you may wish to even enumerate specific hazards that contractors must control. Also, it is critical to emphasize to bidders that evaluation of their written safety and health program will be part of the buyer's evaluation of the bid proposal. Because the buyer is still responsible for what occurs on his property and under his supervision, it is important to ensure that contractors and subcontractors have adequate safety programs.

Some of the aspects you should study in evaluating a bidder's safety program are the following:

- Past safety record—Examine contractors' workers' compensation and OSHA experience.
- SMS—Does the contractor even have one, is it formal, and what does it look like?
- Safety management structure—Who is responsible for safety and health, and how are policies communicated and enforced?
- Work procedures used and how safety is incorporated into the procedures.
- System for reporting accidents.
- Process used to identify and control workplace hazards.
- Safety requirements for subcontractors.
- On-site safety representative—The individual who will enforce compliance of safety procedures.
- When and how employee safety meetings are conducted—Contractors should hold a safety briefing before each hazardous operation starts.
- Formal safety program—Description of the contractor's safety program (this should be their SMS description).

There are almost endless ways to evaluate contractor safety—of course, evaluation of contractor safety programs is totally dependent on the actual operations. Every industry (e.g., construction of petrochemical refineries or operations of subways) does it a little differently, but almost all have similar content. If the contract is large and long term, the company should ensure that the contractor develops and implements a comprehensive safety program similar to the SMS described in Section 4.3.2. If the contractor is transient, contracting for less than 1 year, then an adapted program can be used. What is the most important is that the company designs their contractor/subcontractor safety program evaluation process specifically to company operations and in compliance with the company SMS and of course local and federal laws, regulations, and best practices. Though a contractor safety program may not be as involved as a company SMS, it must not violate any company SMS safety policies. Table 4.7 shows how NASA evaluates outside transient industrial contractors (National Aeronautics and Space Administration, 1985). Though the form is old and is specific for occupational safety and health (not to be confused with managing the operations of NASA launches), it is an example of some of the concepts that should be considered.

TABLE 4.7
Evaluation and Effectiveness Criteria Checklist for Safety Programs

	Satisfactory/ Unsatisfactory	Comments

Program Elements
1. Written policy, purpose, scope
2. Statement of objectives
3. Standards compliance
4. Responsibility/accountability assigned
5. Implementation activities
   A. Training/certification
   B. Written procedures
   C. Hazardous operations
   D. Hazardous materials
   E. Hazardous operations permits
   F. Personal protective program
   G. Mishap/near-miss investigations
   H. Mishap/near-miss records and reports
   I. Fire/explosion safety
   J. Emergency preparedness
   K. Procurement and subcontract safety
   L. Safety representatives on-site
6. Implementation of supplemental requirements
   A. _____
   B. _____
   C. _____

Program Implementation
1. Hazard analyses on record
2. Risk priorities stated
3. Measurable results defined
4. Self-evaluation performed
5. Modifications effected

Program Accomplishments
1. Hazard analyses and surveys
   A. Near-miss investigations, causes corrected
   B. Hazardous operations, materials, permits surveyed and documented
   C. Regular inspections documented
2. Changes and deviations analyzed and modifications documented
3. Trend analysis
   A. Uncorrected vs. corrected discrepancies for the period
   B. Lost-time injuries and illnesses
   C. Severity of injuries and illnesses
   D. Performance compared to other periods
   E. Performance compared to other companies, same industry, or similar operations
4. Employee involvement
   A. Safety meetings and on-site safety briefings
   B. Inspections, evaluations, job hazard analyses
   C. Training and certification

(Continued)

TABLE 4.7 (*Continued*)
Evaluation and Effectiveness Criteria Checklist for Safety Programs

	Satisfactory/ Unsatisfactory	Comments

5. Motivational and promotional projects/participation
   A. Awards and recognition
   B. Seasonal campaigns
   C. Bulletins and newsletters
   D. Management communications
6. OSHA compliance
   A. OSHA citations, serious violations, penalties
   B. OSHA log, poster

Source: National Aeronautics and Space Administration, Johnson Space Center safety manual, JSCM 1700D, National Aeronautics and Space Administration, Houston, TX, 1985, 1-13-4.

4.4.4 Emergency Preparedness Programs

Even though we hope we never have to respond to accidents or other emergencies, it is important (and mandated by OSHA and EPA regulations and local laws in the United States) to have emergency preparedness plans. The emergency preparedness plan is an outgrowth of the system safety process. After conducting hazard analyses, the company-planning group (in concert with local authorities, such as police, fire, and emergency medical services) should take steps to prevent accidents, communicate risks to the public, and develop an emergency plan. Once the system has been well analyzed, the company can take that information and develop emergency response plans to mitigate any mishaps that could create hazardous situations to people and the environment. The emergency preparedness plan is exactly what it states—a plan of action developed a priori to any emergency to detail exactly what is to be done in the event of different accident scenarios. The plan is a stand-alone document, but its process is part of the system safety process. If the system safety process has changed, then the plan should be reviewed and appropriately updated.

A typical emergency response plan should include

1. Introduction
2. Policies and interorganizational agreements
3. Emergency telephone list
4. Response functions
5. Containment and cleanup
6. Documentation
7. Procedures for testing and updating plan
8. Summary of safety analyses
9. Resources and references
10. Emergency procedures
    a. Preparation
    b. Initial response

    c. Assessment/response by emergency response personnel
    d. Decision-making aids
    e. Hazard control
    f. Support operations
    g. Emergency care
    h. Extrication
    i. Evacuation
    j. Debriefing
    k. Restoring normal operations

Usually, companies publish under separate cover the actual emergency procedures that employees are to follow in the event of an accident. These are step-by-step instructions that employees are to follow in order to respond safely to the emergency. The emergency procedures are located at the worksite and should be practiced at least annually, and the entire emergency program should be reviewed every 3 years.

PRACTICAL TIPS AND BEST PRACTICE

Some important ideas to remember when developing an emergency preparedness plan are the following:

- Include not only plant engineers but also public officials and local community representatives in your planning.
- Document the entire emergency planning process.
- Review and update periodically.
- Contact local equipment rental agencies to have the necessary type of cleanup equipment accessible at a moment's notice.
- Develop a clear line of authority for handling the emergency response, cleanup, and return to normal operations.
- Develop a quick phone list of emergency contact phone numbers (including home phone numbers) for emergency response units, company officials, state and federal regulatory agencies (e.g., EPA), and plant engineering managers.
- Practice and test the emergency procedures and conduct self-critiques on a regular basis.
- Contact other plants near you that may be affected by any accident you may precipitate and, likewise, any accident they may cause that could impact you.
- Develop your emergency response plans together, sharing resources whenever possible.
- Review emergency power systems and communication systems to ensure that if utility services are lost you are still in a safe situation.
- Include public information and community relations in your planning.
- Review the plan to ensure that there are adequate protective actions in place for the public.

Remember, if you have not conducted a methodical safety analysis of your process, it is impossible to be sure that you have an adequate response to the most likely types of emergencies.

It is important to mention that many companies have started to implement business continuity management programs. These programs are to help the company prepare, respond, and recover through significant business disruptions including accidents. The ISO 22301, *Business Continuity Management* Standard, is designed to help companies recover vital business functions during or after a crisis. Obviously, emergency response plans for safety would be a subset of business continuity. ISO 22320:2011, *Societal Security—Emergency Management—Requirements for Incident Response*, is an international standard of how to respond to any kind of emergency—terrorist, natural disaster, or accidental.

4.4.5   CASE STUDY: HOW A LEADING GLOBAL PERSONAL CARE PRODUCTS COMPANY CREATED A BEST PRACTICE SAFETY PROGRAM

Share price was high, the company was respected, the corporation took safety seriously, and had a decent safety record. Then an accident occurred in one of the factories, and corporate leadership was shocked that it could even occur. Luckily, no one was killed, though several were severely injured. The global company took the wake-up call seriously and decided to act upon it. In brief, this is their story.

4.4.5.1   So What Happened?

Safety inspections were frequent in this factory, in fact across all factories globally. The corporate office had corporate safety standards that were promulgated worldwide and audited regularly. Factory safety managers, though not high on the *importance ladder*, did have basic respect in their factories, and in many cases (but not all), safety managers worked closely with production managers. There was a strong management commitment to safety, mostly post accident but not preventative in nature. Internal, forensic investigation found the cause of the accident to be operator error in filling a mixing tank of various products. Seemingly, they seemed to be doing all the right things, so what went wrong? The corporation decided to find out and then go and fix it.

4.4.5.2   Safety Management System Review, Not Accident Investigation

The company quickly looked internally and found a few things that shocked them:

- Upon review, it turned out that there had been other, *apparently* unrelated accidents at other factories around the world.
- But the factory audit records said that these factories were in safety program compliance and met local laws and standards.
- Also, many of the safety programs were not fully implemented, but it was unclear why.

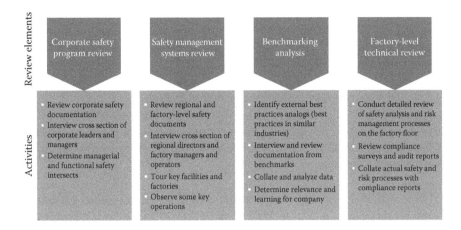

FIGURE 4.4   Understanding how my SMS is working.

The corporate office decided to take a four-step process to understand why the safety program seemed to be misaligned to corporate goals and not performing at the appropriate level. Though it is a four-step process, most of the steps were conducted concurrently (see Figure 4.4).

The review results fall into three key areas: organizational design and performance, safety processes and communications, and risk-based management system.

1. *Organizational design and performance*: The safety program was buried in the corporate structure and at the factory level. It didn't have the importance to generate adequate safety policies at the corporate level or ensure their implementation on the factory floor. There was a lack of safety expertise at all levels of the company, and there was insufficient demarcation and communication of safety performance accountability. The safety metrics measured the wrong performance indicators to be of use.

2. *Safety processes and communications*: Formal safety processes were inadequate with insufficient regular safety planning. Safety goals were not regularly set and when they were set frequently weren't met. Insufficient budget was allocated for safety, and the safety team didn't have the visibility to advocate for more. There was insufficient safety information sharing between factories or even with the corporate office.

3. *Risk-based management system*: There was no structured risk-based approach to safety management. Frequently, policies and procedures were confused and unclear. Only *known* risks were managed, and the approach did not allow for *unknown* risks to be identified. There was nominal corporate support at the factory level to improve implementation and to ensure implementation through appropriate audits and reviews (of both policies and implementation).

According to outside sources, the company had a good safety record. The review found pockets of best-in-class safety performance. But the challenge was that the company did not have an integrated approach, was not unified in its management of safety, and was not sufficiently self-reflective of its performance and self-learning. But probably of most concern, the company did not have a formal, methodical, and uniform safety risk assessment process.

4.4.5.3 Designing a New Safety Management System

The company decided to leverage its internal best practices, and the results of its external benchmarking to capitalize on its existing successful programs yet build attributes toward a more proactive SMS and safety culture. They focused on the following actions:

Commitment: The company was already committed and everyone knew it. What they had to do was to build on this goodwill and use it as a platform to promulgate their new SMS.

Policies and procedures: Needed to be clearly segregated and roles and responsibilities clearly articulated. Corporate documentation was redesigned to more appropriately communicate safety policies and give guidance on how to design safety procedures at the local level to meet corporate requirements.

Communications: Informal networks were strengthened, and new communications channels were created at all organizational levels.

Planning and goal setting: Review found many great plans and goals but results not always accrued. The corporate office led teams to coordinate at the regional level to develop realistic plans and goals, performance tracking metrics, and forums to share results and lessons learned.

Safety culture: Because it is a high-hazard industry, safety culture was already good, but the corporate office wanted to improve it and make it sustainable. Awareness programs were expanded and factory-level staff was invited to corporate headquarters to brief senior leaders on how to improve safety culture worldwide.

Technical expertise: A safety career track was created, and the importance of the safety manager was elevated, including a new executive safety director reporting to corporate senior leadership. Other safety managers were regularly trained, and even frontline factory workers were given targeted safety training.

Organizational design: The corporate safety office was strengthened, a safety division created within the production vertical, and regional safety directors created worldwide.

Risk assessment: A formal, structured, and methodical safety risk assessment process was developed and embedded on the factory floor. The safety risk assessment process was designed to be proactive and identify leading safety risk indicators that would point to new safety problems, not just based on past history (lagging safety indicators). Results were then used to create leading safety risk indicators—metrics that will identify where I will have future safety problems.

4.4.5.4 Another Word on Developing a Risk Assessment Process

One of the key findings of the initial review after the accident was that there was no formal, methodical safety risk assessment process. So the company set out to develop one. They decided that it needed to be a uniform and consistent methodology that could be used worldwide, across all product lines. Though product lines vary greatly, how their safety risks should be assessed needed to be very methodical and repeatable. Also, it was important that safety hazards and their associated risks must be communicated to the corporate level in a uniform and consistent manner so that risks could be compared across all operations.

The company decided to have two levels of safety risk assessments: the first is a general or high level that would look at top level risks across the factory or region and the second is a detailed safety risk assessment process that would dive deep into the causal factors of the individual safety hazards. The second would be applied based on the highest-level risks identified by the high-level risk assessment. This is a very useful approach because it allows you to first identify the primary hazards and risks across all operations but then dig deeply into the ones of highest concern—spend the time and money to understand those risks and how to best manage them.

The corporate office wanted to be sure that the new safety risk assessment process would work and was sustainable. So they piloted it on various product lines in different countries. This gave the company the ability to see if similar safety risks would be identified, determine if risks were evaluated consistently, and test the applicability and ease of the tool to actual safety hazards. It was a resounding success and quickly led to the promulgation of the methodology across all operations worldwide.

4.4.5.5 Corporate Results

Though this case study is a true story, the details need to be kept sparse due to the confidentiality. However, some general corporate results can be reported:

- The company reduced an already low incident rate by another 75% over the following decade.
- Many factories have not had any lost-time injuries for the last few years.
- The company has set annual aggressive safety goals based on the prior years' experience. For example, an additional double-digit safety performance improvement goal has been publicly set to be met over the next 10 years.
- Incorporated safety and safety performance into their corporate social responsibility program—they are not just saying that they will be safe but also putting their reputation on the line, daily, and in the mass media.
- Very active internationally at safety conferences to share lessons learned and promote better safety results for all. In fact, they benchmark their safety performance and share best practices at various international forums.

And most interestingly, on the financial side:

- Company sales continue to rise (even during a downturn in the global economy and in their market).
- Profitability reached record levels.
- Even though the sector slowed, the company outperformed the market.

NOTES FROM NICK'S FILE

I picked up three lessons learned from working this project that I have carried with me for many years:

- Even if you think you have a good safety program you may not really have one, you have to prove it.
- A robust, thorough, systematic, and repeatable safety risk assessment process is core to any safety program. If you can't identify hazards, assess them, and then figure out how to control them and track to closure; then you will never, ever have a good safety program.
- An SMS is the superstructure that all companies *must* have to adequately control their safety risks and do it to realize true benefits and not just save lives but also save money.

A lesson learned for all of us from this case study is that safety does not have to negatively impact corporate financials—in fact, the opposite is true. Safety saves money! Safety is not just common sense—it is also good business.

4.5   COMMON MISTAKES IN IMPLEMENTING SAFETY PROGRAMS

The following are some common mistakes found in implementing SMSs:

- Lack of follow-through in the safety process.
- No closed-loop safety process.
- No verification that hazard controls are adequate and in place.
- Failure to document the hazard control and safety processes.
- Accident investigations don't go all the way to determine root causes.
- Inadequate safety analysis.
- Wrong hazard controls.
- Did not follow the hazard reduction precedence.
- Lack of communication with all parties, including management and the various technical people.
- Poor safety culture.
- Lack of staff and management competency in safety.
- SMS does not follow how safety really is managed.
- Trying to transfer the safety risk to wrong parties.

- Lack of senior management engagement and support.
- Not enough money in the budget to do the job correctly.

4.6 CLOSED-LOOP PROCESS

As you may remember from Chapter 2, the system safety process is a closed-loop process. If there is no feedback in the system, it can never regulate itself nor fine-tune its processes. This closed-loop process is twofold: tracking individual hazards in each system or subsystem and periodic review of the entire system safety process. Not only are you able to verify that safety controls are truly working, you can also optimize the safety process and make the entire system as cost-effective as possible.

4.6.1 Hazard Tracking and Resolution

One area where safety programs often fail miserably is verifying that the problem has been fixed. All the safety analyses in the world will not prevent accidents if you are not certain that hazard controls are in place. Many accident investigations have found that safety concerns were identified but never corrected. The safety control got lost in myriad safety reports. With a safety program set up, mountains of data will start stacking up. It becomes very easy to lose sight of what is important—identifying and correcting the hazard.

That is exactly what has been going on at Northeast Utilities' Millstone Unit 1 nuclear power plant in Waterford, Connecticut. According to Pooley (1996) of *Time* magazine, numerous safety hazards were identified, but controls were regularly circumnavigated. "By sidestepping the safety requirements Millstone saved about 2 weeks of downtime for each refueling [of the nuclear reactor]—during which Northeast Utilities had to pay $500,000 a day for replacement power." Plant safety engineers discovered the problem and tried to use the safety process to remedy the situation. The Nuclear Regulatory Commission (NRC) knew of the problem but failed to take action. What the system needed was a method that would not only track safety problems but also ensure that they were adequately resolved—either accept the risk or remedy the problem. According to *Time*, the problem was just ignored, and safety rules were waived without serious engineering studies. Luckily, no accident occurred. And through strong public pressure, the NRC not only investigated the problem itself, it also tightened up its commercial nuclear power plant–licensing requirements. However, the negative publicity for Northeast was tremendous: cover stories in *Time* magazine and numerous articles in local and national newspapers. Unit 1 was shut down permanently in 1998, though the two other units are operational as of this writing. Considerable revenue was lost, Standard & Poor's downgraded the operator's debt rating from stable to negative, the public was even more jittery about the commercial nuclear power industry (and its federal regulators), and jobs were lost at the plant due to the shutdown. All this resulted from the lack of follow-through in ensuring that safety controls were in place and adequate. Remember, this happened without even an accident or near miss occurring, and this was years before the 2011 Fukushima Daiichi nuclear disaster in Japan.

Probably, the best way to handle this type of situation is to develop safety data tracking systems. The purpose is to track and document the status of each identified hazard and control. This is something you should show to any auditors who may visit your facility. Different safety analysis tools can be used to apply different methods of hazard tracking. In general, however, each identified hazard should have a control number. You should be able to sort by hazard type, hazard causes and effects, hazard severity, probability of occurrence, subsystem or process, location of hazard, hazard controls, hazard control verification methods, and status of hazard resolution. (See Chapter 5 for more details.) Your system should apply a hazard hierarchy, ranking the risks in order of importance, and then automatically print out reports and letters to responsible engineers. This now becomes an automated system that documents the status of each hazard, reports it to the responsible party (i.e., plant engineering, design group, maintenance, production line, quality control, or project management), and provides a database for trend analysis.

Many companies also create verification tracking logs. The verification tracking log is a computerized system that tracks each safety control implementation process and verifies that the control is in place. Typical hazard control verifications are inspection, design review, engineering analyses and calculations, and system testing. See Chapter 5 for more details on hazard tracking.

4.6.2 System Safety Reviews and Audits

The other piece of the closed loop is periodic review of the system safety process. This is the fine-tuning of the system safety program and review at the macroscopic level and should be conducted by an independent safety group to avoid conflict of interest. This is precisely why the system safety office should report directly to the corporate office. The purpose of the review and audit is to verify that the system safety program is in place, performs its intended function, and adequately protects employees, consumers, and the company against accidents and other losses. It also serves as a documentation trail for OSHA inspectors, showing that your organization takes safety seriously.

PRACTICAL TIPS AND BEST PRACTICE

Every time a major change in the system or operational mode occurs, the system SMS should be reviewed to ensure that risks are properly addressed and identified. And make sure that through the closed-loop process, all lessons learned from accident investigations, hazard control verifications, and other learning are fed back into the SMS—thus ensuring a self-learning SMS.

The audit should match system safety performance with the SSPP and SMS, verify the success with which risk to people and property is identified and controlled, and verify to what depth the system safety process influences project and corporate decisions.

Audits should be conducted by a team of specialists from a variety of fields, such as process engineering, safety, environmental, maintenance, operations and

Audit Description					Findings		Corrective Action		Tracking	
ID No.	Date	Auditor	Area Evaluated	Require- ment Evaluated	Finding	Finding Impact	Correc- tive Action	Verification of Corrective Action	Comments	Resolution Status

FIGURE 4.5 Sample audit tracking spreadsheet.

production, design, and management. The entire system safety program should be reviewed once every 3 years. However, other special surveys can be performed annually for selected high-risk systems.

From Section 4.2.2, the diagnostic process can be followed for other internal safety audits. See Figure 4.2, for details. Figure 4.5 illustrates a sample audit-tracking format.

The Audit Description section will track each audited item through a unique tracking number, the date that the audit was conducted, area evaluated (e.g., production unit), and the requirement (i.e., regulation, standard, norm, internal SMS policy). The Findings describes the findings from that particular audit item and its impact on safety or how it propagates to other operational or business units. The Corrective Action describes the specific corrective action to be implemented and how it is verified to be in place. Typically, this is a citation of the inspection report, nondestructive evaluation testing, sampling, training, or other similar types of verification schemes. Tracking gives the auditor the ability to add any relevant comments and the resolution status or closure of the item. Some quick items to remember in conducting a safety audit are the following:

- Don't just audit according to written requirements but also consider if any new worrisome safety trends are popping up that should be considered and any other hot topics of concern.
- Spend time gathering and reading the background data before the audit.
- During the audit, focus on causes rather than symptoms.
- Make sure to not just review documents but also interview a cross section of staff (from senior executives to line workers) to see what they think and do, and observe line operations and work practices to see what people really do.

Also, many organizations have found it useful to include system safety reviews at every major project milestone. The safety review process is progressive with the development of the product or program. It is strongly recommended that a gated system safety review occur at each major decision point of the project, such as preliminary design review, critical design review, test and operational readiness reviews, design certification review, and the first article or customer acceptance configuration inspection. It is considerably cheaper to fix safety problems early in the program, rather than later. The system safety review should ensure that hazards have been identified and controlled adequately and that there is verification that the controls are in place. Table 4.1 indicates where the various project safety control points are located.

4.6.3 CASE STUDY: SPECIAL COMMISSION OF INQUIRY, WATERFALL RAIL ACCIDENT SAFETY MANAGEMENT SYSTEM AUDIT

On January 31, 2003, a train derailed, near Waterfall, New South Wales, Australia; seven people were killed, including the driver. A few days later on February 3, 2003, the Honorable Peter Aloysius McInerney QC was appointed, and he immediately established the Special Commission of Inquiry (SCOI) into the Waterfall rail accident. Through deliberate investigation, testing, and testimony, the SCOI found in January 2004 that the forensic cause of the accident was that the train driver had a heart attack and died while the train was in motion. However, the *deadman* pedal, a fail-safe mechanism (the train driver maintains constant pressure on the foot pedal during operations; if the driver takes his foot off the pedal, it will force the train to immediately halt) designed specifically for these types of incidents failed to disengage and bring the train to a controlled stop. The train was heading at overspeed, almost double the rated speed, through a curve and derailed (McInerney, 2005a). After completing the forensic, technical investigation into what technical failures led to the accident, the Honorable Commissioner McInerney immediately commenced a Stage 2 of the inquiry to investigate the managerial, organizational, and systemic issues that led up to the accident. He investigated from the government regulator's side as well as the rail operator and infrastructure owner's side (see Picture 4.3).

For Stage 2, the SCOI established the Safety Management Systems Expert Panel (SMSEP) to oversee and analyze data from an SMS review of the organizations. It is interesting to look at the SMS audit as a case study of how to review both the regulator and industry operator involved in an accident. The Waterfall accident had occurred only a few short years after the Glenbrook rail accident in December 1999, and the SCOI found that most of those accident investigation recommendations had not been adequately implemented. The intent of the SMS audit was to determine what were the state of the operator's SMS and the government's oversight of that program. The purpose of the SMS audit was (McInerney, 2005b)

1. To determine the adequacy of the SMSs applicable to the circumstances of the railway accident
2. To recommend any safety improvements to rail operations that the commissioner considers necessary

4.6.3.1 Safety Management System Audit Design

It was critical for the success of the SMS audit that the process not just identify the critical issues that led up to the accident but also document them in a thorough fashion that could be presented as evidence in a very high-profile investigation and in court. To do this, an SMS review tool was created. The tool was based on a review of international best practices in SMS application and sourced from numerous programs such as

- Qantas Airways system safety audit process
- American Public Transit Association safety audit programs

(a)

(b)

PICTURE 4.3   (See color insert.) Waterfall rail accident pictures. (From Independent Transport NSW Files. Used with permission.)

- BlueScope Steel SMSs programs
- U.S. Department of Defense standard for system safety programs
- NASA system safety analysis techniques
- The Center for Chemical Process Safety's safety programs, among other sources

The selection and review criteria (Bahr, 2005a) were based on

- Best SMS practice
- Applicability to the New South Wales rail industry
- Stage 1 findings (the forensic investigation of Waterfall rail accident)
- Repeatability of data collection findings
- Uniformity of results
- Ability to roll data up into key findings
- Ease of use

Figure 4.6 illustrates the SMS audit tool format, and Tables 4.8 and 4.9 list the SMS elements reviewed of the rail safety regulator, Independent Transport Safety and

ID No.	Program Element	Protocol	Findings	Audit Data	Rating
1.0	Management Commitment				
1.13		• There is an effective means of making senior managers accountable for safety issues.	• Executive-level job descriptions do not include safety responsibilities.	• GM position description • Director of Ops PD • GM interview March 2, 2004 • Director of Ops interview March 2, 2004	1

0	Element not identified	1	Some key aspect of the element is missing and not integrated.	2	Element partially present and partially integrated	3	Element present but not fully integrated	4	Element present and fully integrated

FIGURE 4.6   Example Waterfall SMS audit tool.

TABLE 4.8
ITSRR SMS Key Elements

Safety Elements

1.0 Regulatory independence
2.0 Regulatory mandate
3.0 Policies and objectives
4.0 Organization and function
5.0 Data analysis
6.0 Transition
7.0 Safety enforcement over rail authority
8.0 ITSRR accident/incident investigation
9.0 ITSRR audits
10.0 Safety accreditation
11.0 Partnership with the rail authority

TABLE 4.9
SRA/RailCorp and RIC/RailCorp SMS Key Elements

SMS Key Elements

1.0 Management commitment	17.0 Customer feedback
2.0 Policy and objectives	18.0 Contracted goods and services
3.0 Safety representative and personnel	19.0 Traceability of goods and services
4.0 Safety committee	20.0 Measuring equipment and calibration system
5.0 Management review	21.0 Procurement of goods and services
6.0 Training and education	22.0 Equipment maintenance
7.0 Hazard identification and risk management	23.0 Design and development
8.0 Document control	24.0 Management and staff recruitment
9.0 Record control	25.0 Medical issues
10.0 Internal audit	26.0 Human factors
11.0 Incident/accident reporting system	27.0 Safety organization
12.0 Incident/accident investigation	28.0 Safety awareness
13.0 Analysis and monitoring	29.0 System safety program plan
14.0 Emergency response procedures	
15.0 Change management	
16.0 System for managing requirements and changes	

Reliability Regulator (ITSRR), and the railway organizations State Rail Authority (operator) and Rail Infrastructure Corporation (infrastructure owner), respectively (Bahr, 2005b). The operator and infrastructure owner were renamed RailCorp in 2004. As you can see, the 29 key SMS elements described earlier in this chapter were originally based on the RailCorp SMS audit. Each of the 11 ITSRR and 29 RailCorp SMS elements was then subdivided into numerous detailed safety protocols that were then evaluated.

4.6.3.2 Safety Management System Audit Process

Once the audit tool was developed, it was important to select and train the audit team. The team was comprised of experienced SMS auditors from various industries including aviation, rail, process industry, and aerospace/military. This was purposely done so that there would be a diverse team that could look at the problem from multiple perspectives and bring best practices from their own industries. The SMS audit team collected data from various sources and then, as a team, evaluated the compendium of information and inputted the results into each of the 11 ITSRR and 29 RailCorp SMS elements for review. The sources of input for the audit were

- Numerous design, management, safety, and operational documents
- Interviews with a cross section of leadership, management, and staff
- Tour of facilities and observation of some key operations

To ensure maximum success, the team was trained on interview techniques and practiced through role playing before interviews started. Over 125 RailCorp employees and 23 ITSRR staff were interviewed. Over 500 documents were reviewed as part of the SMS audit. At the end of each audit day, the team was gathered together for a *hot wash* session in which team members discussed findings and filled in the SMS audit tool with a consensus rating. The team met with the SMSEP weekly to review progress and ensure quality control.

4.6.3.3 Safety Management System Audit Findings

The SMS audit findings were divided between the railway operator and the regulator (McInerney, 2005c). Some of the major findings for the rail operator were the following:

- The SMS was ineffective, missing key elements (e.g., system safety engineering, management of change, and requirements assurance), not integrated nor fully implemented.
- There were major deficiencies in the hazard identification, risk assessment, risk management, training, and internal and external audits.
- Inadequate investigation of incidents and more implementation of corrective actions.
- No effective information management system for handling and sharing key safety data.
- Strong bias toward occupational safety at the expense of system safety.
- Too much instability in senior management positions (five CEOs and five safety managers between Glenbrook accident and 2004).
- Overly focused on following rules without understanding how they impacted safety.

As you can see, these are very serious deficiencies in the SMS, and it is not surprising that there have been very serious accidents. But looking at the SMS review of the regulator is interesting too. The SMS audit evaluated how well ITSRR regulated RailCorp. You saw that there were very serious system safety gaps in the SMS at RailCorp, but the regulator did not pick up on them in any meaningful way. At the time of the accident, RailCorp was mandated to submit an annual safety case for regulatory review and approval. The safety case review process became perfunctory and as a result did not identify key signals that something was very wrong in the operator's SMS. Some of the findings of the regulator were (McInerney, 2005c) as follows:

- Key individuals within the regulatory body lacked essential qualifications.
- No processes in place to measure the effectiveness of the regulations.
- Insufficient resources to perform adequate safety regulatory functions.
- Little to no policy or frameworks to guide a consistent regulatory review or provide adequate guidance to the railway operator—this was particularly a problem because ITSRR uses a coregulatory model in which industry and government design the implementation of the regulation together (see Chapter 12 for more on safety regulatory models).
- Unfounded confidence in accreditation baselines conducted earlier.
- Insufficient staff qualifications, training, and experience in system safety and risk assessment.

- Potential conflicts of interest.
- Lack of formal and detailed processes to verify compliance.

A series of very good recommendations were developed and can be found in the investigation report cited at the end of this chapter. One additional recommendation was made that is quite interesting and unique in many systems. The commissioner demanded that an independent body be established to oversee the review of corrective action progress and implementation efficacy of both the railway operator and the regulator. This did not happen after Glenbrook and the commissioner did not want to see a repeat after Waterfall.

4.7 SOME WORDS ON SAFETY GOVERNANCE

Safety governance is a part of the company's corporate governance in which the board directs and controls safety risks to employees, the public, products, and company assets that have been created by the company's own enterprise. As mentioned earlier, safety governance is the business infrastructure, set of processes, customs, and mechanisms that control how safety is managed. The reason it is called out specifically in the SMS is because of its importance; a strong SSPP can't be sustainable if the governance structure is weak. Corporate governance must include safety management, and it must be part of the primary governance structure from the board level through to the factory floor.

NOTES FROM NICK'S FILE

An interesting challenge I once had looking at safety governance was convincing a board member that even though safety was managed at the subsidiary company level, he and the board were still ultimately responsible for safety—in fact, ignoring safety and thinking the subsidiary would own the risk was actually a liability.

In very simple terms, the governance structure must ensure that the SMS is followed and maintained. One could say that all accidents ultimately are due to a safety governance failure. If the safety governance system worked well, then theoretically at least, the accident should have been avoided. We all know that corporate governance sets the strategic vision and direction to manage the business—safety must be inherent to that process. This is typically done through the development of the governance, policy, and objective framework. But it also must create and manage the appropriate accountabilities and authorities for leaders, managers, and staff to do their safety duty. Essentially, the items listed in the SSPP of the SMS are what the safety governance structure must manage and perform the appropriate level of oversight and control. Of particular import is safety culture and change management.

Change is inevitable, and how the governance structure operates as external and internal changes occur will impact the efficacy of the SMS. With significant external pressures in the marketplace impacting company operations, it is very easy to cut or downsize safety or the SMS during hard times. During these times, safety engineers are always faced with the dilemma of how to prove the importance of current safety activities and the SMS when the corporation has a great safety record. How do you prove a negative? But it is precisely because of a strong SMS that safety performance is good. Cutting back on this at the wrong time could have devastating effects. It is much easier to break something than it is to create or build it.

The SMS operating within an adaptive corporate governance framework gives companies the way to flow with the uncontrolled changes externally and internally rather than allow change to govern them. And even during change, or especially during change, safety performance and safety accountabilities must be strictly adhered to. Safety boundaries should not be abrogated just because change is occurring or the business environment is very unpredictable. During times of change is when your corporate values are put to the true test, and the SMS must be core to corporate values. If safety and the SMS are not a corporate value, then it is not valuable to the corporation. Recently, Mohammed Al-Mady, CEO of Sabic said, "Effective Process Safety Governance and Culture is not a choice but a must for survival in our industry" (OECD, 2012a). The Organization for Economic Co-operation and Development Document, "The Corporate Governance for Process Safety Guidance for Senior Leaders in High Hazard Industries," (OECD, 2012b) asks "Do you know what impact your business decisions have on the level of risk of your site—and not just now, but several years into the future? Analysis of past incidents reveals that inadequate leadership and poor organizational culture have been recurrent features with:

- A failure to recognize things were out of control (or potentially out of control), often due to lack of competence at different levels of the organization;
- An absence of, or inadequate, information on which to base strategic decisions—including the monitoring of safety performance indicators at Board level;
- A failure to understand the full consequences of changes, including organizational ones; and,
- A failure to manage process safety effectively and take the necessary actions."

The document further emphasizes that "Leaders need to understand the risks posed by their organization's activities, and balance major accident risks alongside the other business threats. Even though major accidents occur infrequently, the potential consequences are so high that leaders need to recognize:

- Major accidents as credible business risks;
- The integrated nature of many major hazard businesses – including the potential for supply chain disruption;
- Management of process safety risks should have equal focus with other business processes including financial governance, markets, and investment decisions, etc."

Safety governance and the SMS address all these items.

4.8 VOLUNTARY PROTECTION PROGRAM

There are many sources of information on good corporate safety governance. One interesting program to investigate is the Voluntary Protection Program (VPP), which is a U.S. government program that encourages companies to good SMS and safety governance. It is an OSHA program created in 1982 to recognize worksites with outstanding, comprehensive safety and health programs and strong partnerships between labor and management. It is the OSHA equivalent of the prestigious Malcolm Baldrige Award for quality. OSHA says (U.S. OSHA, 1996a), "Participants are not subject to routine OSHA inspections, because OSHA's VPP on-site reviews ensure that their safety and health programs provide superior protection. Establishing and maintaining safety and health programs on the VPP model are reflected in substantially lower than average worker injury rates at VPP worksites."

OSHA reviews OSHA 200 reports over a 3-year period, performs a 2-day worksite walk-through (reviewing workplace hazards, controls, and safety management programs), and interviews employees. Plant unions have to sign on together with plant management; where plants are not unionized, employees still must be part of the program together with their local management. The purpose is to recognize plants that are far superior to mere OSHA compliance. It is tough to get in to and stay within the VPP; there are only 2000 participants today. The VPP recognition program has three levels (in order of importance):

Star: Designed for exemplary worksites that have implemented comprehensive, successful safety and health management systems and achieved injury/illness rates below their industry's national average

Merit: Designed for worksites with the potential and commitment to achieve Star quality within 3 years

Demonstration: Designed for worksites with Star quality safety and health protection that want to test alternatives to current Star eligibility and performance requirements

The costs for implementing the program initially are high, as for the Baldrige Award; however, for a long-term investment, the payback is significant. Companies must already have a very good safety program in place to be competitive to win the recognition, but once a good safety program is in place, plant operating costs are reduced dramatically. Edwin G. Foulke, Jr., assistant secretary of Labor HR Florida State Council in 2007 dramatically illustrated the advantage (U.S. OSHA, 2014a). He explained that in 1999 in Milwaukee, Wisconsin, United States, an OSHA inspector responded to calls from employees working during very heavy winds with *Big Blue*, one of the largest cranes in the world used to build a baseball stadium. Unfortunately, the crane collapsed during very high winds and killed three. Not only were people killed, but also the out-of-court settlement paid to the victim's families was $60 million, and property damage reached $100 million. The stadium was a year late, and the builders had to compensate the team owners an additional

$20.5 million for lost revenue. He compares that disaster to the construction of the Paul Brown Stadium in 2000 in Cincinnati, Ohio, United States, that was completed on time with a safety record better than industry averages, reduced workers' compensation claims, and reduced liability costs with an overall savings of $4.6 million. The difference between the two stadium projects was that the company building the Cincinnati stadium designed a strong SMS with its contractors, local OSHA office, and state and county safety officials.

Some of the statistics OSHA have gathered and verified (U.S. OSHA, 2014b) are quite remarkable:

- A 2012 study concluded that inspections conducted by California's Division of Occupational Safety and Health (Cal/OSHA) reduce injuries with no job loss. The study showed a 9.4% drop in injury claims and a 26% average savings on workers' compensation costs in the 4 years after a Cal/OSHA inspection compared to a similar set of uninspected workplaces. On average, inspected firms saved an estimated $355,000 in injury claims and compensation paid for lost work over that period. There was no evidence that these improvements came at the expense of employment, sales, credit rating, or firm survival.
- According to Goldman Sachs, companies that did not adequately manage workplace safety and health performed worse financially than those who did from November 2004 to October 2007. Investors could have increased their returns during this period had they accounted for workplace safety and health performance in their investment strategy.
- There is a direct positive correlation between investment in safety, health, and environmental performance and its subsequent return on investment.
- Over 60% of chief financial officers in one survey reported that each $1 invested in injury prevention returns $2 or more. Over 40% of chief financial officers cited productivity as the top benefit of an effective workplace safety program.
- A forest products company saved over $1 million in workers' compensation and other costs from 2001 to 2006 by investing approximately $50,000 in safety improvements and employee training costs. The company has participated in OSHA's Safety and Health Achievement Recognition Program (SHARP) since 1998.
- The average worksite in OSHA's VPPs has a days away, restricted, or transferred (DART) case rate of 52% below the average for its industry. Fewer injuries and illnesses mean greater profits as workers' compensation premiums and other costs plummet. Entire industries benefit as VPP sites evolve into models of excellence and influence practices industry wide.
- An OSHA strategic partnership covering construction of a power plant in Wisconsin resulted in injury and illness rates significantly below the construction industry rates in Wisconsin. In 2006, employees worked over 1.7 million man hours at the site with zero fatalities. The 2006 TCIR was

69% below the Wisconsin average, and the 2006 days away, restricted, and time away (DART) rate for the site was 75% below the Wisconsin average.

4.9 SAFETY CULTURE

The word culture, in general, is difficult to define. We seem to all recognize that it exists but have trouble defining it. Safety culture seems to us to be something that we know when we don't have it. Section 4.2 defines it "is the complete suite of enterprise employee, management, and leadership attitudes towards safety risk." In essence, it means what all employees from the chief executive to frontline staff think, feel, and act in regard to safety. An organization has a good safety culture if they have set up an appropriate governance structure that supports and nurtures a strong positive attitude toward safety. Safety must be designed into everyone's job and everyone's thinking. It is not a separate add-on to any management or corporate culture. It is part of the DNA of a corporate culture. And the SMS is the infrastructure of a strong organizational safety culture.

Much has been written on safety culture: how to define it, measure it, and manage it. This section will briefly discuss some of these elements so that your SMS appropriately manages and nurtures your organization's safety culture. The Further Reading section at the end of this chapter gives suggestions of where to seek out more information.

4.9.1 What Is Safety Culture?

Figure 4.7 illustrates the primary components of the safety culture puzzle. Certainly, many may break it down into more constituent parts, but for ease of understanding, shown in the following are the four key areas:

1. *Just culture*: Is how all employees view the level of trust shown in an organization. If what leadership and managers say doesn't comport with what they do, the trust factor is minimal or zero. Also, staff must feel that they are not unfairly viewed or penalized for safety issues. But it is not only the safety issues that ensure trust; leadership and management must also demonstrate in general that they operate in a fairly transparent and open way. If not, critical issues that could impact safety will be hidden. An unbalanced blame culture can make how appropriate and true faults are attributed. The disciplinary process must also be transparent and fair. To encourage staff to be the organization's *eyes and ears* to safety problems, then they must be encouraged to speak up and not feel that they can be unduly blamed.

2. *Leadership, management, and staff involvement*: Is imperative for a strong safety culture. The safety vision and mission must be clearly articulated

Just Culture
▪ Trust
▪ Appropriate fault
  attribution
▪ Disciplinary process
▪ Encouragement to voice
  safety concerns

Leadership, management,
and staff involvement
▪ Safety vision, mission
▪ Demonstrated leadership and
  management commitment
▪ Actions to support safety
  message
▪ Staff involvement in safety
  process
▪ Staff participation in changes
▪ Commonly understood goals

Continuous improvement
and learning
▪ Organizational responsiveness to change
▪ Internal monitoring, continuous safety evaluation
▪ External learning of new tools and techniques
▪ Appropriate safety prioritization
▪ Encouragement to developing skills and knowledge

Communications and
information flow
▪ Reporting behavior
▪ Organizational reporting
  systems
▪ Feedback loop
▪ Effective communications

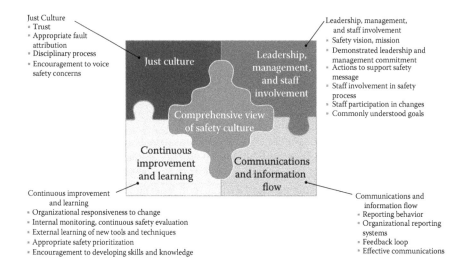

FIGURE 4.7   The safety culture puzzle.

from top leadership, but more importantly, it must be demonstrated on a daily basis through active safety support, actions, and visibility. A key metric of this is time and resource commitments to safety. Leadership decisions must show that safety is prioritized above all other decisions. Staff must also be part of the process and involved in managing change. They should be part of the safety improvements solution design, implementation, and monitoring for continuous improvement.

3. *Communications and information flow*: Are central to a living and positive safety culture. How leadership, management, and staff behave as it relates to safety, fault, and quality reporting is critical for a strong safety culture. Of course, reporting systems must be in place (and where possible they should be anonymous to encourage unfettered reporting). The feedback loop must be closed; in other words, when someone offers input, they should get feedback on its evaluation and implementation (where appropriate). Poor communications will negatively impact the safety culture, and the organization must always be vigilant that communications are clear, succinct, and truthful. It is imperative that all employees are communicated of the safety risks and understand how to manage them.

4. *Continuous improvement and learning*: Are key to a sustainable safety culture. How organizations respond to change and continuously improve is important. This is done through internal monitoring and continuously evaluating and questioning the safety process—ensuring that it is a *learning process*. But reaching out externally to the organization and finding best practices help an organization improve. All of this learning helps leadership prioritization actions that will continuously improve safety. Also, all staff,

managers, and leaders must be encouraged to develop their professional skills and knowledge and also their safety understanding to the hazards the organization faces.

4.9.2 Measuring Safety Culture

So, if defining safety culture is difficult, measuring it is harder. However, it is not impossible, and it should be done on a regular schedule. Chapter 2 discussed the safety maturity model. Safety culture and its measurement can be viewed in a similar fashion. Figure 4.8 illustrates, from worst to best, a safety culture maturity model and is purposely almost exactly the same as the safety maturity model in Chapter 2; the two are totally intertwined. A strong SMS must have a strong safety culture, one cannot happen without the other.

Clearly the left end of the diagram, the "Pathological" step, is not just the worst; it is also illegal. Unfortunately, as this book has shown you, there are too many times that we can find a pathological safety culture in place. Certainly up to the last few years, Level 2, reactive safety culture, had been the norm. Thankfully, that has changed through better regulations and oversight programs. Level 3 is probably one of the most common levels of where safety culture lies in the organization. There is a strong desire to be safe but many do not know how to do it. These organizations probably don't have an SMS in place; all of their safety activities are disparate and not coordinated or systematic. Level 4 is what many companies aspire to and are working to have an SMS in place to embed safety culture but that is not sufficient. The last level is really the best. Safety is on everyone's mind (from the top leadership to frontline staff) and thought about daily and is continuously improved. A strong and vibrant SMS is the organizational DNA that keeps the safety culture alive and growing.

The measurement of safety culture in a company or organization should include the safety maturity model as its basis. The purpose is to first determine where the company is on the safety culture maturity model and then what needs to be done to close the gaps and move to the next level. Figure 4.9 illustrates the process.

Step 1: Determine Safety Culture Review Assessment Framework
As with all assessments, the assessment goals and objectives should be discussed, defined, and agreed with senior leaders. Then the evaluation criteria should be defined into a safety culture review assessment framework. Each of the safety culture puzzle elements defined earlier can be used as the baseline of the safety culture review assessment framework. You can subdivide each of the elements into detailed evaluation criteria.

Step 2: Review SMS and Observe Operations
The SMS is central to any healthy safety culture in an organization. The discussion of how to review an SMS is discussed earlier in Section 4.2, and that material can

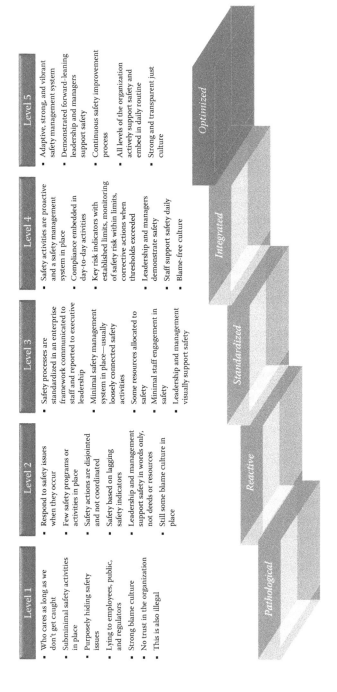

Level 1

- Who cares as long as we don't get caught
- Subminimal safety activities in place
- Purposely hiding safety issues
- Lying to employees, public, and regulators
- Strong blame culture
- No trust in the organization
- This is also illegal

Level 2

- Respond to safety issues when they occur
- Few safety programs or activities in place
- Safety actions are disjointed and not coordinated
- Safety based on lagging safety indicators
- Leadership and management support safety in words only, not deeds or resources
- Still some blame culture in place

Level 3

- Safety processes are standardized in an enterprise framework communicated to staff and reported to executive leadership
- Minimal safety management system in place—usually loosely connected safety activities
- Some resources allocated to safety
- Minimal staff engagement in safety
- Leadership and management visually support safety

Level 4

- Safety activities are proactive and a safety management system in place
- Compliance embedded in day-to-day activities
- Key risk indicators with established limits, monitoring of safety risk within limits, corrective actions when thresholds exceeded
- Leadership and managers demonstrate safety
- Staff support safety daily
- Blame-free culture

Level 5

- Adaptive, strong, and vibrant safety management system
- Demonstrated forward-leaning leadership and managers support safety
- Continuous safety improvement process
- All levels of the organization actively support safety and embed in daily routine
- Strong and transparent just culture

Pathological *Reactive* *Standardized* *Integrated* *Optimized*

FIGURE 4.8 Safety culture maturity model.

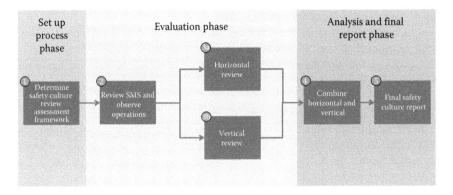

FIGURE 4.9   Safety culture review methodology.

be extracted to serve as the SMS review. The SMS material can be truncated to make the SMS review faster.

It is important to also review selected operations to see how well staff actually implements the SMS and what the general safety culture mood is on the shop floor.

One of the problems of determining the safety culture of an organization is that it is not always measured directly. The first two steps focused on understanding what the safety processes are and if they are implemented. The next two steps will focus on determining people's attitude toward safety. This is done through two types of discussions with the organization's staff. Of course, all of the information gathered in the next two steps must be through nonattribution. If employees can't speak freely, then you won't get as much valuable information.

Step 3a: Horizontal Review

The first is a horizontal review. This review is comprised of a safety culture survey that is sent out to staff with a set of top level questions that determines how they view safety and how they think the company views safety (in words and in deeds). This can be thought of as a 60,000 foot view of the safety of an organization. It will give overall attitudes toward safety illustrating a *topological map of safety culture*, a lot of general views but not sufficiently detailed for further action. That is done in the next step.

Step 3b: Vertical Review

This step will focus on very detailed, open-ended questions (unlike the safety survey in Step 3a) to drill down into better understanding motivations for safety behavior. This can be done through very small focus groups or targeted interviews across the leadership, management, and frontline staff pyramid. The information gathered in this step will help determine a much deeper meaning of safety culture on individuals in the organization.

Step 4: Combine Horizontal and Vertical

Now, the data from horizontal and vertical gathering should be combined. The horizontal, high-level view of the safety culture is very useful to frame the safety health of the organization. But the vertical data added to the horizontal view will give real color to the analysis.

Step 5: Final Safety Culture Report

All the data culled from the previous steps and analyzed in the previous step are now summarized in a final report. Of course, just finding what is wrong isn't enough. Recommendations for each of the safety culture findings should be included, discussed, and prioritized for senior leadership to act upon.

4.9.3 DESIGNING SUSTAINABLE SAFETY CULTURE: FITTING A GLOBAL COMPANY'S CORPORATE SAFETY PROGRAM INTO THE LOCAL CULTURE

We all know that the world is a much smaller place than even just a few years ago. More and more companies are going global and operating in not just different countries but very different cultures. Many will say that some countries value safety higher than others. But really this isn't so. People are people wherever they live. We all love our families and want a stable life and to succeed in life. What is true is that not all countries have the resources or tools to best protect life, property, and the environment. Having said this, it is important to realize that how different cultures manifest a strong safety culture can be different and should be taken into consideration.

NOTES FROM NICK'S FILE

I've worked a large part of my life outside the United States and have encountered many challenges in designing safety systems to fit a local culture. With one client in the Middle East, I set up an international board to oversee design and implementation. I brought in experts from the usual suspects—the United States, the United Kingdom, Australia, and Canada. But because the system would have to operate in the Middle East, it was imperative to bring in other Middle Eastern experts who could challenge our western thinking. I brought in experts from Egypt and Saudi Arabia. They gave us healthy pushback on some of our ideas and forced us to think about how to implement world-class best practices in a local environment in a sustainable way. We not only learned more, but the client truly benefited.

Here are a few things to consider in designing a world-class SMS to operate in a local environment.

PRACTICAL TIPS AND BEST PRACTICE

Spend some time and cycles on understanding the culture that you or your company will be operating in or servicing; in the end, it will save you money and headache:

- Make trips out early in the process, and get to know locals well before designing or implementing any systems. Research where you are going before you go.
- Always, always remember people are people no matter where they live. They have the same desires that we have, and you should always respect and honor that. Don't be condescending even if you know more or are setting the operating standard.
- Seek out advice with locals and ask them detailed questions about your SMS and how they think that it will work. Seriously consider their advice and try to figure out ways to implement it.
- Wherever practical, translate policies and procedures into the local language, and make sure that you translate ideas into local culture.
- Test your programs out before you implement them. You want to know about unintended consequences before they happen.
- Be sensitive to religion and politics.
- Be sensitive to local traditions, what they mean, and potential misunderstandings; remember you are a guest in their country.
- Benchmark with like cultures, countries, and others that are similar to see how well your SMS will work and be viewed on the ground.
- Be aware of culture shock! It follows this pattern: initial euphoria and excitement and interest in the area, then confusion and irritability because you don't understand the cultural norms and hidden signals, and finally acceptance (of what's good and bad in the country) and adaptation.
- Understand that different cultures do business differently, but in the end, both parties want a deal. Spend time understanding the business operational norms.
- Where possible, hire locally and use local staff to help you understand their culture. Make local staff part of your process.
- Send local staff abroad or bring in outside experts for training to demonstrate other ways of thinking.
- Be patient, flexible, and persistent in all that you do in the local community. It always takes longer than you think, but in the end, the payoff will be high.

Most accidents are the result of a poor SMS, but it is not that difficult, or overly bureaucratic, to develop a good, strong SMS. This chapter has detailed the major elements of a good SMS and practical steps toward implementing them. It is now up to you to take these sample program plans and tailor them to your organization.

REFERENCES

American Institute of Chemical Engineers. 1994. *Process Safety Management with Case Studies: Flixborough, Pasadena and Other Incidents*. New York: American Institute of Chemical Engineers.

Bahr, N. J. 1988. The Johnson Space Center Test Safety Program. 88-WA/SAF-l. Chicago, IL: American Society of Mechanical Engineers Winter Annual Meeting.

Bahr, N. J. 2005a, January. Special commission of inquiry into the waterfall rail accident, final report, vol. 2, Appendix F: SMS review methodology, New South Wales, Australia, dated May 12, 2004, p. 5.

Bahr, N. J. 2005b, January. Special commission of inquiry into the waterfall rail accident, final report, vol. 2, Appendix F: SMS review methodology, New South Wales, Australia, dated May 12, 2004, pp. 11, 19, 20.

Barab, J. 2012. Remarks by, Jordan Barab, Deputy Assistant Secretary of Labor for Occupational Safety and Health, Voluntary Protection Programs Participants' Association, Annual Meeting, Anaheim, CA, August 20, 2012. https://www.osha.gov/pls/oshaweb/owadisp.show_document?p_table=SPEECHES&p_id=2860, downloaded May 16, 2014.

Barr, S. 1995, October 26. Foundation awards honor 15 creative government programs. *The Washington Post*.

Bunn, W. 1986, April 17. *Commission Interview Transcripts*. Washington, DC, pp. 42–43.

Cunliffe, J. 1994, May. Eurotunnel channel update. *Professional Safety*, 27–31.

Donaldson, K. and Edkins, G. 2004. A case study of systemic failure in rail safety: The waterfall accident. Presented at *the International Rail Safety Conference*, Perth, Western Australia, Australia, p. 8.

Gallagher, V. 1993, January. Liability, OSHA and the safety of outside contractors. *Professional Safety*, 27.

Gibbs, D. and Lahale, E. 2014. OSHA's voluntary protection programs: A model of safety and health excellence that works! Washington, DC: OSHA, Department of Labor. https://www.osha.gov/dcsp/vpp/articles/modelthatworks_2009.html, downloaded May 16, 2014.

Groupe Eurotunnel. 2012. Annual review. http://www.eurotunnelgroup.com/uploaded-Files/assets-uk/Shareholders-Investors/Publication/Annual-Review/RA2012-UK-Eurotunnel-Group.pdf, downloaded May 16, 2014.

Kitzes, W. F. 1991, April. Safety management and the consumer product safety commission. *Professional Safety*, 25–30.

McInerney, P. A. 2005a, January. *Special Commission of Inquiry into the Waterfall Rail Accident, Final Report*, vol. 1. Sydney, New South Wales, Australia, p. i.

McInerney, P. A. 2005b, January. *Special Commission of Inquiry into the Waterfall Rail Accident, Final Report*, vol. 2. Sydney, New South Wales, Australia, p. xiii.

McInerney, P. A. 2005c, January. *Special Commission of Inquiry into the Waterfall Rail Accident, Final Report*, vol. 2. Sydney, New South Wales, Australia, pp. xiv–xviii.

National Aeronautics and Space Administration. 1985. Johnson Space Center safety manual, JSCM 1700D. Houston, TX: National Aeronautics and Space Administration, 1-13-4.

Organization for Economic Co-Operation and Development (OECD). 2012a. Corporate governance for process safety guidance for senior leaders in high hazard industries. Organization for Economic Co-operation and Development. OECD Environment, Health and Safety Chemical Accidents Programme, June 2012, p. 6. http://www.tukes.fi/Tiedostot/vaaralli-set_aineet/ohjeet/OECD_guidance_hazard_indust.pdf, downloaded on January 18, 2014.

Organization for Economic Co-Operation and Development (OECD). 2012b. Corporate governance for process safety guidance for senior leaders in high hazard industries. Organization for Economic Co-operation and Development. OECD Environment, Health and Safety Chemical Accidents Programme, June 2012, p. 8. http://www.tukes.fi/Tiedostot/vaaralli-set_aineet/ohjeet/OECD_guidance_hazard_indust.pdf, downloaded on January 18, 2014.

Parker, R. J., Pope, J. A., Davidson, J. F., and Simpson, W. J. 1975. *The Flixborough Disaster: Report of the Court of Inquiry*. London, U.K.: Her Majesty's Stationery Office.

Pooley, E. 1996, March 4. Nuclear warriors. *Time*, pp. 46–54.

Report. n.d. *Developing a workplace safety and health program*. Hunt Valley, MD: Maryland Occupational Safety and Health.

Report. 1985. Erosion of solid rocket motor pressure seal updated from August 19, 1985. Revised February 10, 1986. Thiokol, TWR-15150, PC 000769.

Report of the Presidential Commission on the Space Shuttle Challenger Accident. 1986a. Washington, DC: Presidential Commission on the Space Shuttle Challenger Accident, p. 1.

Report of the Presidential Commission on the Space Shuttle Challenger Accident. 1986b. Washington, DC: Presidential Commission on the Space Shuttle Challenger Accident, p.19.

Report of the Presidential Commission on the Space Shuttle Challenger Accident. 1986c. Washington, DC: Presidential Commission on the Space Shuttle Challenger Accident, p.72.

Report of the Presidential Commission on the Space Shuttle Challenger Accident. 1986d. Washington, DC: Presidential Commission on the Space Shuttle Challenger Accident, pp. 129–131.

Report of the Presidential Commission on the Space Shuttle Challenger Accident. 1986e. Washington, DC: Presidential Commission on the Space Shuttle Challenger Accident, p. 148.

Roughton, J. E. 1995, January. Contractor safety. *Professional Safety*, 31.

Sadee, C., Samuels, D. E., and O'Brien, T. P. 1976. The characteristics of the explosion of cyclohexane at the Nypro (UK) Flixborough plant on June 1, 1974. *Journal of Occupational Accidents*, 1:203.

Senate and House of Representatives of the United States of America in Congress. 1970. Occupational Safety and Health Act of 1970. Public Law 91-596. 91st Congress, S. 2193.

The Institution of Engineering and Technology. 2012, August. Do accidents and ill health really cost me money? Health and Safety Briefing No. 48. London, UK: Health and Safety Policy Advisory Group. www.theiet.org, downloaded May 16, 2014.

U.S. Occupational Safety and Health Administration. 1992. Process safety management of highly hazardous chemicals. 29 CFR Part 1919.119. Washington, DC: Bureau of National Affairs.

U.S. Occupational Safety and Health Administration. 1996a, February 23. The benefits of participating in the VPP. OSHA web page. http://www.osha.gov, downloaded May 16, 2014.

U.S. Occupational Safety and Health Administration. 1996b, February 22. The new OSHA—Reinventing worker safety and health. OSHA web page. http://www.osha.gov, downloaded May 16, 2014.

U.S. Occupational Safety and Health Administration. 2013. What is an OSHA partnership. https://www.osha.gov/dcsp/partnerships/what_is.html, downloaded on January 10, 2014.

U.S. Occupational Safety and Health Administration. 2014a. Voluntary protection programs for federal worksites. http://www.osha.gov/dcsp/vpp/vppflyer.pdf, downloaded May 16, 2014.

U.S. Occupational Safety and Health Administration. 2014b. Business case for safety and health. https://www.osha.gov/dcsp/products/topics/businesscase/index.html, downloaded May 16, 2014.

FURTHER READING

Balog, J. N. and Baird, M. C. 1989. Safety planning information directed to emergency response (SPIDER). West Virginia resource manual for the safe operation of transportation systems (set of instructor guide, participant guide, videotapes, and slides). Malvern, PA: Ketron, Inc.

Center for Chemical Process Safety. 2011. *Guidelines for Auditing Process Safety Management Systems*, 2nd edn. New York: Wiley.

Coble, D. F. 1995, May. Complying with OSHA: Are you ready? *Professional Safety*, pp. 44–47.

Health and Safety Executive. 2014. Health and safety management systems. http://www.hse. gov.uk/managing/health.htm, downloaded May 16, 2014.

International Standards Organization ISO 22000:2005. Food safety management systems. http://www.iso.org/iso/home/standards/management-standards/iso22000.htm, downloaded May 16, 2014.

International Standards Organization ISO 22301:2012. Business continuity management standard. http://www.iso.org/iso/news.htm?refid=Ref1587, downloaded May 16, 2014.

International Standards Organization ISO 31000:2009. Risk management. http://www.iso.org/ iso/home/standards/iso31000.htm, downloaded May 16, 2014.

Kausek, J. 2007. OHSAS 18001: Designing and implementing effective health and safety management systems. New York: Government Institutes.

Levitt, R. E. and Samelson, N. M. 1987. *Construction Safety Management*. New York: McGraw-Hill.

McInerney, P. A. 2005, January. Special commission of inquiry into the waterfall rail accident, final report, vols. 1 and 2, Sydney, New South Wales, Australia.

Ministry of Business Innovation and Employment, Institute of Directors of New Zealand. 2013, May. Good governance practices guideline for managing health and safety risks. https://www.iod.org.nz/Publications/Healthandsafety.aspx, downloaded May 16, 2014.

National Safety Council. 1992. *Accident Prevention Manual for Business & Industry, Engineering and Technology*, 10th edn. Washington, DC: National Safety Council.

Perrow, C. 1999. *Normal Accidents Living with High-Risk Technologies*. Princeton, NJ: Princeton University Press.

Report. 1987, January 12. *System Safety Handbook for the Acquisition Manager*. SDP 127-1. Los Angeles, CA: U.S. Department of the Air Force, Chapters 1–6, 13.

Report. 1993, December 7. Beyond enforcement: OSHA's voluntary protection programs. Business and legal reporter: OSHA compliance advisor, pp. 3–7.

Roland, H. E. and Moriarty, B. 1990. *System Safety Engineering and Management*, 2nd edn. New York: John Wiley, pp. 62–69.

Shirely, G. 1993. How to control contractor safety. *Industrial Safety and Hygiene News*.

Thomas, M. 2012. A systematic review of the effectiveness of safety management systems. Australian transport safety report. Cross-modal research investigation XR-2011-002, final. Canberra, Australia Capital Territory, Australia: Australian Transport Safety Bureau. http://www.atsb.gov.au/media/4053559/xr2011002_final.pdf.

Transport Safety Victoria. 2014. Safety culture elements and sub-elements organizational safety culture appraisal tool. Transport Safety Victoria. http://www.transportsafety. vic.gov.au/rail-safety/safety-improvement/organisational-safety-culture-appraisal-tool/ using-oscat/safety-culture-elements, downloaded February 28, 2014.

UK Health and Safety Commission. n.d. Reduce risks-cut costs: The real cost of accidents and ill health at work. HSE booklet INDG 355. http://www.rbkc.gov.uk/pdf/Reduce%20 risks%20and%20cut%20costs%20in%20the%20workplace.pdf.

U.S. Department of Defense. 2012. Military standard, system safety program requirements. Mil-Std-882E. Washington, DC: U.S. Department of Defense.

5 Hazard Analysis

No profit grows where is no pleasure ta'en;
In brief, sir, study what you most affect.

The Taming of the Shrew, 1623
William Shakespeare

We fooled ourselves into thinking this thing wouldn't crash. When I was in astronaut training I asked, "What is the likelihood of another accident?" The answer I got was: one in 10,000, with an asterisk. The asterisk meant, "We don't know."

Space News interview, 1996
Bryan O'Connor
Former Astronaut and Deputy Associate Administrator for the Space Shuttle

Please know that I am aware of the hazards. I want to do it because I want to do it.

Letter to her husband on the eve of her last flight, 1937
Amelia Earhart

Not everything that can be counted counts, and not everything that counts can be counted.

Informal Sociology: A Casual Introduction to Sociological Thinking, 1963
William Bruce Cameron

As Chapter 3 demonstrates, there are numerous safety analysis methods, almost as many different styles as there are industries. However, they all seem to have one aspect in common, the identification of hazards and recommended controls. After understanding the macroscopic view of safety—safety management system—the next step is to add depth to the understanding of what safety is and how to achieve it. The purpose of this chapter is to detail a few of the most commonly used techniques in hazard analysis and to demonstrate their practical applications. It is hoped that you will walk away with a good understanding of how and when to apply these tools. The safety techniques are not difficult to learn and actually are quite easy to pick up and apply immediately.

Chapter 5 walks you through the hazard analysis procedure, explaining how the most frequently used methods work. This hazard analysis procedure is the foundation for many of the other safety techniques described in this book. Succeeding chapters and sections work through real examples of safety analyses, taken from various industries. In this, you will see how the safety technique is actually applied, along with some practical tips on implementation.

It is important to remember that hazard analyses are not to be done in a vacuum. Hazard analysis (as well as other safety analyses) is an umbrella under which an engineer performs other standard engineering analyses, such as fluid flow calculations and dynamic and static analyses. The purpose of the hazard analysis is to identify hazards to the system, evaluate the hazards by determining their impact severity and the probability of occurrence, rank those risks in a prioritized order, and then implement controls to those hazard risks.

Charles Perrow, in his deftly written book, *Normal Accidents: Living with High-Risk Technologies* (1984), argues that with the advent of high complexity and interactivity in our technological systems, even small perturbations in the system—nuances usually not identified or taken seriously by the engineer—can cause horrendous consequences. Perrow goes on to say that complex systems do not necessarily imply highly sophisticated technology, only that its interactions are not readily apparent to the engineer. Many times these hidden interactions can have undesired consequences. Perrow calls this *tight coupling*. Interactions are not always the linear responses that we expect. Their nonlinearity and propagations through the system can cause disaster.

A good example is the demise of the *Dauntless Colocotronis* (Perrow, 1984), an oil tanker slowly steaming up the Mississippi. The tanker grazed the top of a submerged wreck, taking out a large piece of hull in the bottom of the ship next to the tanker's pump room. Oil gushed into the Mississippi. Also, some of it slowly seeped into the pump room. From the pump room, it seeped into the engine room through a packing around a shaft. The heat of the engine room quickly caused the oil to vaporize, and a stray spark in the vicinity caused an explosion. Unfortunately, the *tight coupling* of the tanker did not allow the system to recover easily and safely from the mishap.

So what is important in a hazard analysis? The answer is that *the hazard analysis must be systematic and comprehensive*. In other words, it must be a method that can be applied to the entire system. It must also identify interactions and their consequences and the linkages in a system. Here is where the concept, *the devil is in the details*, should be memorized. If the safety analysis cannot discover and analyze these hidden interactions, then we are only applying safety by luck—the worst possible thing to do.

The very first thing you need to understand about the hazard analysis methodology is what a system is. We all know that a system is composed of many interacting subsystems: procedures, support equipment, people, software, hardware systems, facilities, operating environment, and other interfaces. All of these subsystems operate in the overall natural environment. Figure 5.1 illustrates how all of these subsystem parts interact with each other and how a perturbation in one can propagate into another. A safety analysis must look at all of these parts together and their interrelationships. The tighter the constituent parts are linked together, the stronger the interactions will be, and when something changes in one part, it will most likely affect other parts.

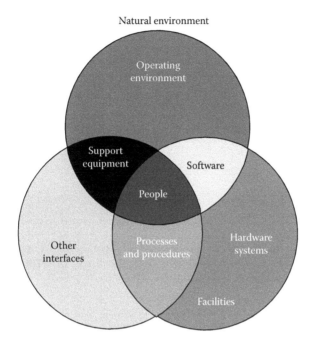

FIGURE 5.1 **(See color insert.)** The system.

5.1 HAZARD ANALYSIS METHODOLOGY

The hazard analysis process is a systematic, comprehensive method to identify, evaluate, and control hazards in a system. Figure 5.2 shows how easy it is to apply.

The first step is to define the analysis criteria and parameters. Then it is important to define and understand the physical and functional characteristics of the system under study. It is important to look not only at the major subsystems but also at their functions and interrelationships. Understanding the subsystem and system interfaces is critical to identifying hazards. Many engineers fail at this stage because they feel they adequately understand how the system works and do not need to spend time accurately defining it. What is important is not just how the system works but also its operating conditions and environment. Remember to look at the system and its elements in context to their surroundings. This means that it is critical to define the people, processes, and technologies that make up the system. The elements of Figure 5.1 are a pictorial representation of the kinds of things that should be viewed in defining the system.

Next, hazards and their root causes must be identified. You should go through the system step by step and postulate what the associated hazards with this system under all operating conditions are (including abnormal conditions). It is important to

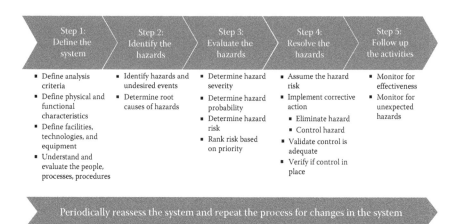

FIGURE 5.2 Hazard analysis process. (Revised from Hathaway, W.T. and Markos, S.H., *Passenger Carrying Submersibles: System Safety Analysis*, DOT-TSC-CG-89-2, U.S. Department of Transportation, Washington, DC, 1989.)

study the system through all phases of the life cycle—remember that there may be very different hazards at particular stages in the life cycle.

Once the hazards, including the causal factors, have been identified, it becomes important to evaluate the hazards and their effects. Most hazard analysis methodologies apply some type of severity classification. This classification is used as a marker to compare the consequences of one hazard to another. Usually some engineering analysis is done so that you can understand what the effects of the hazard would be if an accident did occur.

Just stating worst-case scenarios is not sufficient in hazard analysis. To control hazards, you must understand the probability or likelihood of the hazard event actually occurring. Most hazard analysis techniques use a qualitative probability ranking. Other safety analyses such as fault tree analysis or probabilistic risk assessment use quantitative analysis. Either way, the event probability must be determined before management can decide whether the risk should be eliminated, controlled, or accepted. If the hazard analysis does not address *both* hazard severity and likelihood of occurrence, then the analysis is not very useful. Hazard analysis methodologies employ a qualitative ranking system, merging the severity of the accident with the probability of the event occurring to give a hazard risk. This is then rank ordered.

This risk ranking is then used to decide whether the hazard risk should be accepted or not. This disposition or resolving of hazards requires the acceptance of the risk or implementation of a corrective action system to eliminate or control the hazard. The hazard reduction precedence described in Chapter 2 is then applied to determine how best to eliminate or control the hazard. It is imperative to validate that the control is adequate—that is, it actually does control the hazard—and to verify that the control is physically in place.

NOTES FROM NICK'S FILE

Once, working on a two-phase flow system, we needed to make sure that it was two-fault tolerant to an accident—in other words, after two failures, the system was still safe. We put in two relief valves to handle overpressure. The problem was they both had to operate in tandem to handle the flow and pressure profiles. That was an important example of validating that the hazard control is adequate. Obviously, the two relief valves weren't independent in their operations and therefore didn't independently control the hazard. Luckily, we *validated* our control was *NOT* adequate during testing and resized the relief valves.

The last step is to conduct follow-up activities. It is important to monitor the system to ensure the effectiveness of the hazard controls and to check for new or unexpected hazards. Things change so it is important to periodically reassess the system and determine if hazards are still adequately identified and control mechanisms still work. This is especially important if the system is modified, expanded, or reconfigured or operating conditions change. If there are any material changes to the system, then the hazard analysis should be updated to reflect the changes and their impacts to the system.

All of the hazard analysis techniques discussed in this chapter follow this hazard analysis process. The remainder of this chapter describes the most commonly used hazard analysis techniques and gives some concrete examples of their applications. However, the first step in using hazard analysis is to create a preliminary hazard list (PHL).

5.2 PRELIMINARY HAZARD LIST

Looking back at Figure 5.2, you can see that once the system is well understood, the next step is to identify hazards. A PHL is a brainstorming tool to identify as many hazards as possible in a system and provide input for the hazard analysis. The idea is to develop a list of all possibilities, without regard to the likelihood of the event actually occurring (this will come later on in the hazard analysis process). Once the list is completed, then you can go through it and cross out what is not credible.

It should now be obvious why the first step in hazard analysis is to fully understand the system: this is precisely where you will find hazards. The PHL should be performed as early in the design process as possible. It can also be conducted at various stages of the design as a *sanity check*. It can be very helpful in evaluating existing designs as well. So to identify hazards, you should do the following:

- Use a *team approach*. Gather engineering and management representatives from each major discipline (mechanical, electrical, structural, operations, maintenance, etc.) and brainstorm together.
- Review any prior safety data (i.e., Occupational Safety and Health Administration [OSHA] injury rates, National Safety Council data, safety analyses, company records, and safety trend analyses).
- Review data from previous accidents, designs, and operating experience (in the company and the industry).

- Examine and inspect similar designs (including identification of the kinds of hazards and how they were controlled); actually go to the sites and look them over.
- Talk to current or intended users and operators (compare written procedures to the way work is actually done; pay close attention to shift operations); witness operations.
- Study system specifications and expectations.
- Review applicable local, state, and international codes, standards, and regulations (see Section 2.9 for more details).
- Review detailed design data (electrical, mechanical, structural, materials handling, fluid flow schematics, etc.) and detailed engineering analyses (stress analysis, thermal analysis, mechanical design, etc.).
- Review test data (remember that much retrofitting and design modifications are made either during or immediately after in-house tests).
- Study preventative, scheduled, and unscheduled maintenance records (a lot of information can be gleaned from these often-overlooked sources).
- Consider all life-cycle phases of the system or product. Different hazards can occur at different stages of development or operation.
- Consider all system elements (as shown in Figure 5.1, including human factors, local weather, and organizational systems).
- Identify all energy sources and trace their propagation through the system (Appendix A gives an example list of some typical energy sources).
- Review generic hazard lists (Appendix B is an example hazard list).

From this information, a PHL is created. The idea is to list all the possible hazards without regard to likelihood or severity. This can be an arduous task for the design of a new petrochemical plant, or it can be fairly easy for a small retrofit or modification of a cooling tower. The PHL is then divided into hazard category sublists. This will help you to manage the mountain of data generated. Typical hazard categories found in all industries are

- Collision
- Contamination
- Corrosion
- Electrical
- Cyber controlled
- Explosion
- Fire
- Human factors
- Physiological factors
- Loss of capability
- Ionizing radiation
- Nonionizing radiation
- Temperature extremes
- Mechanical
- Pressure

5.3 PASSENGER-CARRYING SUBMERSIBLE EXAMPLE

The U.S. Department of Transportation and U.S. Coast Guard are responsible for enforcing a minimum level of safety for all U.S. flag vessels. A number of private, commercial companies have been offering underwater sightseeing tours in the Caribbean Sea and the Pacific Ocean. These tours allow tourists to explore underwater shipwrecks, coral reefs, and other sea life. Manufacturers need to demonstrate to the Coast Guard that commercially provided passenger submersibles are safe. Briefly walking through part of the Coast Guard's hazard identification and hazard analysis process is a good way to see how it actually works.

An example PHL is shown in Table 5.1. The Coast Guard followed the same process described earlier (Hathaway, 1989). Notice that the hazard list varies somewhat from the hazard category list. The reason is that each system is unique, and therefore, hazards and their categories could be different. Also, the list really is a compilation of hazard sources. Each item is actually a compact description of many more individual hazards, which are too numerous to show here.

Once the hazard boundary conditions have been created, the system can be divided into a system functional organizational chart. The purpose of the system functional organizational chart is to help you order the information in such a way as to ensure that the entire system—hardware, software, facilities, support equipment, operating environment, etc.—is studied. Figure 5.3 is an example of how the submersible system was subdivided. It is important to remember that it is not critical how the system is functionally parceled; it all depends on what makes the most sense. Also, this step is very important; someone reviewing the final analysis will be able to use this functional diagram to step through the process and note any logic errors.

The functional tree is obviously a hierarchical structure, going from the more general top item—or total system name—to the more specific subelements (i.e., life support systems). You walk through each item in the functional tree and develop a PHL.

TABLE 5.1
Passenger-Carrying Submersible PHL

Collision (underwater or surface)
Entanglement
Fire
Flooding
Loss of power
Passenger illness
Loss of air in ballast/trim system
Stranded on bottom
Emergency or uncontrolled ascent
Inability to rescue submersible
Oxygen leak/CO_2 removal system failure
Loss of communication

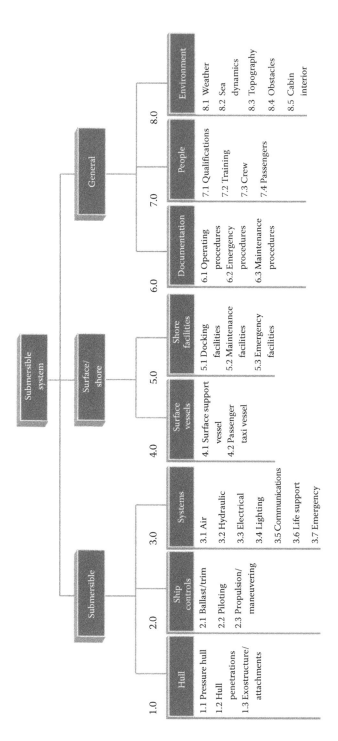

FIGURE 5.3   Passenger-carrying submersible functional tree. (From Hathaway, W.T. and Markos, S.H., *Passenger Carrying Submersibles: System Safety Analysis*, DOT-TSC-CG-89-2, U.S. Department of Transportation, Washington, DC, 1989.)

As you go through a system, it very rapidly becomes apparent that hazards are not always well confined within one subsystem but actually reach across various barriers, both physical and functional.

NOTES FROM NICK'S FILE

A lot of people who are doing hazard analysis for the first time ask me how do I take all these disparate data and make sense of them. This looks a little crazy but is something that I've picked up over the years doing lots and lots of hazard analyses. I call my approach the 3D approach. Here it is:

Look at your system from three perspectives or dimensions: (1) Find patterns in your data. I call it a static view (like a 1D view). (2) Identify relationships—people and info—between the patterns (2D). And then (3) look at the information from a *temporal* perspective—temporal movement or changes, in the patterns and relationships (3D). In summary, I'm looking at the data from a perspective of patterns that the data fall into, relationships between the identified patterns, and then how the patterns and relationships change over time.

5.4 HAZARD ANALYSIS: PRELIMINARY, SUBSYSTEM, AND SYSTEM

For many years, system safety engineers have made quite a bit of noise about the differences between preliminary, subsystem, and system hazard analyses. In reality, there is really only one hazard analysis. The primary objective of a hazard analysis is to *identify all possible hazards*, then it must *categorize the hazards in terms of severity of consequences (catastrophic, critical, marginal)*, and then *it must evaluate the probability of the hazard occurring.*

So if we say that the hazard analysis is the baseline document, then *a preliminary hazard analysis (PHA) is the initial hazard assessment conducted on the system.* It identifies safety-critical areas within the system and starts evaluating hazards and identifying safety design criteria and applicable safety requirements. Also, *a subsystem hazard analysis (SSHA) examines each major subsystem* (such as shown on the functional organizational tree in Figure 5.3) *and identifies specific hazards and safety concerns including failures, faults, processes, or procedures and human errors.* An SSHA also should address hazard controls and how those controls are verified.

A *system hazard analysis (SHA) does the same thing as an SSHA except that it identifies hazards across subsystem boundaries and interfaces.* It looks at system-level hazards. An SHA also should list hazard controls and verifications.

Figure 5.4 shows pictorially how the system, subsystems, interfaces, and other systems interconnect. In conducting a hazard analysis, you must look at each

FIGURE 5.4   (**See color insert.**) A car in traffic as a system.

subsystem, its interfaces, the entire system and how it operates, and other systems and their influences:

System: The car in traffic (with all the other cars, traffic lights, weather, maintenance schedule, etc.)

Subsystems (*some*): Car electrical system, climate control subsystem, audio system, car warning and instrumentation subsystems, steering subsystem, engine cooling subsystem, drive train, emission control subsystem, driver (physiological, psychological factors), safety systems (e.g., air bags, seat belts, antilock brakes, hazard signals, and electrical circuit protection)

Components in a tire subsystem: Tires, spare tire and accessories, brakes, tire treads, tire maintenance schedule, and tire operating parameters (i.e., tire pressure, driving style, driving terrain, tire rotation, and alignment)

Some interfaces: Tires on the road, air bags to passengers, car chassis to electrical system

Some other systems: Traffic lights, traffic flow pattern, accidents on other side of the highway, feeder roads, etc.

Returning to the submersible example, potential hazards and causes are identified for each of the systems and subsystems within the functional areas. Once the hazards and their corresponding causes are found, you need to figure a way to rank those hazards. There are literally millions of hazards in any given system. Which ones are most important? Do we need to fix all the hazards? How can we tell which ones to attack first, especially with a limited budget?

The answer is to correlate the hazard severity and probability or likelihood and create a ranking scheme. Different safety analysis techniques handle this question in different ways. The hazard analysis standard is Mil-Std-882 *System Safety Program Requirements* (U.S. Department of Defense, 2012). The ranking method is a very easy, three-step process:

1. First, assign hazard severity categories to each hazard.
2. Then, allocate the probability of occurrence of hazard (this can be either qualitative or quantitative, depending on the confidence level of your data).
3. And finally, correlate the two to rank which hazards will be addressed first.

Tables 5.2 and 5.3 are the most commonly used hazard severity and probability of occurrence classifications used in hazard analysis. You may wish to revise these two tables slightly to reflect your particular system more accurately. There are many other standards and guidelines that have variations of these tables. You do not have to use the ones shown here. If it makes more sense, create ones that better fit your needs.

Hazard severity, in Table 5.2, is a qualitative measure of the worst credible mishap resulting from the particular hazard cause. For more granularity, the hazard severity table can be quantified. For example, the safe distance or exclusion zone of an expanding toxic vapor cloud could be used to categorize hazard severity. You may also wish to expand the severity definitions and customize them to fit your particular operation.

This table is used for separating credible and noncredible hazards through a methodical approach. You can now rate each hazard in the PHL with severity categories.

Hazard probability, in Table 5.3, is the likelihood that the hazard will occur during a given time period. Assigning a quantitative value early in the design process or of a

---

TABLE 5.2
Hazard Severity Categories

Description	Category	Definition
Catastrophic	I	Death, permanent total disability, system loss, irreversible significant environmental impact, or monetary loss equal to or exceeding $10M
Critical	II	Permanent partial disability, injuries, or occupational illness that may result in hospitalization of at least three personnel, reversible significant environmental impact, or monetary loss equal to or exceeding $1M but less than $10M
Marginal	III	Injury or occupational illness resulting in 1 or more lost workdays, reversible moderate environmental impact, or monetary loss equal to or exceeding $100K but less than $1M
Negligible	IV	Injury or occupational illness not resulting in a lost workday, minimal environmental impact, or monetary loss less than $100K

Source: U.S. Department of Defense, Military standard: System safety program requirements, Mil-Std-882E, U.S. Department of Defense, Washington, DC, 2012.

TABLE 5.3

Qualitative Hazard Probability Levels

Description[a]	Level	Specific Individual Item	Fleet or Inventory[b]
Frequent	A	Likely to occur in the life of an item.	Continuously experienced.
Probable	B	Will occur several times in the life of an item.	Will occur frequently.
Occasional	C	Likely to occur sometime in the life of an item.	Will occur several times.
Remote	D	Unlikely, but possible to occur in the life of an item.	Unlikely but can reasonably be expected to occur.
Improbable	E	So unlikely that it can be assumed occurrence may not be experienced in the life of the item.	Unlikely to occur, but possible.
Eliminated	F	Incapable of occurrence. This level is used when potential hazards are identified and later eliminated.	Incapable of occurrence. This level is used when potential hazards are identified and later eliminated.

Source: U.S. Department of Defense, Military standard: System safety program requirements, Mil-Std-882E, U.S. Department of Defense, Washington, DC, 2012.

[a] Definitions of descriptive words may have to be modified based on quantity involved.

[b] The size of the fleet or inventory should be defined.

process where little quantitative information is available is usually difficult. However, that does not obviate the fact that to assess the hazard, you must still take a stab at stating what the likelihood of occurrence is. If there are a lot of data, then numerical probability levels are usually better than just qualitative values. Refer to any traditional probability and statistics handbook for numerical manipulation techniques.

Again, you want to go back through the PHL and rate the likelihood of occurrence of each hazard. And, of course, the table can be customized to fit your particular needs.

The next step is to correlate the severity with the likelihood and make an assessment as to the relative risk of an accident. Chapters 13 and 14 describe a more sophisticated approach to risk assessment. For now, however, the fairly simple risk assessment matrix in Table 5.4 is the most often used in hazard analysis.

The purpose of the matrix is to help you prioritize hazards for corrective action. The categorization of hazards is based on severity and likelihood. Some hazards may be very likely to occur but of very minor consequences. One example is the minute release of nitrogen gas from a flapper valve into a well-ventilated, open area. Even if release is frequent, the severity of the hazard is low because the quantities are so low. However, an explosion at a commercial nuclear power plant may be remote (but obviously not impossible, as demonstrated by Chernobyl, or the remote possibility of an earthquake creating a tsunami wave hits a nuclear power plant and causes a meltdown as demonstrated by the Fukushima Daiichi nuclear disaster), but the consequences are great. These two hazards must be treated differently. Engineers too often treat all hazards equally, either overreacting or underreacting to the risk.

The hazard frequency can be either qualitative or quantitative. The matrix shows both for illustration. If you do wish to quantify the frequency of occurrence, it is

TABLE 5.4
Hazard Risk Assessment Matrix

Hazard Category Frequency	(1) Catastrophic	(2) Critical	(3) Marginal	(4) Negligible
(A) Frequent ($x > 10^{-1}$)	1A	2A	3A	4A
(B) Probable ($10^{-1} > x > 10^{-2}$)	1B	2B	3B	4B
(C) Occasional ($10^{-1} > x\ 10^{-3}$)	1C	2C	3C	4C
(D) Remote ($10^{-3} > x > 10^{-6}$)	1D	2D	3D	4D
(E) Improbable ($10^{-6} > x$)	1E	2E	3E	4E

HRI	Risk Decision Criteria
1A, 1B, 1C, 2A, 2B, 3A	Unacceptable; stop operations and rectify immediately.
1D, 2C, 2D, 3B, 3C	Undesirable; upper-management decision to accept or reject risk.
1E, 2E, 3D, 3E, 4A, 4B	Acceptable with management review.
4C, 4D, 4E	Acceptable without review.

Source: Modified from U.S. Department of Defense, System safety program requirements, Mil-Std-882C, Washington, DC, 1993, p. A.5.

Note: This table does not include Category F, eliminated as shown in Table 5.3. Also, this table was taken from an earlier Mil-Std version than the one cited in the preceding tables.

very important that you understand clearly how you arrived at that frequency range. Once an event is quantified, people tend to view it as a discrete function and not as a continuous one. NASA suffered this problem early in the Apollo program. The quantitative risk values indicated that going to the moon was too risky. A reanalysis of the mission in qualitative terms demonstrated that it was feasible. This is a good example of how numbers can be misleading. Because NASA was embarking on something never before attempted, it was extremely difficult to quantify the risk involved.

Equally important to assessing the probability and severity of the hazardous event is knowing what to do with that information. The hazard risk index (HRI), coupled with the risk decision criteria, indicates how the hazard should be treated. Obviously, if the hazard is catastrophic and frequent, it must be controlled immediately. So, an HRI of 1E, 2E, 3D, 3E, 4A, or 4B would have a lower priority for corrective action than one of 1A, 1B, 1C, 2A, 2B, or 3A.

NOTES FROM NICK'S FILE

I was working with a global manufacturing company, and we were designing their hazard analysis process. We tried to convince them to use standard hazard severity, probability, and hazard risk assessment matrices. But they wanted their own that fit their processes. Interestingly enough, we spent little time with the severity and probability categories. But we spent a lot of time agreeing across the company on the risk decision criteria—what is acceptable and unacceptable. This had major implications because it determined when the production line would be shut down.

This step in the hazard analysis process is important; as mentioned in Chapter 4, the need to track hazards and verify that either they have been controlled or the risk accepted is very important. A closed-loop tracking system prevents something from slipping through the cracks.

Now, all of this information must be put into some sort of format that makes review and tracking of hazards, causes, and corrective actions easy. A hazard analysis worksheet is prepared, taking into account

- Hazard description
- Potential causes
- Potential effects (including propagation throughout the system or to other systems)
- HRI
- Recommended corrective action
- Effect of corrective action implementation
- Hazard control references

Returning to the submersible example (Hathaway, 1989), Table 5.5 shows how this worksheet provides a good hazard tracking system and documents to present in court or to OSHA inspectors if your sincerity in maintaining a safe system is ever called into question. Many people call the hazard analysis tracking worksheet the risk register. This term is very common. Don't be put off by the term if you have not seen it before. It is the same as the hazard tracking worksheet. The reason they are the same is that the hazard tracking worksheet also contains the risk assessment and risk prioritization.

At the top of the worksheet, of course, is the title of the hazard analysis, the date of the analysis, the name of the analyst, and a page number. The left-hand side of the sheet, indicating the element, system, and subsystem, tracks directly back to the passenger-carrying submersible functional tree in Figure 5.3. In fact, the *control number* correlates to the particular components of a tree branch (e.g., 1.1 pressure hull). Remember that even as we do safety analysis, we must also be sufficiently organized to be able to demonstrate to regulatory inspectors that we are tracking all items and have disposition of all these items.

The next column, *Hazard Description*, is very important. Though it should be written in succinct language, it should still be sufficiently detailed to allow the reviewer to understand what the hazard is. You may wish to try to use some sort of standard words, thus giving you the maximum possibility of sorting this information in a database.

Potential Causal Factors is the third column. Obviously, identifying the hazard is not enough. You must study the hazard sufficiently to be able to identify the potential causes. It is these potential causes (or the following column, *Potential Effects*) that you want to eliminate or control. Notice that each cause is entered separately. Again, this is done for tracking purposes. Many times engineers think that by controlling one cause, they have controlled the hazard. Whether the causes are independent, dependent, or a combination of the two, they must still be controlled. Remember that you do not need to have a failure to create a hazard. If you are at the wrong end of a gun and are killed, no failure occurred: the gun operated as designed, but there still was a hazard.

TABLE 5.5

Sample Passenger-Carrying Submersible Hazard Analysis Worksheet

Element: Submersible

System: Hull

Subsystem: Pressure Hull

Date: 03/11/2014

Analyst: John Doe

Page: 45

Control Number	Hazard Description	Potential Causal Factors	Potential Effects	HRI	Hazard Control Recommendation	Effect of Recommendation on HRI	Hazard Control References	Verification of Control	Status of Control	Notes
1.1.01A	Implosion or failure of hull	Improper design of hull	Internal flooding	1D	Follow CFR, MTS, ASME, ABS, and Navy for design of pressure hull.	1E	46 CFR 54, 197.328. MTS II. Sec. B.2.0 and B.4.0. ASME PVHO-1A. Sec. 1.3 ABS. Sec. 9. NAVMAT P-9290, Appendix B. USCG, May 87, P. 3	Conduct formal design review program per hazard control references.	Open. First design review scheduled for April 2015	
1.1.01B	Implosion or failure of pressure hull	Improper material selection for pressure	Internal flooding	1C	Follow CFT, ASME, ABS, and Navy for pressure boundary material specifications and testing.	1E	46 CFR 176.05, 176.10, 177.10-1. ASME PVHO-1A, Sec. 1.2. ABS, Sec. 3. NAVMAT P-9290, Appendix A.	Conduct formal material selection review and workmanship testing program per hazard control references.	Open. Ongoing	

(Continued)

TABLE 5.5 (*Continued*)
Sample Passenger-Carrying Submersible Hazard Analysis Worksheet

Control Number	Hazard Description	Potential Causal Factors	Potential Effects	HRI	Hazard Control Recommendation	Effect of Recommendation on HRI	Hazard Control References	Verification of Control	Status of Control	Notes
1.1.01C	Implosion or failure of pressure hull	Improper fabrication of pressure hull	Internal flooding	1C	Follow CFR, ASME, ABS, and Navy for fabrication inspection during manufacturing.	1E	46 CFR 177.10-1. ASME PVHO-1A, Sec. 1.3. ABS, Sec. 4. NAVMAT P-9290, Chpt. 4 & Sec. B.3. Inspection: 46 CFR 176.05, 176-10, ABS, Sec. C.17.	Implement formal manufacturing quality assurance review and inspection per hazard control references.	Open. Ongoing	

Source: Modified from Hathaway, W.T. and Markos, S.H., *Passenger Carrying Submersibles: System Safety Analysis*, DOT-TSC-CG-89-2, U.S. Department of Transportation, Washington, DC, 1989.

The fourth column, *Potential Effects*, describes the expected results if the hazard is not controlled. Here, you can see the propagation effects of a hazard into a system and how tight coupling can have disastrous effects. You should also have separate *control numbers* for each *potential effect* listed. One causal factor may have various effects. Some hazard analyses show the potential effects at the different indent levels: element, system, and subsystem.

The next entry is *HRI*. This is the severity and probability as taken from Table 5.4. You have to analyze the hazard and make an assessment of its risk. This risk index is particularly useful in database management. You can sort and print out all HRIs that are IA to IC, for example. This then allows you to prioritize which hazards will be rectified first and also help track those hazard closures.

Hazard Control Recommendation is one of the most important entries. Criticizing a design without offering solutions is unacceptable. Of course, this column is probably the most debated within an organization. There may be various recommendations: some expensive and possibly some not. In this column, the information should be listed briefly. After the worksheets are completed and the safety information is analyzed, then the recommendations need to be greatly expanded, listing in detail how the hazard is to be controlled, including design reviews, tests, and analysis.

Effect of Recommendation on HRI gives you the understanding of how effective the recommendation is to controlling the hazard. This helps you determine if the control is really worth the effort.

Hazard Control References cite any standards or norms that relate to the control of the hazard. The references will assist you in identifying which standards are the most applicable.

The next column, *Verification of Control*, is very important. This information is a method to ensure that the control is sufficient and in place. Many times wonderful controls are created but not implemented. The purpose of this column is to verify that the control is really there and that it does its job. This is an important point to demonstrate to safety regulators and inspectors.

Status of Control indicates whether the hazard control is open (not controlled), closed (controls are in place and verified), or in progress. Hazards must be tracked until closed.

The *Notes* column allows you to include any additional information that should be noted.

What has just been described is the heart of the hazard analysis, a methodical approach to reviewing the entire system. Hazard analysis worksheets can be designed into any commercially available software so that you may more easily manage the data.

With the hazard information located in a database, you can easily sort by anything you want. For example, if you wish to sort all hazards in the system related to failure of the pressure hull, you can do that. Or you may wish to sort all hazards that are still open, listing due dates and status. Sorting the recommendations column can give you insight about how significant the design or operational changes to the system need to be.

The safety analysis worksheet database is the foundation for creating an in-house safety database. Chapter 10 explains further how you can take advantage of these databases and gives examples of safety databases you can use. With this database, some of the information you can glean includes

- Safety lessons learned, creating corporate memory
- A quick review of how you may have controlled a similar hazard on an earlier system
- A listing of all design, safety, or operational requirements, standards, or norms levied on a system or how certain requirements have been implemented
- A listing of *typical* hazards and their effects on particular systems and sub-systems (this is very useful, saving money by not rediscovering the kinds of hazards that may exist)
- A quick sanity check for other systems to help ensure that you have not missed any other hazards
- A listing of all open items and their status, avoiding the *dropping out of data* and something falling through the cracks (especially important for large or highly dynamic systems or programs)
- Sorting common causal factors, helping to identify trends and patterns
- Sorting of potential effects, bouncing off other hazard lists, again ensuring that all potential effects have been identified
- Helping in trending data of how a hazard may propagate through a tightly coupled system, by sorting combinations of causal factors and potential effects
- Printing out the worksheets, and showing off the database, demonstrates to OSHA or other safety inspectors that you are serious about safety and controlling hazards

As mentioned earlier, the hazard analysis worksheet is not the end. You have gone through and looked at all potential hazards with your PHL. From the PHL, you have identified typical hazard categories and used them to generate a functional tree that describes your system. Methodically going through every function in the tree, bouncing off the PHL each hazard on the list, you now have listed all the hazards in your system. Once again you have gone back and looked at each hazard in your subsystem, system, and interfaces, assigning a probability or likelihood of occurrence. In your hazard analysis, you have assessed each hazard for severity and probability and have evaluated the risk associated with that particular hazard. Your hazard analysis not only identifies the hazard, causes, and effects, it also recommends corrective action, cites appropriate standards or norms, and indicates how controls are verified to actually be in place and adequate. So what is left to do?

The last step in the hazard analysis methodology is the *safety assessment*. The safety assessment is just like any research paper in that it discusses the results of the analysis. If you think of the hazard analysis worksheets as the analysis results, then the safety assessment is the discussion of those results.

The purpose of the safety assessment is to sort through the mountain of data in the worksheets and perform a comprehensive evaluation of the hazards and risks and controls. Depending on your initial safety criteria of what constitutes an acceptable level of risk, the safety assessment will discuss the information in the worksheets and put it into perspective with the overall system.

The safety assessment summarizes the methods used to eliminate or control the hazards identified on the worksheet. It should summarize the following:

- Safety criteria and definition of risk (especially acceptable and unacceptable risk).
- Any assumptions used in the hazard analysis.
- Definition of credible hazards and evaluation criteria.
- Discussion of hazards and how they will be controlled or the effects will be mitigated.
- Discussion of how hazard controls are validated to work and verified to be in place (describe test and analysis results and inspection reports).
- Which hazard control references were cited and why and how the controls meet the requirements.
- Status of hazard controls; those hazards that are still open (or that are controlled on a continuous basis, such as by procedure) should be addressed.
- The safety assessment should explain to the reader how the hazard analysis plugs into the safety management system (see Chapter 4 for more details on safety management system).

Continuing our passenger-carrying submersible example, the safety assessment would look something like the one shown in Table 5.6.

PRACTICAL TIPS AND BEST PRACTICE

- Assembling such a large array of data may seem pretty intimidating. Remember that most of the time you already have these data somewhere in your office or in the company. Many times you will only have to do extensive analysis of a few areas that are not already covered somewhere else.
- Also, tailor the analysis to your needs. Use everything listed here, but decide what level of detail is necessary. Remember that this is *system* safety engineering. You are actually doing a systems analysis. If you already do that, then you already have most of the data in place. If you are not, then this will serve two purposes: incorporate safety into your system and give you an overall view (systems analysis) of how well your total system operates. Two for the price of one.
- Keep the worksheets in the database updated. If you can do that, then the next time you do a safety analysis, use the database as a baseline document and just change what needs to be modified. Be careful not to assume that two systems are exactly alike. We all know that they are not.

- If you add cost data with each hazard control and then trend the success of those data, you are well on your way to modeling the cost of upgrading the system. You can take the entire hazard analysis process and change out the hazards for design or operational trade-off ideas, associate costs to each trade-off, trend the data with any typical statistical package, and have a pretty elaborate efficiency and productivity analysis of your system.

TABLE 5.6

System Safety Assessment Summary

System safety methodology
    Brief description of hazard analysis methodology
    Definition of risk
    Safety criteria used in the analysis
    Assumptions
    References
    Sources of data
Discussion of hazard analysis
    Description of hazard analysis worksheets
    Summary of hazards
    Discussion of how hazards are controlled and verified[a]
        Analysis (stress analysis, thermal analysis, etc.)
        Test (vibration testing, acceptance testing, life testing, coupon testing, etc.)
        Inspection (nondestructive evaluation, quality control inspection, etc.)
        Design reviews (critical design review, safety review, operational reviews, mechanical systems group review, electrical subsystem manager review, etc.)
        Training (general training of personnel, emergency procedures)
        Emergency preparedness
        Procedures (operations, emergency procedures, maintenance, preventive maintenance, emergency maintenance, etc.)
Status of safety control verifications (a tabular form of which controls are in place and therefore closed and which are not, along with expected closure dates and responsible individuals)
Cost of implementation
The hazard analysis in the system safety program
Appendices (hazard analysis worksheets, analysis report summaries, test report summaries, inspection summaries, etc.)

---

[a] Each major hazard or hazard category (electrical shock to personnel, flooding, fire, air contamination, etc.) should be discussed separately, addressing how the hazard is controlled and addressing how each of the items listed in the "Discussion of Hazards" section verifies that controls are adequate and in place.

5.5 FACILITY HAZARD ANALYSIS

Various extractions have been taken from the basic hazard analysis format. The SSHA looks only at hazards within a subsystem. Likewise, the SHA concentrates on system-level hazards. PHA looks only at the initial design of the system. Fault hazard analysis emphasizes faults in a system that can create hazards. As stated previously, these hazard analyses really are all the same thing.

There is one that is sufficiently different that it should be treated separately—the facility hazard analysis. As the name implies, it focuses on hazards in a facility. The same hazard analysis procedure is used, but emphasizing facilities and the facility acquisition process. The purpose of the facility hazard analysis is to apply hazard analysis techniques to a facility and its operations for the entire facility life cycle—from concept through disposal. The *facility* is construed to mean actual buildings, the area around the buildings, and the operations into, out of, and inside the buildings. Specifically, it looks at items such as the following:

- The facilities themselves, including building structure electrical (power) systems, lighting, heating, ventilation, and air-conditioning
- Facility siting, including where the facility is located geographically, other facilities located near the facility of interest, and other outside environmental or operational issues that could impact the facility
- Facility modifications
- Fire protection systems, including fire suppression; fire detection; alarms, monitoring, and communications; and fire department operations
- Facility operations, including manufacturing lines, office areas, and management
- Pressurized systems, such as pneumatic, hydraulic, and two-phase flow systems
- Materials handling, including storage; transfer, delivery, and movement; disposal or waste; and bulk systems
- Handling of hazardous materials
- Unique operations, such as laboratories, computer rooms, and testing facilities
- Operator training (normal and emergency operations)

It is difficult to list all, or even the majority, of the items that fit into the description of a facility. The purpose of the preceding list is to give a flavor of what kinds of things you need to look at in a facility hazard analysis. Appendix C offers a very general outline of some of the items that should be considered in a facility hazard analysis. Obviously, you should start with this list and expand it to include the unique features of your facility. The functional divisions in Appendix C also can help you set up your functional tree.

The facility hazard analysis follows the same path and sequence of order as the hazard analysis process (review Figure 5.2 to refresh your memory). As mentioned in Section 2.6, the safety management system in a facility should also use the hazard reduction precedence: design out the hazard, use safety devices, use warning devices, and finally, use special procedures and training.

Again, a PHL is developed (use Appendix C as a starting point). The PHL is divided into hazard categories. The functional tree is created. Then the actual facility hazard analysis is started. Each hazard is assigned a severity and probability level, and the other portions of the hazard analysis worksheet are completed. Then a system safety assessment is performed and the worksheet results are analyzed.

A few deviations from the hazard analysis procedure are unique to facility hazard analysis and need to be discussed.

The purpose of a facility hazard analysis is to identify and evaluate hazards and make recommendations for the elimination and control of hazards. The major safety concerns that facility managers should address are loss of life or serious injury to personnel, reportable (to the Environmental Protection Agency and/or state and local agencies) hazardous materials discharge to the environment, serious damage to facilities or equipment resulting in large dollar loss, and hazards that could have serious adverse effects on the plant or company mission capability, operability, or public opinion.

As with other safety analysis tools, the facility hazard analysis should be performed as early in the program as feasible. The same hazard analysis format can be used, with the columns filled in with preliminary or available information. The facility hazard analysis should be revisited at the 30%, 60%, 90%, and 100% design review stages, updating the worksheets as information becomes available. When the acceptance inspection and operations review is conducted, the facility hazard analysis should again be updated. The same holds for facilities that are modified or retrofitted.

Because the cost of constructing or modifying a facility is so great, it is critical to focus your energy only on what is important. NASA uses a very good system (National Aeronautics and Space Administration, 1998), the facility risk indicator (FRI). The FRI is different from the HRI on the worksheet in that it ranks the risk of individual facilities in relation to one another and the overall mission of the organization. Obviously, much more effort should be given to a laboratory that uses high-pressure hazardous gases than to an office building or guard shack. Briefly, the FRI definitions, per NASA, are as follows:

FRI 1 (high risk): There is a *high* probability that hazards in this facility can cause loss of life. Hazards may result in loss of life, permanent disability, or serious occupational illnesses to one or more persons, three or more lost-time injuries, loss of facility operational capability for 1 month or greater, or damage to equipment or property in excess of $500,000.

FRI 2 (medium risk): There is a *medium* probability that hazards in this facility can cause loss of life. Hazards may result in permanent disability to one or more persons, hospitalization (associated with illness or injury) of three or more persons, up to two lost-time injuries, loss of facility operational capability from 2 to 4 weeks, or damage to equipment or property from $250,000 to $500,000.

FRI 3 (low risk): There is a *low* probability that hazards in this facility can cause loss of life. Hazards may result in hospitalization of one or two persons, occupational injury or illness resulting in a lost workday or restricted-duty case, loss of facility operational capability from 1 day to 2 weeks, or damage to equipment or property from $25,000 to $250,000.

There is a *low* probability that hazards in this facility can cause loss of life. Hazards may result in permanent disability to one or more persons, occupational injury or illness resulting in a lost workday or restricted-duty case, loss of facility operational capability for from 1 day to 2 weeks, or damage to equipment or property from $25,000 to $250,000.

FRI 4 (*acceptable risk*): Loss of life as a result of hazards in this facility is unlikely. Hazards may result in no lost workday injuries or no restricted-duty cases, loss of facility operational capability of less than 1 day, or damage to equipment or property less than $25,000.

Table 5.7 indicates how one company might rank its buildings. The ranking could be very different from one plant to another. Also, the current plant layout is very critical for ranking the buildings. In the example in Table 5.7, the day care center is fairly isolated from the rest of the plant operations. In other companies, the day care center might be located close to hazardous operations and therefore would receive an FRI of 1.

In Figure 5.5, you can clearly see that the functional tree is set up in exactly the same format as the passenger-carrying submersible. Also, Table 5.8 is almost the same as the passenger-carrying submersible SHA.

Table 5.8 shows what a typical worksheet might look like for a laboratory that handles hazardous gases. Note that this worksheet includes a position for the facility risk index. As you can see, if the control is implemented, then the FRI changes. Again, it is very important to include verification of control. This is what ensures that your control is adequate to control the hazard, and it is something that can be verified to be in place.

TABLE 5.7
Example FRIs for Various Buildings

Building	FRI
Cafeteria	3
Research projects lab	1
Gas cylinder storage building	3
Day care center	4
Maintenance building	3
Heating and refrigeration plant	1
Microchip processing facility	1
Information management and computer facility	2
Shipping and receiving	3
Main administrative building	4
Security gate house	3
Plant operations building	1
Tool and equipment storage	3

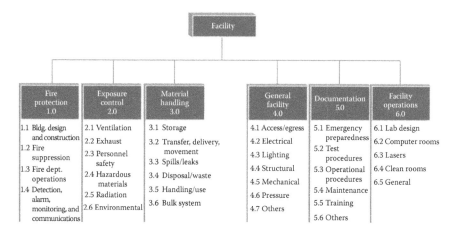

FIGURE 5.5 Sample facility functional tree.

5.6 OPERATIONS AND SUPPORT HAZARD ANALYSIS

One last transformation of the hazard analysis technique that is worth investigating is the operations and support hazard analysis (O&SHA). Most hazard analyses (and safety analyses, in general) are directed toward uncovering hardware design problems; however, this is not the intent of an O&SHA. Simply put, an O&SHA *identifies and evaluates the hazards associated with the operations of a system.* As with all hazard analyses, it looks at hardware systems, software, facilities, support equipment, procedures, personnel, operating environment, natural environment, human–machine interfaces, and other interfaces, but with the telling difference of how all of these factors relate to the operation of the system by people. The O&SHA is a very useful technique to understand how operations-focused hazards impact the system. It is not a human factors analysis. See Chapter 8 for more on human factors analysis.

Many engineers perform an O&SHA instead of a hardware hazard analysis. This is not a particularly good idea since the operations of a system or plant are intrinsically related to the hardware design. You may wish to use the O&SHA as a separate, more in-depth look at operational aspects of the system, but not in lieu of other analyses. Actually, the best idea is to combine the O&SHA with the hardware hazard analyses. Many times the human–machine interface is a very ambiguous area, and it is unclear which affects which.

Remember that the hazard reduction precedence indicates that using training or procedures to control hazards is the least effective method of hazard control. Unfortunately, this method may be the only viable option, due to the high cost of redesign. The O&SHA is used many times during the modification or upgrade of a plant or process. But the best time to conduct an O&SHA is during the original design.

TABLE 5.8

Sample Facility Hazard Analysis

Element: Chemical Facility

System: Processing Lab

Subsystem: Exposure Control—Hazardous Materials

Control Number	Hazard Description	Potential Causal Factors	Potential Effects	HRI	Facility Risk Index	Hazard Control Recommendation
2.4.01 or personnel asphyxiation	High hydrogen concentration leads to a fire devices in facility.	Lack of hydrogen gas detection	Personnel death, fire, or explosion	ID	2	Provide hydrogen gas detectors in lab areas that store and use hydrogen gas. Provide emergency power to the gas detection system. The alarm should sound both locally and on the emergency console in the process control center.
2.4.02	Release of toxic or highly toxic gases.	Lack of means to detect toxic and highly toxic gases	Personnel death or illness	IIA	1	Provide a continuous gas detection system to detect the presence of gas at or below the permissible exposure limit or ceiling limit in lab areas that store and use toxic and highly toxic gases. The detection system shall initiate a local alarm and an alarm in the emergency console in the process control center. The alarm shall be both visible and audible. The system shall be provided with emergency power.
2.4.03 toxic gases	Undetected buildup of hydrogen gas due to failure to accurately calibrate and maintain gas detection system.	Gas detection system illness	Fire/explosion, personnel death	IB	1	Provide accurate calibration and follow written maintenance plan.

(Continued)

TABLE 5.8 (Continued)
Sample Facility Hazard Analysis

Element: Chemical Facility
System: Processing Lab
Subsystem: Exposure Control—Hazardous Materials

Date: 05/05/2014
Analyst: John Doe
Page: 45

Effect of Recommendation on HRI	Effect of Recommendation on Facility Risk Index	Hazard Control References	Verification of Control	Status of Control	Notes
IE	3	UFC 80.303 (a)(9); UFC 80.303 (a)(7)	Review of Drwg. E607 and Design Spec. Section 16723. Verify as-built drawings to actual hardware with facility walkdown. Conduct operational tests.	OPEN. Currently in design phase. Review scheduled for 5/14. OPEN. Inspection to be completed during acceptance inspection TBD date	
IIE	3	UFC 80.303 (a)(9); UFC 80.303 (a)(7)	Review of Drwg. E607 and Design Spec. Section 16723. Verify as-built drawings to actual hardware with facility walkdown. Conduct operational tests.	OPEN. Currently in design phase. Review scheduled for 5/14. OPEN. Inspection to be completed during acceptance inspection TBD date	
IE	3		Plant quality assurance office to verify calibration and maintenance procedures are followed.	OPEN. Completion date TBD	

Perform the O&SHA as you would conduct any other hazard analysis. Follow the same procedure as outlined earlier in this chapter. However, this time, concentrate more on

- Operation or task sequence
- Concurrent task effects and limitations
- Planned system configuration at each phase of activity
- Human–machine–environment interfaces
- Planned and unplanned operations in the system (and its subsystems)
- Hazardous operations

The typical operational sequences you should assess are

- Normal operations
- Testing
- Installation
- Modification
- Support operations
- Maintenance
- Transportation
- Storage operations
- Servicing operations
- Contingency operations
- Emergency operations
- Activation and decommissioning
- Postaccident operations
- Training

The O&SHA should review a host of plant activities and documentation. Review the various operational and maintenance procedures; look at the mental and physical demands placed on the operators. Verify that the timing of procedures and operations is realistic. Section 8.2 goes into more detail about how to model human–machine interactions.

PRACTICAL TIPS AND BEST PRACTICE

When engineers conduct an O&SHA or some other analysis of operations, they many times assume that operators follow the written procedures precisely. Reality is vastly different. In fact, more times than not, operators take shortcuts and do not do exactly what the procedure states. Be sure to observe how the operators actually do their work. You will find a lot of surprises.

Review the written procedures. Verify the actual work performed (on all work shifts). Make sure that the various tasks do not lead to an accident. Some items to focus on include the following:

- Study each step of the operation and make sure that each individual step in the procedure is necessary, clearly understood, and conducted in a safe manner.
- Examine how human error (operator, maintenance, etc.) can alter the desired effects of the operation.
- If human error can affect a significant hazard, look for ways to control the hazard using the hazard reduction precedence.
- Any safety-critical operations must be clearly identified and assured of operation.

Table 5.9 shows a typical O&SHA worksheet. Of course, the worksheet can be in a tabular form as with the first two types of hazard analysis, but many engineers like to put each operational hazard on one worksheet. The figure shows such a format.

The only difference (besides visual) between the O&SHA and the other hazard analyses is that the O&SHA includes task descriptions and events leading up to hazard state.

Task descriptions describe the purpose of tasks or activities. It is important to remember to write the task or activity *intended*. The actual analysis of the task will verify if the operation really accomplishes what is intended.

Events leading up to the hazard state are an addition to the hazard analysis format. The events can occur either sequentially or in parallel or both. In developing controls to prevent the development of the hazard state, remember that you have the opportunity to interrupt the hazard event sequence at various points. It is critical that you spend some time in deciding where you wish to intervene. A poor decision can be very expensive.

Note that the O&SHA addresses only human errors or operator errors—not hardware failures. This is its strength and its weakness. Because this hazard analysis technique focuses on the operations of a system, it is very good at identifying the kinds of operational hazards that are often obscure to the engineer. However, as the aforementioned example shows, this worksheet does not identify any structural or design inadequacies of the crane or other lifting hardware. It is precisely for this reason that the O&SHA should never be used alone, but only in tandem with a hardware hazard analysis or a subset of the overall SHA.

5.7 EXAMPLES OF HAZARD ANALYSES

This section gives detailed, real-life examples of how hazard analysis, facility hazard analysis, and operating and support hazard analysis are applied. It is a difficult exercise to devise examples that are not overly complex, and therefore too long to be included here, and yet not make them trivial and nonrepresentative. The examples presented are actual engineering problems; however, the presentation has been truncated while retaining the important points that must be emphasized. It should be noted that these are not complete analyses but rather pieces of analyses that demonstrate how the technique is applied.

TABLE 5.9
Sample O&SHA Worksheet

Operating and Support Hazard Analysis

Date: 3/11/2014
Analyst: John Doe
Control Number: 1.1.01

Task description:

Nozzle parts loaded and unloaded into hydroclave for curing. Only trained and certified personnel are allowed to perform the operation (including the crane operations). Operations are performed with written procedures. All lifting and handling equipment is regularly tested and maintained. The quality assurance department is responsible for maintaining all hardware and personnel certification records.

Hazard description:

Mandrel and billet damaged while loading billet into hydroclave, affecting program-critical hardware and schedule. Damaged hardware causes possible injury or death to operations personnel.

Events leading up to hazard state:

1. Suspended billet is lowered onto mandrel in the hydroclave.
2. Operator of 15 ton crane attempts to lower and position billet load into hydroclave by himself, without attention or presence of an activity director.
3. Bottom of billet hits top of mandrel at excessive speed.
4. Structural damage inflicted on mandrel and billet.

Potential causal factors:

1. Operator of 15 ton crane attempted to lower and position billet onto mandrel in hydroclave without the assistance of an activity director, inattentive during operation.
2. Poorly written procedures or procedures not followed.
3. Adverse environmental conditions (i.e., lighting, air conditioning, noise) during operations.
4. Vision obstructed during operation.

Potential effects:

Operations personnel injured or killed due to loss of control of loading or lifting hardware. Major hardware damage forces shutdown of operations and major cost and schedule effects.

HRI: 1C

Hazard control recommendations:

1.1 Add activity director approval signature for Hydroclave Operations Proc. Traveler A. 1.2SS.
1.2 Conduct basic crane operations and hydroclave operations recertification courses every 24 months instead of current time period.
2.1 Conduct test operations with dummy loads to verify that written procedures are sufficient.
2.2 Add mandatory safety briefing to procedures.
2.3 Add to procedures verification step that three operators are present during task: activity director, spotter, and crane operator.
3.1 Measure average luminance, ambient temperature and humidity, and noise levels. Compare to ASHRAE standards.
4.1 Conduct test operations with dummy loads to identify blind spots. If equipment layout design cannot be modified, install necessary mirrors and/or video cameras to increase visibility.

(Continued)

TABLE 5.9 (*Continued*)
Sample O&SHA Worksheet

Verification of control:

  1.1.1 QA to verify all signatures on travelers before work commences
  1.2.1 Activity director responsible for assuring all personnel certified
  2.1.1 Lead engineer to review and approve new procedures
  2.2.1 Lead engineer to review and approve new procedures
  2.3.1 Lead engineer to review and approve new procedures
  3.1.1 Team leader to conduct analysis. Engineering office to approve
  4.1.1 Team leader to conduct analysis. Engineering office to approve

Effect of recommendations on HRI: 1E

Hazard control references:
ASHRAE *Handbook of Fundamentals*
ASHRAE *Handbook & Product Directory—Systems and Applications*
ASHRAE *Guide and Data Book—Equipment*
Status of Control: CLOSED
Notes:

5.7.1 EXAMPLE HAZARD ANALYSIS OF NASA LASER

NASA has been a leader in monitoring the changing environment in the Arctic. One of the major issues concerning polar ozone loss is the monitoring of the developing Arctic ozone hole. Monitoring the Arctic ozone hole and studying the increase or decrease in the Arctic ozone hole is extremely important from scientific and environmental viewpoints. NASA flies an airborne laboratory into the polar vortex air, sampling ozone loss. To conduct these experiments, NASA flies on a DC-8 aircraft, a methane light detection and ranging (LIDAR) system consisting of two excimer lasers, gas. Replacement and cooling subsystems, laser optics, laser electrical power and control subsystem, ground and airborne support, and support structure. NASA conducted a detailed hazard analysis (NASA, no date) to ensure that laser and other hazardous operations did not injure people or adversely affect the environment.

The LIDAR experiment uses two excimer lasers. Passing an electric current through the gas excites the gas lasers; the discharge runs the length of the tube. Then the excited gas emits light, which resonates within the laser cavity and emerges to form the laser beam. The laser tubes are designed to be periodically purged and refilled with fresh laser gas. Replacement gas is provided using two gas cylinders and a fluorine generator. Approximately 1.3 L of fluorine is generated and contained within the unit. Figure 5.6 is a general layout of the laser.

Replacement gas is provided using two gas cylinders and a fluorine generator. Figure 5.7 shows the mechanical system for refilling the lasers. One gas cylinder provides the helium and the other provides the premix of other gases. The fluorine source is initially in the solid form of potassium fluoride. During filling, the laser's controller activates heaters that raise the temperature of the solid compound, which

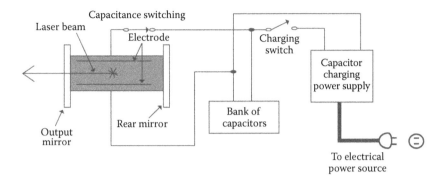

FIGURE 5.6 General *LASER* layout.

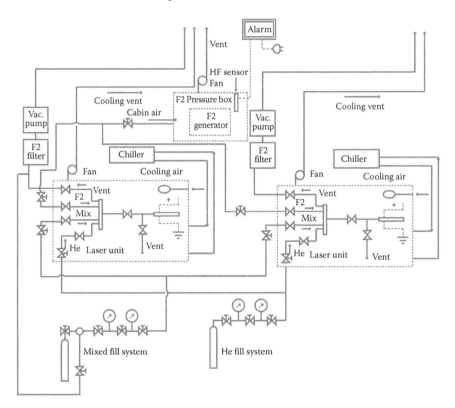

FIGURE 5.7 Mechanical system schematic.

decomposes to generate fluorine gas. A separate cooling system with a water chiller is used to control the excimer lasers.

A power supply includes a dc motor, which charges a capacitor, and a fast discharge circuit applies the electrical pulse across the laser to excite the gas-laser media. Two Class 2 lasers are used in an oscillator–amplifier configuration, emitting laser light out through a 16 in. diameter Newtonian telescope out from the top of the aircraft into the environment. Light is reflected back through the telescope and into a set of filters and an instrumentation system.

Various equipment supports the system. Instrumentation racks include computers, laser synchronization units, sensors, and other equipment used to control the experiment. The entire LIDAR setup is mounted on high-grade aluminum beams.

A detailed hazard analysis was performed, following the same methodology as described in Section 5.4. The format used by NASA is slightly different from the one shown in Section 5.4. A column titled "Hazard Risk Acceptance (HRA)" prioritizes which hazards must be dispositioned first:

- HRA 1—Unacceptable risk, must be resolved before flight.
- HRA 2—Undesirable risk, must be resolved or accept residual risk prior to flight.
- HRA 3—Acceptable (with review) Risk, resolution is desirable.
- HRA 4—Acceptable (w/o review) Risk, resolution is not required.

Figure 5.8 indicates how the functional areas were divided for the hazard analysis. Table 5.10 is a brief excerpt of the hazard analysis.

There are many more hazards in an experiment such as this one. However, it is worthwhile to focus on what is unique about the LIDAR experiment. In studying this experiment, it was important for NASA to go back and study in detail certain aspects of the design. Other safety analyses in this book will demonstrate how to conduct a detailed safety analysis of other systems, such as pressure or hazardous fluids and processes.

In reviewing the hazard analysis worksheets, you can see that gas detection is important. The experiment is flying in a closed-in environment, so laser gas leakage is a very real concern. Many times engineers forget to add appropriate monitoring systems. Of course, designing out the hazard would have been the best solution if it were possible.

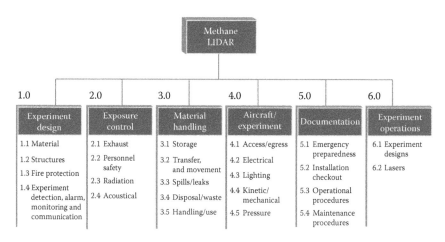

FIGURE 5.8   Function/concern organization chart. (From National Aeronautics and Space Administration, Goddard Space Flight Center, *Methane LIDAR Experiment, Preliminary Hazard Analysis*, Prepared by Hernandez Engineering, Inc., under NASA Contract NAS5-31825, Task S-1 1-94, no date. With permission.)

TABLE 5.10

LIDAR Hazard Analysis Worksheet

Control Number	Hazard Description	Potential Causal Factors	Potential Effects	HRI/ HRA	Hazard Control Recommendation	Effect of Recommendation on HRI and HRA	Hazard Control References	Status of Control
1.4.12	Lack of means to detect toxic or flammable gas within facility or aircraft	Improper selection or lack of gas detector systems	Air contamination, fire, or explosion.	IIA/1	Provide means to monitor throughout experiment ground and flight operations.	ID/3	AMES DC-8 *Experimenter's Handbook*	CLOSED. Two continuous monitoring Sensidyne F2/ detectors are provided, one installed in the exhaust manifold and the other used as req'd also. Second monitor available for *confined space entry* of aircraft.
3.4.04	Discharge of waste (particulates, toxins, etc.) into environment	Environmental effects not considered in design of discharge systems (drainage, sewers, ventilation/ exhaust, etc.)	Environmental contamination and/or pollution. Increased danger of disease in local population.	IIB/1	Include environmental quality considerations in planning and design of discharge systems.	IID/3	AMES DC-8 *Experimenter's Handbook.*	CLOSED. Halogen filters ensure compliance with F2 discharge ordinance. Exhaust concentration below 13 ppm.
4.2.02	Continued operation of equipment under dangerous conditions	Inability to easily shut off power to equipment in the event of an accident	Possible death or severe injury to personnel. Equipment damage.	IC/1	Provide readily accessible power shutoff as required.	ID/3	OSHA 1910.305 NFPA 70. AMES DC-8 *Experimenter's Handbook.*	CLOSED. All shutoff control electronics and power switching are easily accessible.

(Continued)

TABLE 5.10 (*Continued*)
LIDAR Hazard Analysis Worksheet

Control Number	Hazard Description	Potential Causal Factors	Potential Effects	HRI/ HRA	Hazard Control Recommendation	Effect of Recommendation on HRI and HRA	Hazard Control References	Status of Control
4.2.05	Electric motors, outlets and any other spark generating sources in hazardous areas	1. Not using explosion-proof motors. 2. Incompatible outlets/ connectors. 3. Static electricity. 4. Anything that will generate a spark that may cause an explosion of undetected accumulated vapors due to leak	Injury or death. Equipment damage or loss due to explosion of vapors.	IC/1	Select, design, install, and maintain all electrical system components in accordance with industry and NASA guidelines.	ID/3	NFPA 30, NFPA 70. *AMES DC-8 Experimenter's Handbook.*	CLOSED. All equipment selected for compliance with aircraft req. Vacuum pumps used for filling are brushless and/or operated on the ground only. Chiller compressor electronics modified to use solid-state relays to eliminate any potential for sparking.
5.3.05	Increase in hazards due to changes in design or operations	Lack of or inadequate review and safety consideration during planning/design changes of experiment processes	Personnel injury. Equipment damage.	IIB/1	Ensure that all proposed changes are adequately reviewed, especially for any implications on hazards and safety.	IID/3	*AMES DC-8 Experimenter's Handbook*	CLOSED. NASA AMES safety and flightworthiness reviews conducted for LIDAR experiment mission.

	Hazard	Cause	Effect		Recommendation		Reference	Status
6.2.02	High energy release	Unprotected capacitors (laser power supply)	Fire, injury, death.	IB/1	Isolate laser power supply capacitors with screens, shields, barriers, or uninhabited rooms. Doors and covers should be interlocked.	ID/3	ANSI Z136.1 Sec. 4.6.3, 4.6.4. AMES DC-8 Experimenter's Handbook.	CLOSED. All circuitry with high-voltage potential is enclosed and protected against inadvertent contact. Appropriate interlocks, grounding, and isolation are provided for lasers and other equipment.
6.2.08	Undesired exposure to laser beam	Reflected laser beam. Laser beam directed at other aircraft, vehicles, or buildings	Injury, property damage.	IIC2	Ensure beam path is monitored when laser is operational.	IID/3	21 CFR 1040; NSC Fundamentals of Industrial Hygiene Chapter 11, AMES DC-8 Experimenter's Handbook.	CLOSED. Laser emission and optical path completely enclosed. There is no possibility of personnel exposure to laser emission during in-flight operations. All laser ops are monitored.

Source: National Aeronautics and Space Administration, Goddard Space Flight Center, *Methane LIDAR Experiment, Preliminary Hazard Analysis*, Prepared by Hernandez Engineering, Inc., under NASA Contract NAS5-31825, Task S-1 1-94, no date.

Discharging fluorine gas is illegal in many municipalities in the United States. Halogen filters were added to the flight system so that a negligible amount would be exhausted to the environment.

Operating the system under unusual or dangerous conditions is another important point that was engendered from this analysis. It is critical always to have a way to shut off the system quickly, without causing more hazards. Item 4.2.05 uses explosion-proof electrical equipment to mitigate any danger from spark sources. If there is an explosive environment in the aircraft, being able to shut off power will not necessarily prevent an explosion. In this case, NASA has provided very strong hazard controls to prevent a spark from reaching an explosive environment.

Item 5.3.05 is a very interesting entry. Many engineers feel that once a system has been accepted, no other safety concerns should ever crop up in the future. This item is a good control for any modifications made to the system or the operation of the experiment.

The last two entries are laser specific. The first demonstrates the importance of system interlocks so that it is physically impossible to operate the laser while the high-voltage system is accessible to personnel. The second shows that sometimes only procedures and training can be used to control the hazard. As you remember, the least desirable method of hazard control is relying on personnel actions and training. However, in this case, there is no other way to control where and how the laser beam is directed.

There are a few additional points to ponder: NASA purposely used a fluorine generator, thus eliminating the need to carry large quantities of gas on board the aircraft. If all of the available fluorine were released into the aircraft cabin, it would result in a concentration of 5.1 ppm, approximately five times below the IDLH value—effectively eliminating the hazard. However, gas monitors are still used.

Primarily safety goggles and procedural methods control the Class 2 laser beam hazard. The laser beam is completely enclosed and can only exit the aircraft through the telescope window mounted on the roof of the aircraft. The only other hazard is that the aircraft flies near other aircraft while the laser is operated. That itself presents other, more immediate dangers.

A safety checklist (which will be further described in Chapter 6) was also used in tandem with the operation of the LIDAR experiment. This is a good example of combining various safety methods for maximum safety coverage.

5.7.2 Brief Example of a Hazardous Waste Storage Facility Hazard Analysis

We have seen considerable detail of how to conduct a hazard analysis with an example of shooting lasers into the ozone layer and have also learned how to perform a facility hazard analysis. Though the processes of performing a hazard analysis and a facility hazard analysis are the same, it is still worthwhile to show very briefly the advantages of one application of facility hazard analysis—analyzing a hazardous waste storage facility, something all industries around the world must contend with.

This semiconductor manufacturing plant, located in a seismic zone at the confluence of two navigable rivers, is 5 miles from an elementary school. The finished products are shipped to the United States, Europe, and some Latin American countries. During the manufacturing process, various hazardous chemicals are used and stored on site temporarily, before they are shipped to their disposal sites. Some of the materials stored together are acids, oxidizers, old battery packs, spray paint, acetone, fluorides, lead, ammonia, trichloroethylene, chlorine, and other chemicals. A fire with various toxic by-products that shut the plant down convinced the owners to move the hazardous waste storage facility to another building, separate from the manufacturing process. A facility hazard analysis was conducted of the manufacturing plant, and numerous problems were identified and corrected; then the owners wished to conduct a separate analysis of the hazardous waste facility.

The engineers performing the facility hazard analysis of the hazardous waste facility used the following hazard resolution process:

- Define the physical and functional characteristics of the entire building (or system) as related to the interactions among people, handling and storage procedures, equipment used, and the environment.
- Identify hazards related to all aspects of the operation.
- Assess the hazards to determine the severity and probability and to recommend means for their elimination and control.
- Decide to accept the hazard, or implement corrective measures to eliminate or control the hazard.
- Conduct follow-up analysis to determine the effectiveness of preventive measures and to identify any unexpected hazards.

All of this information was combined and emergency preparedness plans were developed from the analysis results. Further, the local fire department was given information about which chemicals were stored in the facility. Figure 5.9 shows the functional tree used for the analysis.

The facility hazard analysis worksheets will not be reproduced here, but some of the more interesting findings were as follows:

- Inadequate containment and control in case of spill of toxic substances—recommend use of berms.
- Obstructed doorways do not allow rapid evacuation if needed.
- Unsafe practices in workplace, such as smoking (especially a problem with visitors)—need a better employee training program and hazard warning signs.
- Chemicals not properly labeled.
- Continued operation of building during dangerous conditions—recommend a single power shutoff for entire building, operable from a remote location.
- Poor lighting design and labeling caused employees to take out the wrong drum of materials and sometimes creates a hazard in the rest of the plant.
- Poor design of floor could cause inventory to fall and cause a fire.

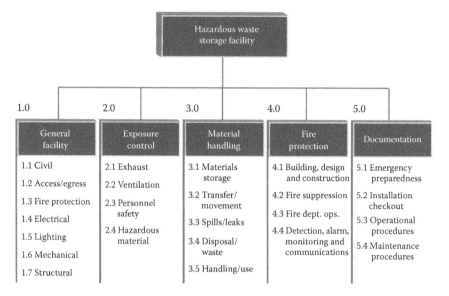

FIGURE 5.9 Hazard waste storage facility functional tree.

- Building not designed for earthquakes. This is particularly important because the plant is not far from a residential neighborhood. Industrial and residential zoning was not followed.
- Runoff of minor spills in facility released untreated into river.
- Drifting of volatile vapors into areas where baseboard heaters are used—recommend installing warning device to indicate failure of HVAC system. An even better solution would be to make all electrical systems explosion proof.
- Building located too close to other plant buildings, thus allowing rapid spread of fire.
- No fire protection system in facility (suppression and detection).
- Local fire fighters not prepared to fight fire (not informed of hazards and hazardous materials in the facility).
- Accidental mixing of combustible materials with incompatible materials (especially oxidizers)—recommend providing separate storage areas for combustible and other reactive materials, physically separated from other material classes.
- Workers not wearing appropriate protective clothing during hazardous materials handling.
- Release of explosive gas due to materials mixing with water; this was possible because of periodic river rising and flooding the building—recommend moving the building to higher ground.
- Arson—install security system.
- Inaccurate list of hazardous materials storage.

- Majority of structural steel does not have fire-resistive encasement.
- Store smaller quantities of materials. A waste minimization analysis indicated that the plant could reduce over 70% of its hazardous materials by simple changes to the manufacturing process.

REFERENCES

Hathaway, W. T. and Markos, S. H. 1989. *Passenger Carrying Submersibles: System Safety Analysis*. DOT-TSC-CG-89-2. Washington, DC: U.S. Department of Transportation.

National Aeronautics and Space Administration. 1998. *NASA Facility System Safety Handbook*. NASA-STD-8719.7, January 1998. Washington, DC: National Aeronautics and Space Administration, pp. 5–7.

National Aeronautics and Space Administration, Goddard Space Flight Center. No date. *Methane LIDAR Experiment, Preliminary Hazard Analysis*. Prepared by Hernandez Engineering, Inc., under NASA Contract NAS5-31825, Task S-1 1-94.

Perrow, C. 1984. *Normal Accidents: Living with High-Risk Technologies*. New York: Basic Books, pp. 72–100.

U.S. Department of Defense. 1993. *Military standard: Systems Safety Program Requirements*. Mil-Std 882C. Washington, DC: U.S. Department of Defense, p. A.5.

U.S. Department of Defense. 2012. *Military Standard: System Safety Program Requirements*. Mil-Std-882E. Washington, DC: U.S. Department of Defense.

FURTHER READING

Ericson, C. A. 2005. *Hazard Analysis Techniques for System Safety*. Hoboken, NJ: Wiley-Interscience.

Goldberg, B. E., Everhar, K., Stevens, R., Babbitt, N., III, Clemens, P., and Stout, L. 1994. *System Engineering "Toolbox" for Design-Oriented Engineers*. NASA Reference Publication 1358. Huntsville, AL: U.S. National Aeronautics and Space Administration.

Gustafson, R. M., Stahr, J. J., and Burke, D. H. 1987. The use of safety and risk assessment procedures in the analysis of biological process systems: A case study of the verax system 2000. *American Society of Mechanical Engineers 1987 Winter Annual Meeting*, Boston, MA, December 13–18, 1987.

Hempseed, J. W. 1987. Hazard risk analysis associated with production and storage of industrial gases. *Eleventh International Symposium on the Prevention of Occupational Accidents and Diseases in the Chemical Industry*, International Social Security Association, Annecy, France.

King, R. and Magid, J. 1979. *Industrial Hazard and Safety Handbook*. London, U.K.: Newnes-Butterworths.

Kletz, T. A. 1999. *Hazop and Hazan: Identifying and Assessing Process Industry Hazards*, 4th edn. Philadelphia, PA: CRC Press.

Lees, F. P. 1980. *Loss Prevention in the Process Industries*. London, U.K.: Butterworths.

National Aeronautics and Space Administration. 1995. *NASA Facility System Safety Handbook* (draft). Washington, DC: National Aeronautics and Space Administration.

Powell, A. E. 1993. An analysis of system safety training requirements for facility acquisition managers, planners and engineers. Masters thesis, University of Maryland, College Park, MD.

Roland, H. E. and Moriarty, B. 1985. *System Safety Engineering and Management*. New York: John Wiley.

Stephans, R. A. and Talso, W. W. 1993. *System Safety Analysis Handbook*. Albuquerque, NM: System Safety Society.

U.S. Department of Transportation. Federal Transit Administration. January 2014. *Hazard Analysis Guidelines for Transit Projects*. Reprint from DOT-FTA-MA-26-5005-00-01, January, 2000. Cambridge, MA: CreateSpace Independent Publishing Platform.

U.S. Environmental Protection Agency, Federal Emergency Management Agency, U.S. Department of Transportation. 1987. *Technical Guidance for Hazard Analysis: Emergency Planning for Extremely Hazardous Substances*. Washington, DC: U.S. Environmental Protection Agency.

World Health Organization. 2010. *Hazard Analysis and Critical Control Point Generic Models for Some Traditional Foods: A Manual for the Eastern Mediterranean Region*. Cairo, Egypt: WHO Regional Publications Eastern Mediterranean Series.

6 Process Safety Analysis

Life is what happens to us while we are planning other things.

Beautiful Boy, 1980
John Lennon

Few enterprises of great labor or hazard would be undertaken if we had not the power of magnifying the advantages we expect from them.

The Rambler, No. 2, March 24, 1750
Samuel Johnson

You can tell whether a man is clever by his answers. You can tell whether a man is wise by his questions.

Naguib Mahfouz
Nobel Prize in Literature winner in 1988

"The employer shall perform an initial process hazard analysis (hazard evaluation) on processes covered by this standard. The process hazard analysis shall be appropriate to the complexity of the process and shall identify, evaluate, and control the hazards involved in the process" (U.S. Occupational Safety and Health Administration, 1992; note, the most recent update is February 2013).

After the Bhopal accident in 1984, OSHA started work on a new standard, *Process Safety Management of Highly Hazardous Chemicals*, 29 CFR Part 1910.119. The U.S. government had determined that the current OSHA standards and the General Duty Clause did not adequately cover the chemical process industry. Much of the earlier work done by the American Institute of Chemical Engineers' Center for Chemical Process Safety (CCPS) formed OSHA's baseline of how to treat process safety hazards. The CCPS published their 1992 landmark publication *Guidelines for Hazard Evaluation Procedures* (3rd edition was published in 2008) and set the new international standard for process safety. And so on May 26, 1992, the OSHA Process Safety Management standard was enacted. The purpose of the standard is to promulgate a proactive and comprehensive approach to safety management in the chemical process industries. Since then, many countries and industries have developed their own regulations that are derivations of the CCPS guidelines and OSHA standard or specified it for the unique applications of a particular industry or industrial activity.

183

6.1 PROCESS HAZARD ANALYSIS

Process hazard analysis is the cornerstone of the OSHA standard. The standard goes on to say that the employer may choose from one or more of the following safety analysis methodologies as a process hazard analysis:

- What-if
- Checklist
- What-if/checklist
- Hazard and operability (HAZOP) study
- Failure mode and effects analysis (FMEA)
- Fault tree analysis
- An appropriate equivalent methodology

The remainder of this chapter will discuss HAZOP and what-if techniques in detail and illustrate specific examples of how they are applied. Chapter 7 will address fault tree analysis and Chapter 8 will discuss failure modes effects and criticality analysis. An excellent reference manual for these techniques is the *Guidelines for Hazard Evaluation Procedures*, published by the American Institute for Chemical Engineers' CCPS (2008).

6.2 HAZOP

A *hazard and operability* (HAZOP) study is a *systematic group approach to identify process hazards and inefficiencies in a system.* A team of engineers methodically analyzes a system and, through the use of guide words, asks how the process could deviate from its intended operation and what the effects would be. The group divides the system into nodes and, using the preestablished guide words (no flow, less flow, high temperature, etc.), ponders the questions of what could occur if the process deviated in some fashion. In other words, a HAZOP is a somewhat controlled *technical brainstorming session.*

The HAZOP is an extremely useful tool in *making sense* of highly complex process flows. The advantage of working in a group is that with the synergy of the individuals, more creativity is applied to identifying possible hazard scenarios. Of course, the HAZOP can be used at any phase of system or plant development; however, the design has to be somewhat mature to truly take advantage of the HAZOP's powers. This does not mean that you cannot perform a HAZOP on a preliminary design that is still not fully defined; HAZOPs are also great for looking at the effects of a modification to an existing design or operations.

The HAZOP is the primary method used in the petrochemical industry to identify, control, and document hazards. The HAZOP report is one of the pieces of information you have available to demonstrate your compliance to OSHA or other government inspectors.

Once you decide that you wish to conduct a HAZOP study, you must

1. Define objectives and scope
2. Select the HAZOP team

3. Conduct the HAZOP analysis
4. Document the results
5. Track the hazard control implementation

To dissect the design, the team should define what a *node* is. A thumbnail definition of *a node is that it is the location (on piping and instrumentation diagrams) where process parameters can change.* Interfaces of functional areas in the plant are good nodes. Significant changes in process parameters are good nodes. Interface points of major pieces of hardware also make good nodes. A pipeline connecting two major plant processes is a node. For example, nodes could be the piping from the feed tank through the feed pump to the feed header; piping carrying coolant into the condenser, through the main expansion valve to the evaporator; or piping from the heat exchanger through the accumulator to the compressor. HAZOP works for both batch and continuous processes. And HAZOP works for normal, omnidirectional flow as well as an emergency reverse flow. In fact, it is quite interesting to see the different types of hazards that are generated from reversing flow.

NOTES FROM NICK'S FILE

While working a HAZOP of a water distribution system, we found that reversing flow for emergency reasons created additional hazards to our process. One of the new hazards was critical safety valves and pumping stations could be blocked from contamination from dirt picked up from reverse flow through our filters.

Now gather your team. The team should obviously have a leader or facilitator, whose job is to gently nudge the team through the entire process so that it does not get *hung up* at one node. Many times the team leader is an engineer who is very familiar with the HAZOP technique. The key is for the team leader to be a very good facilitator. Keep the team on track, but allow creativity to come through. The other team members do not need to be experienced HAZOP team members. What is most important is that the team members be intimately familiar with the design and operation of the plant. In fact, OSHA requires that at least one member of the safety analysis team have a thorough understanding of the process.

One team member should take notes of the proceedings of the HAZOP meetings. The team recorder is probably one of the most important roles in the HAZOP process. The recorder must be able to support the team dynamics and not slow down the process.

The balance of team members (four to eight individuals) should be made up of plant designers, operators, maintainers, and other users. These individuals should be a good mixture of experience and expertise. Mechanical, electrical, chemical, and other process engineers should be part of the team.

With the team assembled, the kinds of information that they will need to review and consider in the HAZOP are

- Process and instrument drawings
- Facility drawings and plant site maps

- Process flow diagrams
- Operation procedures
- Hazard analyses or other safety reports
- Past accident and incident reports
- Interlock descriptions and classifications
- Operating parameters (nominal and emergency)
- Instrumentation set parameters (nominal and emergency)
- Equipment specifications (pressure vessel capacities, maximum design pressures, relief valve set pressures/capacities, flow rates, etc.)
- Any other HAZOPs of similar systems

The HAZOP team follows Figure 6.1, the HAZOP procedure, and Table 6.1, a list of HAZOP guide words, for each node in the system. The team starts with an input into the node by selecting a feed line. Then each of the guide words is applied to the flow line. For example, is more flow possible? If not, then go on to the next deviation. If more flow is possible, then is it hazardous? If not, return to investigating the other causes of more flow.

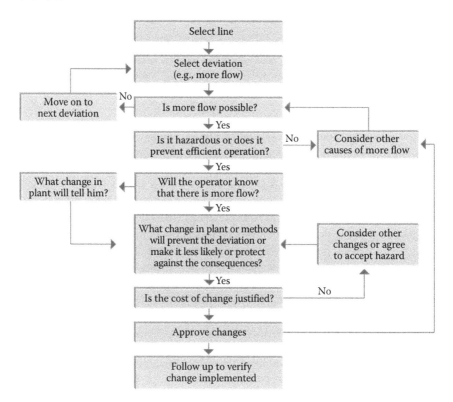

FIGURE 6.1 HAZOP procedure. (From Kletz, T.A., *Hazop and Hazan, Identifying and Assessing Process Industry Hazards*, 3rd edn., Hemisphere, Bristol, PA, p. 9, 1992. With permission.)

TABLE 6.1
HAZOP Guide Words

Guide Word	Deviation Definition/Example
No	*The physical process does not take place.* There is NO forward flow, electricity, mixing, etc.
More	*There is more of some relevant physical property than there should be.* There is MORE pressure than anticipated.
Less	*There is less of some relevant physical property than there should be.* There is LESS (or lower) temperature than anticipated.
As well as	*There are other constituents than those anticipated.* There is gas AS WELL AS fluid in the flow.
Part of	*Composition of process is different than it should be.* PART OF the mixture has particles larger than 200 μm.
Reverse	*The opposite process occurs than is anticipated.* REVERSE flow occurs.
Other than	*Something happens other than normal operations.* OTHER THAN slowing down, it sped up.

Now, if more flow is hazardous, then you must ask: Does the operator know that there is more flow occurring? If he does not realize that more flow is occurring, what will happen to plant operations that will indicate to him that more flow is occurring? How does instrumentation detect this; what are the resultant actions if it does?

Then the team must ascertain what design or operational change will prevent the deviation (more flow) or mitigate its consequences. Will the changes be costly? If the costs are unacceptable, then you must consider other changes to the system to protect against the hazard or formally accept the residual risk.

If changes are made to the system, they must be reviewed and approved by the appropriate engineering disciplines. And lastly, the company must follow up on the changes to verify that the hazard control is implemented and adequate. Remember that when thinking about changes to the system, use the hazard reduction precedence described in Chapter 2 as your guide.

PRACTICAL TIPS AND BEST PRACTICE

- Do not use the HAZOP session to develop hazard controls to the hazards identified in the process. That will only bog you down and it will take forever to complete the original HAZOP. Usually it will not lead to the best solution.
- The best thing to do is to complete the HAZOP identifying the hazards and list out recommendations for further investigation. It may actually take weeks or more to determine the costs and benefits of various design or operational changes to a particular hazard. It is much more efficient to work the solutions outside the HAZOP meeting.

After looking at Table 6.1, one set of questions that the team might generate for a NO FLOW condition could be the following:

- Could there be NO flow to the system?
- If it were to happen, how would it occur? Are the pumps down? Are lines plugged? Are valves closed?
- What are the consequences of NO flow to the process? Does it affect other parts of the system in an adverse way? Does material not reach the right place at the correct time? Do pumps cavitate?
- Are the consequences hazardous? Does this cause loss of critical cooling, therefore resulting in a high-temperature scenario locally or somewhere else in the system? Is there a way for the operator to recognize the hazard?
- If the situation is hazardous, can we change the system in any way to prevent the NO FLOW condition? Or can we do something differently to mitigate the effects of a NO FLOW condition?

Table 6.2 is an example of a typical HAZOP worksheet for a caustic regeneration process. As can be seen in the worksheet, the node is described (along with its components), and then the team walks through the node applying each guide word.

For example, in the first case, the guide word NO FLOW is considered. That generates the question to the group: "What would cause the node not to receive any flow?" Once the causes are determined, then each of the effects of a NO FLOW cause is considered. The *effects* column indicates to the group whether the consequences of NO FLOW are safety related, environmental, operational, etc.

PRACTICAL TIPS AND BEST PRACTICE

Remember that the HAZOP *guide word* and the *process condition* (flow, pressure, temperature, pH level, etc.) is the *process deviation*. And the process deviations that most interest us are the ones with hazardous consequences. For example, a node with MORE PRESSURE can lead to pressure vessel rupture, killing plant operators, spilling toxic wastes to the environment, and forcing the rest of the plant to shut down due to losing a critical process.

The *type* column indicates the category of concern: maintenance, operational, environmental, or safety. This helps the process engineer to sort what is of immediate importance (e.g., safety) but also identifies other areas for process efficiency improvement, such as enhanced maintenance practices.

The team documents the current safeguards in place, and the *risk level before* determines whether that is acceptable or not. If a recommendation is made, then the *after* (or new) *risk level* is indicated. Finally, the status of the process change is presented. Remember that the *recommendation* column is just that—recommendations. The process engineers, managers, and others will need to determine if the

TABLE 6.2
HAZOP Node Summary Report

Project: Caustic regeneration plant

HAZOP Team Members: F. Johnson (lead), C. O'Conner, S. Watt, M. Ramirez, S. Casey, E. George

Date: October 10, 2014

Node design description:

To receive and separate hydrocarbons and caustic materials from the discharge lines 4.3a, 4.4a, 4.4b, and 4.5a. To send the hydrocarbon and caustic to off-site storage

Node components:

Caustic relief drum and associated piping and instrumentation

Guide Word	Cause	Effect	Type	Safeguards	Risk Level Before/After	Recommendation	Status
No/less low	2″ block valve closed when needed	Deadhead the pump, high liquid level in the caustic relief drum, loss of seal, possible personnel exposure to caustic or hydrocarbon	Safety	Operator intervention	IB/ID	Install chemical splash guards on pumps.	OPEN: currently researching vendor prices
No/less low	3″ block valve closed on line 4-4b	Normally closed, not an issue	—	—	—	—	—
More flow	L-33 malfunctions open or bypass open on line 49 from oxidizer vent drum to caustic relief drum	Increased liquid level and carryover of oxygen into vent flare gas header, possible combustible mixture in line, possibility of deflagration	Safety	Two LAHs, PAL-137, flame arrestor on common flare header	ID	None.	CLOSED

(Continued)

TABLE 6.2 (*Continued*)
HAZOP Node Summary Report

Guide Word	Cause	Effect	Type	Safeguards	Risk Level Before/After	Recommendation	Status
More flow	L-35 malfunctions open or bypass open on line 53 from discharge of relief caustic pumps	Not an issue	—	—	—	—	—
More pressure	Tubing compression fittings become loose or tubing and fittings are overpressured from a relief into caustic relief/drain header	Potential personnel exposure to flare gases, injuries, or death	Safety	None currently in place	IC/IE	Replace all tubing with hard piping. Replace tubing compression fittings with welded fittings.	OPEN: currently under design

PICTURE 2.1 Fukushima nuclear accident. (Accessed from http://commons.wikimedia.org/wiki/File:Fukushima_I_by_Digital_Globe_crop.jpg.)

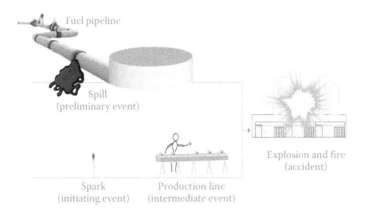

FIGURE 2.1 Events that lead to an accident.

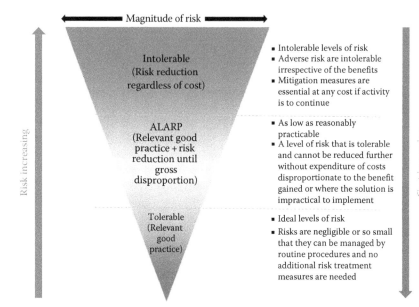

FIGURE 2.2 ALARP principle.

FIGURE 4.1 SMS.

(a)

(b)

PICTURE 4.3 Waterfall rail accident pictures. (From Independent Transport NSW Files. Used with permission.)

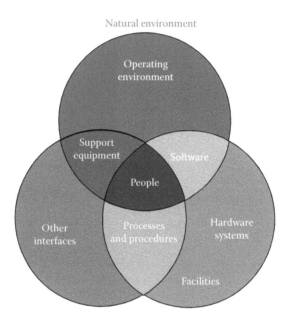

FIGURE 5.1 The system.

FIGURE 5.4 A car in traffic as a system.

FIGURE 9.2 Bow tie example.

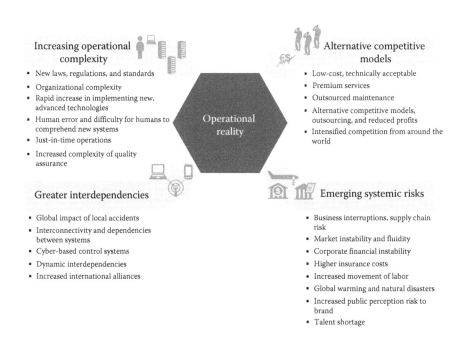

Increasing operational complexity

- New laws, regulations, and standards
- Organizational complexity
- Rapid increase in implementing new, advanced technologies
- Human error and difficulty for humans to comprehend new systems
- Just-in-time operations
- Increased complexity of quality assurance

Alternative competitive models

- Low-cost, technically acceptable
- Premium services
- Outsourced maintenance
- Alternative competitive models, outsourcing, and reduced profits
- Intensified competition from around the world

Operational reality

Greater interdependencies

- Global impact of local accidents
- Interconnectivity and dependencies between systems
- Cyber-based control systems
- Dynamic interdependencies
- Increased international alliances

Emerging systemic risks

- Business interruptions, supply chain risk
- Market instability and fluidity
- Corporate financial instability
- Higher insurance costs
- Increased movement of labor
- Global warming and natural disasters
- Increased public perception risk to brand
- Talent shortage

FIGURE 11.1 Changing operating environment and its challenges.

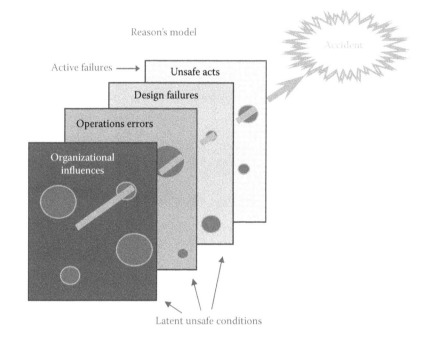

FIGURE 11.2 Sequence of events leading to an accident ("Reason's model").

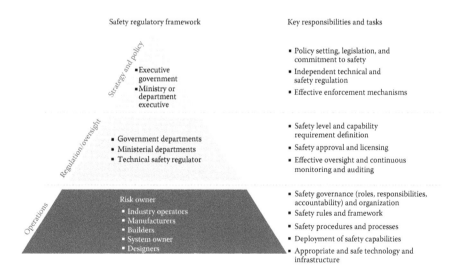

FIGURE 12.1 Government–industry regulatory relationship.

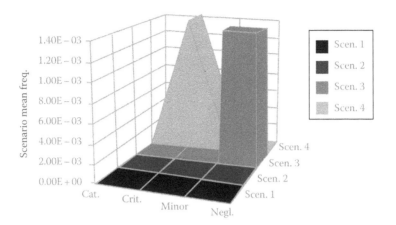

FIGURE 14.2 Sample failure consequences risk profile.

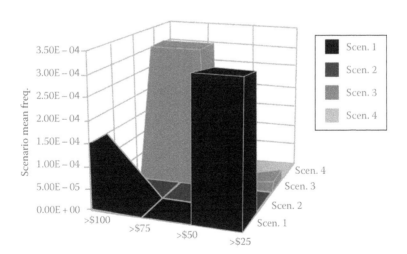

FIGURE 14.3 Sample risk expectation profile.

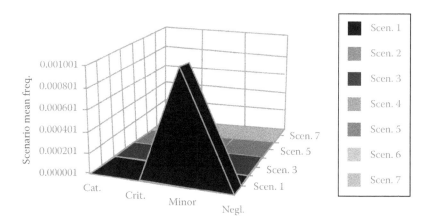

FIGURE 14.7 Operator error (valve 5 open) severity classification risk profile.

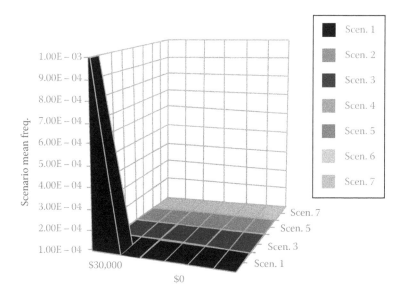

FIGURE 14.8 Operator error (valve 5 open) dollars-at-risk profile.

recommendation makes sense; how to design, operate, and maintain the recommendation; and whether it is even worth the cost and effort to do it. One recommendation could lead to various design or operational options, and the best option will most likely need to be thought through carefully and judiciously.

PRACTICAL TIPS AND BEST PRACTICE

- Do not go overboard with this. As you can see, the worksheet is very detailed; be judicious about the level of detail that you need. You certainly will not want to HAZOP in detail an entire plant. However, the most hazardous operations must be studied in detail.
- HAZOPs are great for plant modifications, even small ones.
- A HAZOP is only as good as the information and the team members.
- Be sure to consider group dynamics.
- Also, as you can see, there is considerable information generated in a HAZOP that can significantly enhance plant efficiency and productivity; do not be afraid to use the HAZOP for that as well.

As you can see, a HAZOP is a very close cousin of the hazard analysis spoken of earlier.

6.3 *WHAT-IF* ANALYSIS AND SAFETY CHECKLISTS

In the process safety industry, the *what-if* and use of safety checklists are really two separate tools. Now, with the advent of OSHA 1910.119, the two tools are more commonly merged into one. This does not mean that you cannot apply them separately if you so choose. Both are accepted safety analysis tools for OSHA compliance. In this section, however, the two will be combined.

The *what-if* is exactly as it sounds: *what if this occurs; what will be the consequences?* The purpose of a *what-if* is to carefully consider the effects of unexpected events on the system. If you can couple this simplistic questioning with individuals who have a good understanding of the operation of the plant and a systematic review of the plant, then it can become a very useful tool.

Concretely, the procedure is to examine the possible deviations (similar to a HAZOP) from the design and operation of a plant or process. The same kind of information required for a HAZOP is needed for the *what-if*. Follow the same process as the HAZOP: define the objectives and scope, select the team, conduct the questioning, document the results, and then track the hazard control implementation.

Methodically go through each of the functional areas defined by the analysis objectives. Start with one functional group or other manageable unit of the plant, and go through a series of *what-if* scenarios. The HAZOP methodology of choosing a node can also be applied here to help you pick the functional area for study. You may wish to use Appendices A, B, and C to help you consider the major hazard areas by taking the safety checklists in the appendices and rephrasing them as questions. As the group asks these questions, you should document the results. Any of the previously mentioned tabular forms can be easily adapted to the *what-if*/safety checklist.

PRACTICAL TIPS AND BEST PRACTICE

- The *what-if*/safety checklist is not as formal as the other safety tools, but it is a very inexpensive way to *estimate* the magnitude of some of the plant's safety concerns.
- This method is especially good in those moments when you do not have a lot of time for decision making. It is very useful for contingency planning.
- However, it should NOT be the primary safety analysis tool. The problem with checklists is that it is only as good as the checklist itself and is not open ended enough to be comprehensive.

Typical *what-if* questions are as follows:

- What happens if mixing is not complete?
- What occurs if the temperature rises above ambient?
- What happens if pump A cavitates?
- What procedures do operators follow if gauge A is out of range?

You can see just from just this short list of questions how difficult it would be to use the technique as a comprehensive approach. It is difficult to really anticipate all the right questions and be sure that it comprehensively looks at the safety of the process.

6.4 BRIEF HAZOP EXAMPLE OF AN AMMONIA FILL STATION

Anhydrous ammonia is one of the most valuable and versatile chemical compounds used in a variety of industries including food production and processing, textile and chemical manufacturing, petroleum refining, refrigeration plants, metal treating, pollution abatement, and the agricultural industry. Figure 6.2 is a simplified piping and instrumentation drawing of one node of a large metallurgical and metal working plant.

A large, 100 ton storage tank holds the bulk of the plant's ammonia capacity. However, to use the ammonia effectively, it is transferred to a holding tank or ready storage. From the ready storage, the ammonia is dispersed to the various parts of the plant. The distance from the storage tank to the ready storage is approximately one mile. A positive-displacement pump and various control and manual valves are used to transfer the ammonia. Both tanks are fitted with relief valves.

A dry nitrogen gas system is used periodically to purge the system of contamination. All venting is connected to the unit scrubbers.

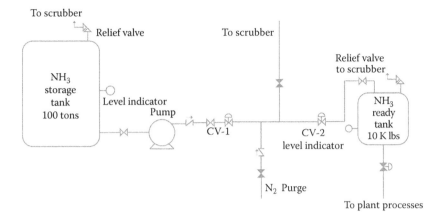

FIGURE 6.2 Ammonia fill station.

It would be impossible to repeat a full-blown HAZOP here, but this ammonia fill station is a good example to demonstrate the important elements in a HAZOP. It does emphasize various inherent problems found in a typical process unit in a chemical plant.

Most HAZOP entries are not quite so detailed as this one. However, it is worthwhile to go into some detail here to note some of the finer points of HAZOPs. Because the HAZOP is a systematic review of the design and operation, it is also fairly repetitive. Most of the repetition has been deleted here for ease of reading. Table 6.3 is the one summary report sheet (of various) of the ammonia fill station HAZOP.

Upon first reading the summary report, you can immediately see that the HAZOP results have yielded not only safety issues but also operational issues. Also, the HAZOP has indicated various entries that are already well controlled. The *safeguards* column demonstrates the current state of safety in the node. This summary report is very useful in demonstrating to inspectors your thoroughness of analysis and current state of hazard control—various items are already under control, and only some of the items must be taken to disposition.

As you can see, the no/less flow entries are numerous. This proves that there are various ways that flow can be interrupted. This is a critical point in any process plant and one area that is almost always ripe for close scrutiny.

The most important no/less flow entry is 1.5. The distance between the two tanks is one mile. If care were not taken, it would be easy to forget to open one of the control valves while the pump is operating. As stated earlier, when presented with a hazard, the best control is to design it out. Currently, operating procedures control the hazard. Installing relief valves that prevent possible overpressurization does not eliminate the danger but does mitigate the hazard even if the operator fails to open the control valve. Further controls could include control feedback to the pump to shut down if a certain pressure is reached downstream (i.e., the valve has not cycled open).

TABLE 6.3
HAZOP Node Summary Report

Project: Ammonia fill station

HAZOP Team Members: F. Johnson (lead), C. O'Conner, S. Watt, M. Ramirez, S. Casey, E. George

Date: October 10, 2014

Node design description:
To store large quantities of anhydrous ammonia and supply to a ready storage facility for various plant operations

Node components:
100 ton ammonia storage tank, 10,000 lbs. ammonia ready storage tank, fluid transfer pump, valving, nitrogen purge system, and system and associated piping and instrumentation

Guide Word	Cause	Effect	Type	Safeguards	Risk Level Before/After	Recommendation	Status
1.1 No/less flow	Ammonia block valves or control valves closed.	No ammonia delivery to ready storage	Maintenance/ operational	Controlled by procedures.	IIIC/IIIC	None.	CLOSED
1.2 No/less flow	N₂ purge open during normal operations.	Same effect	Operational	Controlled by procedures.	IIIC/IIIC	None.	CLOSED
1.3 No/less flow	Line ruptures due to system vibrations.	Ammonia spills in area	Environmental	Safeguards adequate.	IID/IID	Consider adding pressure sensor between CV-1 and CV-2 to shut down pump and sound alarm if loss of pressure is sensed.	OPEN
1.4 No/less flow	Pump fails, seizes, or leaks.	No ammonia delivery to ready storage; ammonia spills in area	Maintenance	Current PM program adequate.	IIIC/IIIC	None.	CLOSED

	Cause	Consequences	Category	Safeguards	Ranking	Recommendations	Status
1.5 No/less flow	CV-2 or CV-1 closed during operation.	Same as the aforementioned; possible system overpressure	Safety	Controlled by procedures.	IIIC/IIID	Verify correct operation (and operator communication) and sequencing of control valves; install relief valves between CV-2, CV-1, and pump.	OPEN
1.6 No/less flow	Manual valve to scrubber open during operation.	Liquid ammonia flow to scrubber; loss of product	Safety/operational	Controlled by procedures.	IIC/IIC	None.	CLOSED
1.7 No/less flow	Valves/flanges leak.	Loss of product	Environmental	Periodic inspection.	IIB/IIID	Change threaded fittings for welded fittings.	OPEN
1.8 No/less flow	Contamination or ice in piping system.	No ammonia delivery to ready storage; possible system overpressurization	Safety/maintenance	Periodic nitrogen purge of system.	IC/IE	Install filters at pump inlet; verify valve packing is ammonia compatible; install pressure-relief valves.	OPEN
2.1 More flow	Pump motor overspeeds.	System overpressurization; possible overfill of ready storage	Safety	Relief valve on ready storage.	IB/IIID	Install electrical limit switches; add excess-flow valve at ready storage entrance; set level indicator to shut off motor if 85% of total liquid level reached; add overpressure relief to pump housing.	OPEN
2.2 More flow	False level indicator signal/reading in ready storage.	Overfill ready storage	Safety/operational	Liquid-level indicator and pressure-relief valve.	IB/IID	Add excess-flow valve to ready storage.	OPEN

(Continued)

TABLE 6.3 (Continued)
HAZOP Node Summary Report

Guide Word	Cause	Effect	Type	Safeguards	Risk Level Before/After	Recommendation	Status
3.1 More pressure	Block valves closed while pump is running.	Overpressurize system	Safety/ operational	Procedures call for block valves to be open.	IB/IID	Install relief valves between all block valves.	OPEN
3.2 More pressure	Thermal expansion in isolated valve section due to trapped ammonia.	Overpressurize system	Safety	None.	IB/IID	Install relief valves between all block valves; add bleed valves to system.	OPEN
3.3 More pressure	Pump motor overspeeds.	Overpressurize system or overfilled ready storage	Safety	Relief valve on ready storage.	IB/IIID	Install electrical limit switches; add excess-flow valve at ready storage entrance; set level indicator to shut off motor if 85% of total liquid level reached.	OPEN
3.4 More pressure	Blockage in lines from contamination or ice, causing trapped ammonia.	Overpressurize system	Safety/ maintenance	Periodic nitrogen purge.	IC/IE	Add filter to pump inlet; add relief valves between all block valves.	OPEN
3.5 More pressure	Valves cycled too quickly causing hydraulic hammer.	Possible system overpressure or line rupture	Safety/ operational	Piping designed for worst case; operators follow well-established procedures.	IIIC/IIIC	None.	CLOSED

3.6 More pressure	Overfill ready storage due to operator inattention or other system failure.	Ready storage overpressurizes	Safety	Liquid-level indicator sounds alarm.	IB/IIID	Add excess-flow valve at ready storage entrance; set level indicator to shut off pump motor if 85% of total liquid level is reached; consider additional pressure sensor in ready storage.	OPEN
4.1 More temperature	Solar effects.	Increases pressure in system	Safety	Not an issue, insufficient thermal expansion to cause problems.	IVE/IVE	None.	CLOSED
5.1 Gas as well as liquid in system	Low liquid level in system.	Cavitates pump: not enough flow to ready storage	Operational	Not an issue.	IVD/IVD	None.	CLOSED
5.2 Gas as well as liquid in system	Nitrogen purge used during process.	Cavitates pump: not enough flow to ready storage	Operational	Procedure controls use of nitrogen purge.	IID/IID	None.	CLOSED

Entry 1.6 looks pretty funny. How could an operator forget to close the line to the scrubber before attempting to fill the ready tank? Yet errors like this occur every day at process and manufacturing plants around the world.

Line contamination, especially icing in ammonia systems, is quite frequent and a typical problem. Adding a filter to the pump inlet will keep lines free from some level of impurities in the ammonia. Breaking into the line, careless workers may allow rain or other liquids to enter the system. This can have disastrous effects. The moisture will freeze upon contact with the ammonia and form an ice plug and could overpressure the line. That is why there is a dry nitrogen purge system. It is important that operators verify valve soft goods are ammonia compatible. Brass and bronze fittings are not appropriate for ammonia systems.

Overspeed of the pump motor creates a more flow and more pressure condition. This would be disastrous to the ready tank, causing overfill of the tank. The recommendations for 2.1 and 3.3 are identical because even though the hazard was identified through different guide words, the cause and effects are the same.

One hazard not identified in the HAZOP is overpressure of the pump housing. Again, the HAZOP is only as good as the team. If the team members do not have good synergy or understanding of how the plant operates, they will not do a good job. Adding a relief valve to the pump housing would be one way to control the hazard.

Entry 3.6 again demonstrates how human error can lead to a catastrophic accident. Of course, the best solution is to design the hazard out. In this case, you cannot take away the potential but have added numerous levels of redundant fail-safe mechanisms to prevent the mishap from occurring. Note that the ready tank already has a liquid-level indicator, but if it is only tied back to an alarm, it does not guarantee that the hazard will be prevented. It only means that the operator will be warned. The overfill hazard is too significant to be left to an alarm indication alone.

NOTES FROM NICK'S FILE

I was looking at a system that was injecting a chemical liquid into a process flow. The tank overfill hazard was controlled by a scale that weighed the contents before, during, and after the injection. Because the liquids were all compatible and nonreactive, no one really thought that there was a real hazard. The big problem was that the hazard was not truly controlled. (1) If the scale was not calibrated correctly, or out of date, it may not sense overfill; and (2) because the overfill alarm was just that—an alarm—it would not stop flow during an overfill situation. An operator had to hear the alarm and then manually stop flow. This means that if we did not hear the alarm, we would not stop the flow. We redesigned it to have a level indicator tied to shutting down the flow to prevent overfill.

Probably the single most important change that can be made to the system is to add adequate relief and bleed capability. Ammonia trapped between block valves is a potential hazard with very grave consequences. Many engineers count on the fact that most of the time, the lines are sufficiently overdesigned to prevent rupture.

The problem is that this is not always the case. Also, what happens when you need to break a line for maintenance? The vapor pressure of ammonia at

- 70°F (21.1°C) is 114.1 psig (786.7 kPa)
- 105°F (40.6°C) is 214.2 psig (1476.8 kPa)
- 115°F (46.1°C) is 251.5 psig (1734.0 kPa)
- 130°F (54.4°C) is 315.6 psig (2176.9 kPa)

These temperatures are not atypical at various locations in processing plants. The vapor pressure rise also brings up the need to leave adequate volume in the ready and storage tanks. Hard-filling liquid ammonia in the tanks is a serious concern and, unfortunately, not an uncommon occurrence. It might be a good idea to add a pressure sensor to the tank. Relief valves must be placed between any parts of the system where pressure could be trapped. Bleed valves should also be installed wherever maintenance workers may wish to break the system and open it up.

It is important to remember that a deviation at the ammonia fill station could have consequences in other parts of the plant. The most obvious is that insufficient product is available to other parts of the plant. If the use of ammonia is critical to the process, then the ammonia fill station could be a single-point failure in the process. If the fill station stops operating, it could bring the entire plant to a halt. Section 8.1 (Failure Modes, Effects, and Criticality Analysis) will discuss a lot more about single-point failures and how they impact safety.

6.5 EXAMPLE *WHAT-IF*/SAFETY CHECKLIST FOR PRESSURE TEST EQUIPMENT

Many times engineers are faced with taking existing hardware and using it in various test configurations. The example shown below is one such case. Figure 6.3 shows an actual gas pressurization system used to proof and leak test a variety of low- and high-pressure systems. Table 6.4 indicates the pressure values of each component. A brief *what-if*/safety checklist analysis is performed to ascertain if this setup is safe.

The plant needs to test both high- and low-pressure systems, so a team of engineers quickly put together this cart to test high-pressure (3000 psig) and low-pressure (65 psig) equipment. The side A high-pressure regulator steps down the incoming 6000 psig pressure to 3000 psig. The side B low-pressure regulator steps down the incoming 2200 to 65 psig. The low-pressure equipment that will be tested at 65 psig is critical for supporting other parts of the plant. To get the job done quickly, an engineer built this system from components found in the shop.

All components in the system have been individually proofed to 1.5 times their maximum operating pressure (MOP). All the high- and low-pressure system components have a design burst pressure of at least four times the MOP. The single manifold valve switches the system from side A to side B so that only one side can be used at a time. This manifold system was taken from a previously used block that needed the ability to switch from one side to the other.

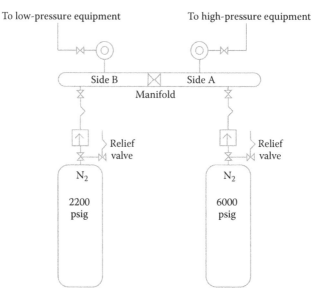

FIGURE 6.3   Pressure tent cart.

TABLE 6.4
Component Pressure Ratings

Item	Max. Operating Pressure (psig)	Component Proof Pressure (psig)	Design Burst Pressure (psig)
Manifold assembly	6000	9,000	24,000
Pressure regulator assembly (side A)	2400	4,000	12,000
Pressure regulator assembly (side B)	6000	9,000	24,000
Manual valves	3000	9,000	24,000
Check valves	3000	9,000	24,000
Flex lines	6000	9,000	24,000
Relief valves	3000	16,000	40,000
Manifold valve	6000	9,000	24,000
Tube assembly (side A)	3000	4,500	24,000
Tube assembly (side B)	65	100	24,000

Note: The gas supply cylinders meet DOT Spec. 3AA6000.

To start the *what-if*/safety checklist, consider Appendix B, *Generic Hazard Checklist*, and form simple questions from the listed hazards. Start with side A and then go to side B, and then look at the entire system.

Obviously, not all the items listed in the generic hazard checklist are applicable to this system, such as electrical, explosives, or radiation. Space does not allow giving a full listing of all the possible hazards; however, some of the more salient ones are shown in Table 6.5.

TABLE 6.5
What-If Results

What-If	Consequence/Hazard	Risk Level	Recommendation
What if the side A regulator fails open or leaks?	Then high-pressure equipment will see 6000 psig and not 3000, as designed.	IC	Install a relief valve set at 3300 psig. Verify adequate relief capacity. Plumb relief away from personnel.
What if the manifold valve leaks?	Then 6000 psig will flow into low-pressure system and cause large-scale rupture.	IB	Take out the manifold valve and permanently *close off* flow communication between side A and side B.
What if the check valve is clogged with contaminate?	Flow will be diminished or blocked. No effect on safety.	IIIC	Clean components before mating.
What if the flex lines come loose during operation?	Whipping effect of flex line could injure personnel.	IC	Clamp flex lines to cart fixture at both interface points.
What if the 6000 psig cylinder is attached to side B?	Possible overpressurization at side B regulator.	IB	Size connector interfaces of side A and side B differently, so it is physically impossible to mismate them.
What if the cylinder or system leaks into the surrounding operating area?	Although nitrogen is inert, it could asphyxiate personnel if operated in a closed environment.	IC	Calculate largest volume or air exchange necessary to avoid asphyxiation. Use an oxygen monitor.
What if cart is operated in high-temperature area?	Supply bottle pressure will rise and be vented through relief valves.	IID	No additional design changes needed. Verify relief valves can handle full flow.
What if the side B regulator fails open or leaks?	Then high-pressure equipment will see 2200 psig and not 65, as designed.	IIC	Install two-step regulation to regulate pressure from 2200 to 100 psig, then to step pressure from 100 to 65 psig. Install a relief valve set at 100 psig between two regulators. Verify adequate relief capacity. Plumb relief away from personnel.
What if pressure is still in system after completing test?	Disconnecting test lines from test article without bleeding pressure could injure personnel.	IB	Include bleed valves to assure pressure has been vented before disassembly.

At first glance, this system would look relatively safe. It has separate high- and low-pressure subsystems. All the components have been proofed to 1.5 times MOP. Even the design burst pressure is significantly higher than what we expect to see in this system.

On closer look, however, we see that there are a number of minor failures—such as leaky valves—that could cause catastrophic events. Also, the danger of using what is lying around in the shop is that we are not always sure of the pedigree of the components. Of course, all the components are good in this system, but their combination is not.

As evidenced by the *what-if*, we can see that a single failure could *blow* our critical low-pressure test equipment. Using two regulators with a relief valve before and after will help assure that we do not jeopardize plant-critical hardware during testing.

However, no matter how fast we need to test the equipment in the plant, it would be best to have two independent test carts.

NOTES FROM NICK'S FILE

Years ago, I was doing an accident investigation of a technician who did not properly understand his work instructions nor the consequences of not following them to the letter. He had to leak test a system. He used a high-pressure air system but did not follow the written procedures and connect a reducer valve between the high-pressure source and the lower-pressure equipment. He was alone and in a hurry, and he thought that if he opened the valve slowly, he could control the pressure. He was very lucky that he was not killed during the resulting explosion and fire from the adiabatic or rapid compression detonation. Very rapid compression pressures can cause local hot spots of several thousand degrees, which, coupled with suspected contamination in the piping system, created the accident. This is precisely why it is imperative to follow the hazard design precedence.

REFERENCES

Center for Chemical Process Safety. 1992. *Guidelines for Hazard Evaluation Procedures.* New York: American Institute for Chemical Engineers.

Center for Chemical Process Safety. 2008. *Guidelines for Hazard Evaluation Procedures*, 3rd edn. New York: Wiley. 2008.

Center for Chemical Process Safety, New York. https://www.aiche.org/ccps.

Kletz, T. A. 1992. *Hazop and Hazan, Identifying and Assessing Process Industry Hazards*, 3rd edn. Bristol, PA: Hemisphere, p. 9.

U.S. Occupational Safety and Health Administration. 1992. *Process Safety Management of Highly Hazardous Chemicals*. 29 CFR Part 1910.119. Washington, DC: Bureau of National Affairs.

FURTHER READING

Carling, N. 1987. *HAZOP Study of BAPCO's FCCU Complex. American Petroleum Institute Committee on Safety and Fire Protection Spring Meeting.* Denver, CO, April 9–11, 1986.

Center for Chemical Process Safety. 2007. *Guidelines for Risk Based Process Safety.* Hoboken, NJ: Wiley.

Center for Chemical Process Safety. https://www.aiche.org/ccps.

Hyatt, N. 2003. *Guidelines for Process Hazards Analysis (PHA, HAZOP), Hazards Identification, and Risk Analysis.* CRC Press.

Kletz, T. A. 1990. *Plant Design for Safety: A User-Friendly Approach.* CRC Press. 1990.

Kletz, T. A. 1999. *Hazop and Hazan, Identifying and Assessing Process Industry Hazards.* 4th edn. CRC Press.

Kletz, T. A. and Amyotte, P. 2010. *Process Plants: A Handbook for Inherently Safer Design,* 2nd edn. Boca Raton, FL: CRC Press.

Knowlton, R. E. 1981. *An Introduction to Hazard and Operability Studies.* Vancouver, BC: Chemetics International.

International Electrotechnical Commission. 2007. IEC 61882 Ed.1.0b:2001, *Hazard and operability studies (HAZOP studies)-Application guide.* International Electrotechnical Commission.

McKelvey, T. C. and Zerafa, M. J. 1990. Vital HAZOP leadership skills and techniques. *American Institute of Chemical Engineers Summer National Meeting.* San Diego, CA, August 19–22, 1990.

7 Fault Tree Analysis

Don't meet troubles halfway.

Sixteenth-century proverb

Dig a well before you are thirsty.

Chinese proverb

Nothing is so easy as to deceive one's self; for what we wish, that we readily believe.

Orations, Vol. 1, 349 BCE
Demosthenes

Fault tree analysis (FTA) is a graphical method commonly used in both reliability engineering and system safety engineering (though it is more well known in reliability circles). It is a deductive approach that is very powerful as a qualitative analysis tool that can be quantified. You postulate a top event—or fault—such as train derailment, then branch down from the top event, listing the faults in the system that must occur for the top event to occur. This top-down method forces you to go through systematically, listing the various sequential and parallel events or combinations of faults that must occur for the undesired top event. Logic gates and standard Boolean algebra allow you to quantify the fault tree with event probabilities and thus determine the probability of the top event.

It is important to understand that this is not a model of all possible system failures or all possible causes, but rather, a model of particular system failure modes and their constituent faults that lead to the top event. Not all system or component failures are listed, only the ones leading to the top event. Like the other safety analysis techniques discussed previously, only credible faults are assessed. The faults can be events associated with component hardware failures, software glitches, human errors, and environmental conditions—in short, any of the elements that make up the complete system.

The fault tree was first developed in 1961 for the U.S. military intercontinental missile program. The U.S. Nuclear Regulatory Commission published a guide in 1981, and since then, FTA has been used in almost every engineering discipline around the world, from mass transit to commercial nuclear power plants, chemical process plants, oil drilling platforms, NASA satellites, and aircraft control centers. Fault trees are used extensively in accident investigation. NASA used fault trees to recreate the events that lead up to the *Challenger* and *Columbia* Space Shuttle accidents. Fault trees have been combined with event trees, and other root cause analyses have been used very effectively in accident investigation, including the investigation of a plutonium spill at a Boulder, Colorado, National Institute of Standards and Technology laboratory.

Dynamic FTA is used more commonly in computer systems fault analysis and involves employing Markov analysis to generate the tree. Dynamic fault trees are also frequently used to model fault-tolerant systems. The challenge is that the size of the tree grows very quickly and can be very cumbersome to manipulate.

NASA succinctly defines (Stamatelatos et al., 2002) the process of conducting an FTA:

1. *Identify the objective for the FTA*—Determines what the engineer wants to know before starting the analysis
2. *Define the top event of the fault tree*—States the end result that is being investigated and should give the information needed to meet the objective defined in Step 1 (it defines the fault mode of the system)
3. *Determine the scope of the FTA*—Bounds how far the analysis should go and determines which faults will be included and their boundary conditions
4. *Define the resolution of the FTA*—Details the level of fault causes that will be followed to reach the top event
5. *Define the ground rules of the FTA*—Determines the naming scheme for the analysis and how the fault tree will be modeled
6. *Construct the fault tree*—Builds the actual fault tree (graphically and logically)
7. *Evaluate the fault tree*—Conducts quantitative and qualitative analysis of the fault tree through cut sets and Boolean algebra
8. *Interpret and present results*—Explains to the reader what all this means (this is the most important part of the analysis; results must be put into a context that makes sense and is understandable)

7.1   FAULT TREE SYMBOLS AND LOGIC

The fault tree uses logic gates to describe graphically how the top event occurs. You read a fault tree from the top event down to the constituent events. The *higher* gates are the outputs from *lower* gates in the tree. Therefore, the top event is the output of all the input faults or events that occur.

First, it is important to understand the difference between fault and failure. Simply put, *failure* means that something has *broken*. *Fault* means that something does not perform the action you desire, even though it operates as designed. For example, a valve closing is a fault if it occurs at the wrong time due to the improper functioning of some upstream component or human error. But the same valve can *fail* closed due to seizing of the poppet. Rupture of a pressure vessel is a component failure. You can say that all failures are faults, though not all faults are failures.

NOTES FROM NICK'S FILE

Spend time clearly understanding the difference between fault and failure. When I review fault trees, that is one of the biggest mistakes that I frequently find. I have seen many fault trees filled with failures but few faults.

The system or subsystem fault is an undesirable state of existence of the system or subsystem. That system fault (the top event) is comprised of component (or subsystem) faults. It depends to what level of detail you wish to delve whether the FTA faults are to the component, black box, or subsystem level. The fault tree of a nuclear plant is very large even at the subsystem level. However, it would be advantageous to go to the component fault level to analyze the plant safety subsystems.

NOTES FROM NICK'S FILE

Fault trees are very diverse and can be used in many ways. They are one of my most favorite safety analysis tools. I have used them for such diverse activities as understanding integrity management of an upstream oil pipeline system, employee and management actions taken during a plutonium spill at a laboratory, and the Sydney, Australia, Waterfall rail accident investigation.

The component fault is the state of existence of that component that contributes to the mechanism that leads to the next-level fault. In understanding what the component fault is, it is important to consider what the component state is in and when it is in that state of existence. Component faults are comprised of primary, secondary, and command faults. However, most primary and secondary faults are comprised of component failures, so they are usually called primary and secondary failures.

A *primary failure* is a failure that occurs under normal operating and environmental conditions. A *secondary failure* is a failure outside of normal conditions. A *command fault* occurs when a component performs as designed but produces the output signal at the wrong time. Roberts et al. (1981) demonstrate command faults with a humorous story from the American Civil War.

It appears that General Beauregard had sent his courier to deliver a message to one of the commanders in the field. The battle situation changed, and sometime later, the general sent another message with updated information. The battle situation changed again, and the general amended the previous messages with a third. The messages all arrived (as designed) to the commander in the field, but in the wrong order. Because the messages arrived in the incorrect order, that *fault* caused the battle commander to take the wrong action—with disastrous results.

Fault tree symbols are divided into four categories: primary event, intermediate event, gate, and transfer. Figure 7.1 defines each of the symbols used in fault tree generation.

Primary events are end events; in other words, for one reason or another, they are not studied further. For example, the circle or *basic event* describes a fault that is an initiating event itself and has no inputs depicted in the fault tree. Some examples are as follows: $K1$ timer contacts inadvertently open; $K2$ relay contacts fail to close; battery 2A is 0 V; or pressure switch contacts fail to open.

An ellipse, or *conditioning event*, is a sort of message bubble that records any conditions or restrictions that apply to any of the logic gates. This symbol is used primarily with INHIBIT and PRIORITY AND gates.

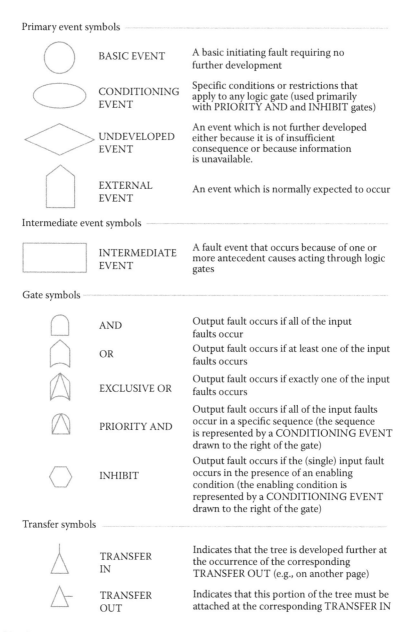

Primary event symbols

	BASIC EVENT	A basic initiating fault requiring no further development
	CONDITIONING EVENT	Specific conditions or restrictions that apply to any logic gate (used primarily with PRIORITY AND and INHIBIT gates)
	UNDEVELOPED EVENT	An event which is not further developed either because it is of insufficient consequence or because information is unavailable.
	EXTERNAL EVENT	An event which is normally expected to occur

Intermediate event symbols

	INTERMEDIATE EVENT	A fault event that occurs because of one or more antecedent causes acting through logic gates

Gate symbols

	AND	Output fault occurs if all of the input faults occur
	OR	Output fault occurs if at least one of the input faults occurs
	EXCLUSIVE OR	Output fault occurs if exactly one of the input faults occurs
	PRIORITY AND	Output fault occurs if all of the input faults occur in a specific sequence (the sequence is represented by a CONDITIONING EVENT drawn to the right of the gate)
	INHIBIT	Output fault occurs if the (single) input fault occurs in the presence of an enabling condition (the enabling condition is represented by a CONDITIONING EVENT drawn to the right of the gate)

Transfer symbols

	TRANSFER IN	Indicates that the tree is developed further at the occurrence of the corresponding TRANSFER OUT (e.g., on another page)
	TRANSFER OUT	Indicates that this portion of the tree must be attached at the corresponding TRANSFER IN

FIGURE 7.1   Fault tree symbols. (From Roberts, N.H. et al., *Fault Tree Handbook*, NUREG-0492, U.S. Nuclear Regulatory Commission, Washington, DC, 1981, p. IV-3.)

A diamond describes an *undeveloped event* for which no further analysis is required or for which information is unavailable to develop the event further. Undeveloped events are de facto boundary conditions to the problem. Any fault can be left undeveloped if you cannot ascertain how it occurs (what its inputs are) or if they are not important to the next event.

An *external event* is a normal event. External events can be thought of as assumptions in the graphical analysis: Gravity as the predominate force, air (with sufficient oxygen to sustain a fire) in a warehouse, etc.

Intermediate events are the basic events described. Various input faults feed into this intermediate event, which itself feeds into the next-level-up fault. A typical intermediate event is that motor 1 fails to start. The various faults that can prevent the motor from starting are the inputs to the intermediate event. The fact that motor 1 does not start is itself an input to the next-up fault that the pump does not operate.

The next group of symbols are the fault tree graphic operators. These are the logic gates. *AND* indicates that all the faults feeding into the AND gate must occur for the output fault to occur. For example, to have an overheated wire (an intermediate event), there must be a 5 mA current in the system, AND power must be applied to the system for $t > 1$ ms.

An OR gate is the opposite of an AND. Any of the input events occurring would result in the event fault. For example, for *P*-2 line not to receive any water, *V*-3 OR *V*-5 OR pump 3 must fail. The other gates are variations of the AND and OR gates and are not used as frequently.

Triangles are used to depict *transfer* gates. The transfer in and transfer out gates really are used to indicate a continuation of the fault tree onto another sheet of paper.

Figure 7.2 shows a high-level fault tree for a maglev (magnetic levitation) train. The top event is as follows: Train comes to a sudden stop. The next tier down states that if

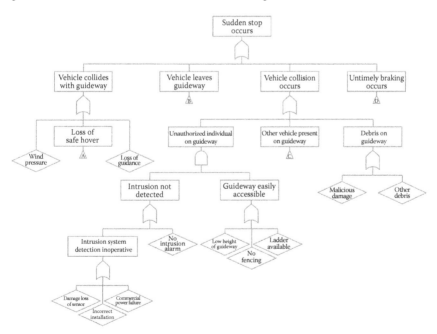

FIGURE 7.2   Fault tree of sudden stop of maglev train. (From Dorer, R.M. and Hathaway, W.T., *Safety of High Speed Magnetic Levitation Transportation Systems: Preliminary Safety Review of the Transrapid Maglev System*, DOT-VNTSC-FRA-90-3, U.S. Department of Transportation, Washington, DC, 1991, p. C-4.)

any of the four intermediate events occur, it will lead to the top event. Transfer gate *A* under the *vehicle leaves guideway* event indicates that that fault is further developed on another page. The diamonds are undeveloped events that the analyst felt did not need to be pursued any further for this study. However, that does not mean that sometime in the future he may wish to take one or more of those diamonds and investigate their faults.

Fault trees are relatively easy to construct. However, there are a few rules that should be followed. The U.S. Nuclear Regulatory Commission's *Fault Tree Handbook,* NUREG-0492 (Roberts et al., 1981), though published many years ago is still a classic, provides some good ground rules:

- Write the statements that are entered in the event boxes as faults; state precisely what the fault is and when it occurs (e.g., motor fails to start when power is applied).
- If the answer to the question, *Can this fault consist of a component failure?* is *Yes,* classify the event as a state-of-component fault, add an OR gate below the event, and look for primary, secondary, and command modes. If the answer is *No,* classify the event as a state-of-system fault, and look for the minimum necessary and sufficient immediate cause or causes. As a general rule, when energy originates outside the component, the event may be classified as state of the system.
- If the normal functioning of a component propagates a fault sequence, then it is assumed that the component functions normally. In other words, no miracles are allowed. If a fault is going to occur, it must occur.
- All the inputs to a particular gate should be completely defined before further analysis of any one of them is undertaken.
- Gate inputs should be properly defined fault events, and gates should not be connected directly to other gates. Many people shortcut the FTA by hooking the outputs of gates directly into another gate without describing the event. Do not do that. It is sloppy.

7.2 FINDING CUT SETS

As stated earlier, the fault tree is a model of the system fault state. There are qualitative and quantitative tools to evaluate the tree. Qualitative analysis of fault trees is conducted through the use of cut sets and simple Boolean algebraic manipulation. Trees are quantified by applying probabilities or frequencies of occurrence of each event fault. The event faults are then combined through Boolean manipulation, and the top-event probability is determined. You may wish to review a math book and become familiar with Boolean algebra and probability theory. The U.S. Nuclear Regulatory Commission's *Fault Tree Handbook* (Roberts et al., 1981) and NASA's *Fault Tree Handbook with Aerospace Applications* (Stamatelatos et al., 2002) are excellent references as well.

In Boolean algebra, the OR gate represents the union of two or more events. The AND gate represents the intersection of two or more events.

In Figure 7.3, for the *no current* example, either *A* OR *B* must occur for event *A* to occur. In Boolean algebra, this can be written as

$$A = B + C$$

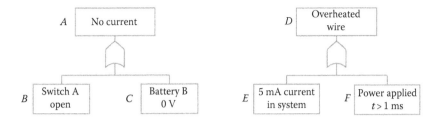

FIGURE 7.3 Fault tree representation.

The same expression in set theory is "A is B union C" ($A = B \cup C$). The *overheated wire* is stated as E AND F must occur for D to occur. Again, in Boolean algebra,

$$D = E \times F$$

In set theory, "$D = E$ intersection F" ($D = E \cap F$).

Table 7.1 is a refresher of the rules for Boolean algebra. Because of space limitations, it is impossible to go into set theory, Venn diagrams, and probability theory; it is strongly recommended that you review a good book on probability before applying any of these rules.

TABLE 7.1
Boolean Manipulation Rules

Algebraic Rule	Set Theory Representation	Engineering Representation
Commutative law	$X \cap Y = Y \cap X$	$X * Y = Y * X$
	$X \cup Y = Y \cup X$	$X + Y = Y + X$
Associative law	$X \cap Y(Y \cap Z) = (X \cap Y) \cap Z$	$X * (Y * Z) = (X * Y) * Z$ or $X(YZ) = (XY)Z$
	$X \cup (Y \cup Z) + (X \cup Y) \cup Z$	$X + (Y + Z) = (X + Y) + Z$
Distributive law	$X \cap (Y \cup Z) = (X \cap Y) \cup (X \cap Z)$	$X(Y + Z) = XY + XZ$
	$X \cup (Y \cap Z) = (X \cup Y) \cap (X \cup Z)$	$X + Y * Z = (X + Y) * (X + Z)$
Idempotent law	$X \cap X = X$	$X * X = X$
	$X \cup X = X$	$X + X = X$
Law of absorption	$X \cap (X \cup Y) = X$	$X * (X = Y) = X$
	$X \cup (X \cap Y) = X$	$X + X * Y = X$
Complementation	$X \cap X' = \varnothing$	$X * X' = \varnothing$
	$X \cup X' = \Omega$	$X + X' = \Omega = 1$
	$(X')' = X$	$(X')' = X$
De Morgan's theorem	$(X \cap Y)' = X' \cup Y'$	$(X * Y)' = X' + Y'$
	$(X \cup Y)' = X' \cap Y'$	$(X + Y) = X' * Y'$
Other operations	$\varnothing \cap X = \varnothing$	$\varnothing * X = \varnothing$
	$\varnothing \cup X = X$	$\varnothing + X = X$
	$\Omega \cap X = X$	$\Omega * X = X$
	$\Omega \cup X = \Omega$	$\Omega + X = \Omega$
	$\varnothing' = \Omega$	$\varnothing' = \Omega$
	$\Omega' = \varnothing$	$\Omega' = \varnothing$

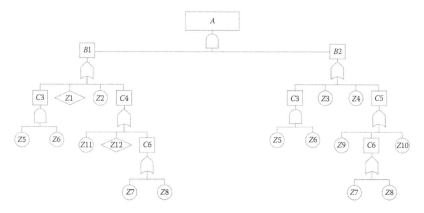

FIGURE 7.4 Example of fault tree.

Figure 7.4 is a typical branch of a large fault tree. There are a number of ways to solve a fault tree: top-down substitution, bottom-up substitution, and even using Monte Carlo simulations (with actual failure data). Also, a number of computer programs can solve (and draw) the tree. It is impossible to keep up to date with the changes in software programs for fault trees. Here are some of the software programs on the market:

- CAFTA (Data Systems and Solutions)
- FaultTree + (Isograph)
- PTC Windchill FTA (formerly Relex Fault Tree)
- Fault Tree Analysis Module of ITEM Tool Kit (Item Software [USA] Inc.)
- SAPHIRE (formerly IRRAS, U.S. Nuclear Regulatory Commission)
- Probabilistic Risk Assessment Workstation (Electric Power Research Institute)
- RiskSpectrum FTA (Lloyds Register)
- FaultrEASE (Arthur D. Little, Inc.)
- Windchill FTA
- LOGAN (RM Consultants)
- FTAnalyzer Lite (SoHaR)
- TDC FTA (TDC)
- ELMAS Fault Tree (Ramentor Oy)
- GRIF-Workshop (TOTAL, STATODEV)
- CARE FTA (BQR)

Using the top-down substitution method—actually writing Boolean equations from the top event down—we can write

$$A = B1 * B2$$

$$B1 = C3 + Z1 + Z3 + Z4$$

$$B2 = C3 + Z3 + Z4 + Z5$$

Start at the top event and then create Boolean equations for each level or branch on the tree. Once the next couple of levels have been written, you can use the various Table 7.1 substitution laws. So, combining $B1$ and $B2$ and through Boolean manipulation,

$$A = C3 + Z1*Z2 + Z3*Z4 + Z2*Z3 + Z2*Z4 + C4*Z3$$

$$+ C4*Z4 + C4*C5 + Z1*C5 + Z2*C5$$

Note that two branches are repeated in the tree, the $C3$ and $C6$ branches. It is not uncommon that the fault scenario is repeated in a large fault tree. If one subsystem feeds various plant units, then that branch will be repeated wherever it occurs. Parallel pumps, dual motors, or even single units (e.g., emergency backup power units) are simple examples of repeat branches. This is a very important point to note: if a repeat branch happens to be failure prone, then its faults will be replicated throughout the fault tree:

$$C3 = Z5*Z6$$

$$C4 = C6 + Z11 + Z12$$

$$C5 = C6 + Z9 + Z10$$

$$C6 = Z9 + Z8$$

Again, using Boolean manipulation, the final fault scenario that leads to the top event, A, can be written as

$$A = (Z7) + (Z8) + (Z5*Z6) + (Z1*Z3) + (Z1*Z4) + (Z2*Z3) + (Z2*Z4)$$

$$+ (Z3*Z11) + (Z11*Z4) + (Z11*Z9) + (Z1*Z9) + (Z5*Z9)$$

A cut set is a collection of basic events that will lead to the top event. A minimal cut set is the smallest combination of component failures, which, if they all occur, will cause the top event to occur. A single-component minimal cut set means that if that single component fails, then the top event will occur. In the aforementioned example, parentheses have been placed around the cut sets. If the components indicated in the parentheses fail, then the system will fail. As can be seen, there are numerous single-point failures.

PRACTICAL TIPS AND BEST PRACTICE

- The more AND gates you use, the safer the system is. AND gates denote a fault tolerance; for example, for a braking subsystem failure, both the primary brake AND the backup (emergency) eddy-current brake must fail.
- Likewise, if you have a lot of OR gates in your system, you are very failure prone. Any of those failures can lead to the event. A string of OR gates leading up to the top event is extremely dangerous. Try to change the system and incorporate more AND gates.

NOTES FROM NICK'S FILE

Once, I was doing a fault tree for a very complicated system that needed to improve its safety performance. I used the fault tree with lots and lots of OR gates to illustrate to senior leadership that we were seriously at risk if we could not find a way to substitute the OR for AND gates and build in more resiliency. The fault tree graphic did the job. Senior leaders immediately understood the risk they were facing, and we changed the design and operations.

Obviously, the bottom-up method of FTA is the exact opposite of what was just demonstrated. You start at the lowest level, substitute the Boolean equations, and solve for the top event.

As stated earlier, the opposite of a fault tree is a success tree. In Boolean algebra, a success tree is the complement of a fault tree. The complement of a cut set is a path set. To solve a success tree, you have two options: either draw a success tree from the start or draw a fault tree and then take the complement of the tree (along with the corresponding Boolean equations).

7.3   FAULT TREE QUANTIFICATION

FTA is not a quantitative analysis; however, the tree can be quantified. The most common method of quantification is to assign failure probabilities to each of the events. Then use the various laws of probability and statistics and solve for the top event. NASA's *Fault Tree Handbook with Aerospace Applications* (Stamatelatos et al., 2002) is a great reference. The U.S. Nuclear Regulatory Commission's *Fault Tree Handbook* (Roberts et al., 1981), again, is an excellent reference. Henley and Kumamoto's *Probabilistic Risk Assessment* book, listed under Further Reading at the end of this chapter, also goes into a lot of detail about how to assign probabilities to fault events. Before assigning failure probabilities to your tree, consult a reliability engineering book to ensure that you are manipulating the data appropriately.

PRACTICAL TIPS AND BEST PRACTICE

A very useful way of demonstrating how your safety system operates is through a success tree. The success tree will demonstrate the *must succeed* events. At times, this can be a very poignant method of demonstrating how difficult it will be to meet an exceedingly success-oriented project.

The fault tree is drawn, and then the Boolean equations and minimal cut sets are derived for the top event. Probability estimates can be generated from hardware failure data, human error estimation, maintenance frequency, etc. Probability estimates are then assigned to the events. Be sure to take into consideration uncertainty limits to your failure data. Through the laws of probability, combine the probabilities to determine the top event. The rare-event approximation is an excellent method to help truncate the math. It is used to facilitate the manipulation of very small probability numbers. Obviously, the smaller the probability, the better will be the approximation.

7.4 EXAMPLE OF A FAULT TREE CONSTRUCTION OF A MOTOR–PUMP PRESSURE SYSTEM

The *Fault Tree Handbook*, Chapter VIII, has an excellent example of how to construct a fault tree of a pump-motor pressure system, as shown in Figure 7.5. The handbook describes the problem (Roberts et al., 1981).

The function of the control system is to regulate the operation of the pump. The latter pumps fluid from an infinitely large reservoir into the tank. We shall assume that it takes 60 s to pressurize the tank. The pressure switch has contacts, which are closed when the tank is empty. When the threshold pressure has been reached, the pressure switch contacts open, de-energizing the coil of relay $K2$ so that relay $K2$ contacts open, removing power from the pump, causing the pump motor to cease operation. The tank is fitted with an outlet valve that drains the entire tank in an essentially negligible time; the outlet valve, however, is not a pressure relief valve. When the tank is empty, the pressure switch contacts close, and the cycle is repeated.

Initially, the system is considered to be in its dormant mode: switch $S1$ contacts open, relay $K1$ contacts open, and relay $K2$ contacts open: that is, the control system is de-energized. In this de-energized state, the contacts of the timer relay are closed. We will also assume that the tank is empty and the pressure switch contacts are therefore closed.

System operation is started by momentarily depressing switch $S1$. This applies power to the coil of relay $K1$, thus closing $K1$ contacts. Relay $K1$ is now electrically self-latched. The closure of relay $K1$ contacts allows power to be applied to the coil of relay $K2$, whose contacts close to start up the pump motor.

The timer relay has been provided to allow emergency shutdown in the event that the pressure switch fails to close. Initially, the timer relay contacts are closed and the

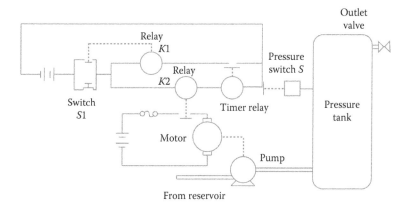

FIGURE 7.5 Pressure tank system. (From Roberts, N.H. et al., *Fault Tree Handbook*, NUREG-0492, U.S. Nuclear Regulatory Commission, Washington, DC, 1981, p. V-III.)

timer relay coil is de-energized. Power is applied to the timer coil as soon as relay $K1$ contacts are closed. This starts a clock in the timer. If the clock registers 60 s of *continuous* power application in the timer relay coil, the timer relay contacts open (and latch in that position), breaking the circuit to the $K1$ relay coil (previously latched closed) and thus producing system shutdown. In normal operation, when the pressure switch contacts open (and consequently relay $K2$ contacts open), the timer resets to 0 s.

Figure 7.6 is the resulting fault tree. In constructing the fault tree from the pressure tank schematic, it is obvious that the top event should be *rupture of pressure tank after the start of pumping*. This is a fairly simplified tree, in which piping, wiring, etc., have been ignored. The *Fault Tree Handbook* makes a good point of emphasizing that the fault must specify what happens and when it occurs.

An OR gate is drawn because the top event can be caused by a component failure. This is a good example of the use of primary and secondary component failures. The circle or primary failure of the tank could be due to things such as material fatigue and poor workmanship. If there is concern that the tank does not meet the minimum necessary design specifications (i.e., ASME Section VIII), then the circle could be another rectangle (or secondary failure). However, in this case, we feel that the tank was designed appropriately. Likewise, the diamond is highly unlikely and would not need to be developed further.

So, now, we concentrate on the secondary failure of tank rupture. The *Fault Tree Handbook* again emphasizes a critical point with primary and secondary faults— namely, that a primary failure is one in which a component fails in the environment for which it is qualified and the secondary failure is one in which it fails in an environment for which it is not qualified—important distinctions.

This secondary failure is composed of component failures, so again, an OR gate is drawn. The same logic as used earlier is applied here to draw the secondary failure and the diamond. The INHIBIT gate documents that the input to the fault is a continuous, $t > 60$ s pump operation. Remember, this is *conditional fault*. The pump must operate longer than 60 s for the failure to occur.

The concept of *state-of-component* and *state-of-system* faults is worth discussing briefly here. If a state-of-component exists—in other words, the fault occurs because of a component failure—then OR gates are used. The use of OR gates connotes that any of the listed fault inputs can cause the event. If a state-of-system fault occurs, that means that something in the system failed that caused the event to occur and thus connotes an AND gate—all the fault inputs must occur for the event to occur.

The fact that two faults are in place without a gate between them is not incorrect; it only indicates that the author wishes to detail the failure sequence. If more detail is needed to understand the process, then a string of rectangles in series can be drawn. It is obvious that for the pump to operate continuously, it must have power for longer than 60 s.

From there, an OR gate is drawn, state-of-component faults; however, the *EMF Applied to K2 Relay Coil for $t > 60$ s* is a state-of-system fault and thus requires an AND gate. This erroneous command signal to the component is due to other faults in the system.

On the left side of the AND gate, all the events end as circles or diamonds. In other fault trees or if the top event is highly significant (such as rupture of the reactor in a

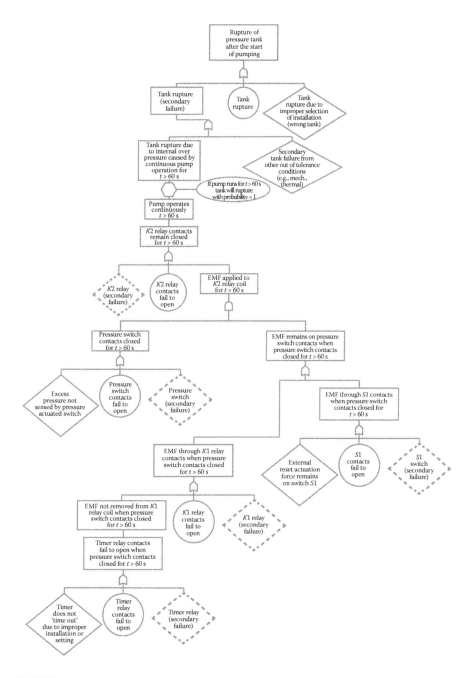

FIGURE 7.6 Pump-motor pressure tank fault tree. (From Roberts, N.H. et al., *Fault Tree Handbook*, NUREG-0492, U.S. Nuclear Regulatory Commission, Washington, DC, 1981, p. V-III.)

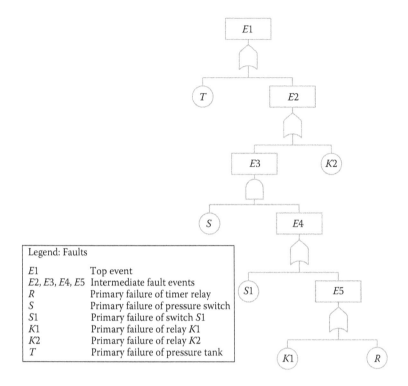

FIGURE 7.7 Fault tree example. (From Roberts, N.H. et al., *Fault Tree Handbook*, NUREG-0492, U.S. Nuclear Regulatory Commission, Washington, DC, 1981, p. VIII.)

nuclear power plant), then these entries may need to be evaluated further. Likewise, the far right side of the fault tree ends with similar faults.

The remainder of the fault tree goes into further detail about how relay circuit can fail. In Figure 7.7, the fault tree has been simplified and the Boolean expressions developed:

$$E1 = T + E2$$

$$= T + (K2 + E3)$$

$$= T + K2 + (S * E4)$$

$$= T + K2 + S * (S1 + E5)$$

$$= T + K2 + (S * S1) + (S * E5)$$

$$= T + K2 + (S * S1) + S * (K1 + R)$$

$$= T + K2 + (S * S1) + (S * K1) + (S * R)$$

The minimal cut sets are $K2$, T, $S * S1$, $S * K1$, and $S * R$.

TABLE 7.2
Failure Probabilities for Pressure Tank Example

Component	Symbol	Failure Probability (Pr)
Pressure tank	T	5×10^{-6}
Relay $K2$	$K2$	3×10^{-5}
Pressure switch	S	1×10^{-4}
Relay $K1$	$K1$	3×10^{-5}
Timer relay	R	1×10^{-4}
Switch $S1$	$S1$	3×10^{-5}

Now, if we assign failure probabilities to system components, we can see how quickly the fault tree can be quantified. Table 7.2 shows the failure probabilities.

It is very important to remember to follow the rules for manipulation of probabilities. In this example, all failures are independent, so they can be easily multiplied together. If they were dependent failures, then the combinations would be different. So the resulting probabilities are

$$Pr(T) = 5 \times 10^{-6}$$

$$Pr(K2) = 3 \times 10^{-5}$$

$$Pr(S*K1) = (1 \times 10^{-4})(3 \times 10^{-5}) = 3 \times 10^{-9}$$

$$Pr(S*R) = (1 \times 10^{-4})(1 \times 10^{-4}) = 1 \times 10^{-8}$$

$$Pr(S*S1) = (1 \times 10^{-4})(3 \times 10^{-5}) = 3 \times 10^{-9}$$

So, by summing the minimal cut sets, the top-event probability of occurrence is

$$Pr(E1) = 3.4 \times 10^{-5}$$

7.5 COMMON MISTAKES IN FAULT TREES

A few mistakes you should try to avoid in constructing, quantifying, and evaluating fault trees are the following:

- Try to model to the highest level possible that you have data; the more the data used, the more uncertainty in the model.
- Do not put too many inputs that have very small probabilities into gates.
- Do not spend too much time on passive components in a system. Remember, the fault tree really looks at functions, not components.
- Do not model human errors of commission because they are very difficult to capture realistically and can skew results.

- Remember, garbage in–garbage out. If the results of the quantified tree do not make sense, do not give them too much weight. It is much better to use quantitative trees for comparison, not as absolute number generators.
- Do not fault tree everything. It is expensive—be judicious.
- Do not try to treat Boolean algebra expressions as regular algebraic equations. Be careful when combining Boolean expressions.
- Look closely at the failure modes to determine if they are independent or dependent. This is very important in probability manipulations.
- Be sure the top event is a high-priority concern.

PRACTICAL TIPS AND BEST PRACTICE

Fault trees are extremely powerful methods to demonstrate your safety systems' fault tolerance to an accident. The next time you want to demonstrate how many things *must* go wrong for an accident, use fault trees. Fault trees are great tools to educate a non engineer (e.g., in a lawsuit) of how hard it is for something to occur.

REFERENCES

Dorer, R. M. and Hathaway, W. T. 1991. *Safety of High Speed Magnetic Levitation Transportation Systems: Preliminary Safety Review of the Transrapid Maglev System.* DOT-VNTSC-FRA-90-3. Washington, DC: U.S. Department of Transportation.

Roberts, N. H., Vesely, W. E., Haasl, D. F., and Goldberg, F. F. 1981. *Fault Tree Handbook.* NUREG-0492. Washington, DC: U.S. Nuclear Regulatory Commission.

Stamatelatos, M., Caraballo, J., and Vesely, W. August 2002. *Fault Tree Handbook with Aerospace Applications.* Version 1.1. Washington, DC: NASA Office of Safety and Mission Assurance NASA Headquarters.

FURTHER READING

Anderson, T. and Lee, P. A. 1981. *Fault Tolerance: Principles and Practice.* Englewood Cliffs, NJ: Prentice-Hall.

Center for Chemical Process Safety. 1999. *Guidelines for Chemical Process Quantitative Risk Analysis*, 2nd edn. Hoboken, NJ: Wiley.

Center for Chemical Process Safety. 2008. *Guidelines for Hazard Evaluation Procedures*, 3rd edn. Hoboken, NJ: Wiley.

Haasl, D. F. 1965. Advanced concepts in fault tree analysis. *System Safety Symposium*, Seattle, WA.

Henley, E. J. and Kumamoto, H. 2000. *Probabilistic Risk Assessment and Management for Engineers and Scientists.* Hoboken, NJ: Wiley-IEEE Press.

International Electrotechnical Commission. 2006. *Fault Tree Analysis.* IEC 61025. Geneva, Switzerland: International Electrotechnical Commission.

Lacey, P. 2011. An application of fault tree analysis to the identification and management of risks in government funded human service delivery. *Proceedings of the Second International Conference on Public Policy and Social Sciences,* Kuching, Sarawak, Malaysia. http://papers.ssrn.com/sol3/papers.cfm?abstract_id=2171117, downloaded July 9, 2013.

Lapp, S. A. and Powers, G. J. 1977. Computer-aided synthesis of fault trees. *IEEE Transactions on Reliability*, R-26: 2–13.

Long, A. Beauty and the beast—Use and abuse of fault tree as a tool. No date. http://www.fault-tree.net/papers/long-beauty-and-beast.pdf downloaded May 17, 2014.

National Institute of Standards and Technology. 2009. *Root Cause Analysis Report of Plutonium Spill at Boulder Laboratory*. Gaithersburg, MD. http://www.nist.gov/public_affairs/releases/upload/root_cause_plutonium_010709.pdf downloaded May 17, 2014.

8 FMECA, Human Factors, and Software Safety

The business of every art is to bring something into existence.

Nicomachean Ethics, 350 BCE
Aristotle

In the land of the blind, the one-eyed man is king.

Adagia, 1500
Erasmus

Know then thyself, presume not God to scan,
the proper study of Mankind is Man.

An Essay on Man, 1734
Alexander Pope

Three nonsafety tools are used in safety analysis: failure modes, effects, and criticality analysis (FMECA); human factors analysis; and software analysis. Because these techniques are extremely helpful in finding equipment failures, human errors, and software mistakes, safety engineers have coupled them to their safety analyses. It is definitely worthwhile to understand how these tools can benefit you.

8.1 FAILURE MODES AND EFFECTS ANALYSIS

Failure modes and effects analysis (FMEA) and its complement, *FMECA*, are used primarily as reliability engineering tools. However, system safety engineers for years have appropriated the FMEA and used it extensively in identifying failures. Reliability engineers have used both FMEA and FMECA since the early 1960s in the U.S. missile program. And, as stated earlier, OSHA recognizes FMEA as a legitimate safety analysis tool. These tools are ubiquitous around the world in most every industry dependent on system design.

Before discussing how to perform an FMEA, it is very important to emphasize that it is a reliability engineering tool and not a primary safety tool. A significant danger in using FMEA is that the engineer will think that by identifying failures he or she has identified hazard causes. As Chapter 4 clearly states, a hazardous situation can occur as part of normal system operation. *A failure does not have to occur for a hazard to be present in the system.* The Space Shuttle used highly volatile propellants which, when operated as designed, will burn profusely and produce very significant thrusts. If all works well, no failures occur, but that does not mean that there are no hazards in operating the Shuttle.

It is strongly recommended that an FMEA be used to investigate further how a particular failure (which leads to a hazard) can come about. The FMEA should not be used as the primary safety analysis tool. A more appropriate application is to hazard and operability (HAZOP) a particular part of the plant. Once the safety-critical operations have been identified, then FMEA can be used very selectively to focus on how particular failure modes might lead to process deviations and thus create a hazard. The primary reason for this is that FMEAs are a very laborious effort and easy to become bogged down. But their strength is going to the piece-part level, as necessary, to determine root causes, and this of course is paramount in understanding how to control a hazard.

Having said all this, FMEAs should not be discounted as not being useful for safety analysis. It is an extremely powerful analytical tool that is applied in virtually all industries, from food processing to aerospace. If FMEA is used as designed, it can be very beneficial to engineers.

8.1.1 CONDUCTING A FAILURE MODES AND EFFECTS ANALYSIS

FMEA is simply *an analysis tool that identifies all the ways a particular component can fail and what its effects would be at the subsystem level and ultimately on the system.* FMEA is vastly different from fault tree analysis. Fault tree analysis is a top-down analysis of faults in a system. FMEA is a bottom-up analysis that identifies failures (not necessarily faults) in the system. The fault tree starts with the top-level or system-level concern (top event) and then works down to the events that lead to that top event. FMEA does exactly the opposite: it starts with the components in the system and analyzes failures and how they impact the subsystem in which it is housed and what are the propagated effects across the system.

FMEA is a very methodical tool that looks at every component in the system (or subsystem) under study and then identifies the ways the component can fail (failure modes). Each failure mode is further analyzed to determine what would be the effects of that particular failure mode on the system. Results are posted in tabular form (very similar to the hazard analysis worksheet).

Return to Figure 5.4, which describes a car in traffic as a system. FMEA starts with the components that make up a car (e.g., wheels, brakes, and steering) and then analyzes how a failure in those particular components would affect the rest of the system. Figure 8.1 shows how a particular system is comprised of various components.

The process to perform an FMEA is straightforward:

1. Define the system and analysis scope and boundaries.
2. Construct functional block diagrams that indicate how the different system indenture levels are related.
3. Assess each functional block (at the block level) and determine if its failure would affect the rest of the system. If it would not, then ignore the block. If its failure would affect the rest of the system, go down another indenture level and perform the following scheme. Continue down to the level of relevance.
4. The real analysis starts here; this is where the bottom-up approach commences. In each functional area where failures could adversely affect the system, look at the component failures. List the *modes* or ways that the component can fail. Be sure to mention what component parts would fail.

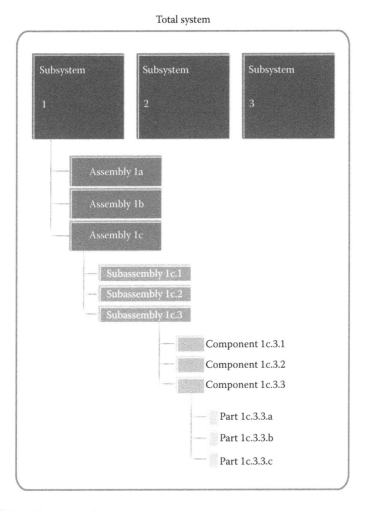

FIGURE 8.1   System breakdown.

5. For each failure mode, assess the failure's *effects*. Usually, engineers assess the worst credible case with consequence severity and probability of occurrence, if possible.

6. Identify whether the failure is a *single-point failure*. This is very important. A single-point failure is the failure of a single component that could bring down the entire system.

7. Determine the failure effects locally and how it propagates to the next system level. Understanding failure propagation is critical so that we can help design our system to be more resilient. If it propagates to loss of system, then it becomes a single-point failure.

8. Determine methods of corrective action. These might take the form of preventing the failure or mitigating its effects.

9. Document on the FMEA worksheet.

NOTES FROM NICK'S FILE

I was doing an FMEA of a blowout preventer, a series of valves that control the back pressure of subsea oil drilling equipment, that had a lot of *designed-in* redundancy for safety. The blowout preventer had multiple redundancy (for safety reasons) methods to operate these critical valves during an emergency. These redundant systems were bundled in redundant hosing down to the seabed, a redundant path to the blowout preventer. Ironically, the single-point failure to the entire safe and reliable operation of the system was a manual valve that switched operations from a side A redundant path to side B. If that valve got stuck in the middle, potentially, neither side would work. When discovered, obviously, this was quickly fixed.

The typical columns shown on an FMEA worksheet are indicated in Table 8.1, though there is no single, correct format for the FMEA worksheet. The *subsystem* is the group of components being analyzed in this functional block. The *assembly* is part of subsystem. The *subassembly* does not have to pertain to a specific hardware system; it could be a functional group of components such as electrical, environmental, mechanical, or structural elements.

The *component number* is the company part number identifier. The *component name* is the common name for the component. The component *function* must state clearly what the component does.

Failure mode describes how the component can fail. All credible failure modes and their causes should be listed. Since a failure mode may have more than one cause, all probable independent causes for each failure mode should be identified and described. Three failure mode causes are shown in the sample worksheet. Some engineers will create a separate *failure cause* column tied to each failure mode so that causes are more easily tracked. Here, the two are combined for ease of understanding. Typical failure modes (conditions) that should be considered are

- Premature operation
- Failure to operate at a prescribed time
- Intermittent operation
- Failure to cease operation at a prescribed time
- Loss of output or failure during operation
- Degraded output or operational capability
- Other unique failure conditions as applicable based on system characteristics and operational requirements or constraints

Mission phase indicates the appropriate phases during the system life cycle. Typical mission phases are installation, operation, maintenance, and repair.

Failure effects locally describe each failure mode effect on the assembly. *Failure propagation to next level* describes how malfunction of the component and/or subassembly will affect other components, assemblies, or the total system.

Single-point failure designates the individual malfunctions that will cause shutdown of the entire system. The true challenge for an engineer is to decide when and

TABLE 8.1
FMEA: Hydraulic Control System

Subsystem: Hydraulic control panel
Assembly: Junction Box A
Subassembly: Mechanical

Date: 10/11/2014
Analyst: John Doe
Page: 31

Component number	Component name	Function	Failure mode	Mission phase	Failure effects locally	Failure propagation next level	Single-point failure	Risk failure class	Control, recommendation
45–341	Solenoid valve	Electropneumatic interface and control of hydraulic panel valves	No pneumatic signal sent from valve due to loss of pressure—fail closed	Ops.	Rendered useless due to loss of working fluid	No pneumatic signal sent to hydraulic valve, resulting in longer response time to control valve 3-A	No	4C	Manually operate hydraulic panel valve. Verify air supply inlet pressure from source. Inspects for leaks.
			Failed valve due to internal spring failure from excessive wear	Ops.	Continuous pneumatic flow through valve	Possible hydraulic valve activation or deactivation due to inappropriate pneumatic pilot signal	No	4C	Inspect and test regularly. Verify correct and smooth spring–plunger–sleeve alignment.
			Failed plunger from excessive wear	Ops.					
45–342	1/4 in., 4-way, 3-position hydraulic valve	Selects side A or B of system	Failed open or closed due to wear, part failure, or leaks	Ops. maint.					

(*Continued*)

TABLE 8.1 (*Continued*)
FMEA: Hydraulic Control System

Component number	Component name	Function	Failure mode	Mission phase	Failure effects locally	Failure propagation next level	Single-point failure	Risk failure class	Control, recommendation
					Incorrect or no pneumatic signal sent	Reduced reliability of hydraulic valve	No	4C	Lubricate well during maintenance phase to avoid excessive wear. Add maintenance procedures.
					Rendered useless	Loss of entire hydraulic control system	Yes	1B	To eliminate single-point failure, install separate uninterrupted supply lines from pumping bank. Lead each line into separate hydraulic panel valves. Add to maintenance procedures to inspect and clean regularly. Add use of anti-seize compound to operating procedures. Add to quarterly inspection to verify all gaskets, O-rings, and proper seals are in correct position. Verify that each gasket fits over shoulder and end caps. Avoid hammering seal palate into valve assembly.

how much money to spend to design out single-point failures. In most cases, it is impossible to design them out entirely. An alternative to designing out single-point failures is to make the component much more robust and less sensitive to failure. That is precisely why FMEAs are conducted.

Risk failure class is the same classification scheme used with all the other safety analyses and described in detail in Chapter 5. It crosses the severity of the consequences with the probability of occurrence. Many times, this column is divided into two columns, clearly indicating the severity and the probability of occurrence. Because FMEAs are really for reliability engineers, the probability column usually contains the actual failure rate of the component.

Control recommendations describe countermeasures that can be taken to eliminate or prevent the failure or reduce the effects.

Sometimes, reliability engineers also add a column that identifies the method of

PRACTICAL TIPS AND BEST PRACTICE

- Be Careful! Do not overuse FMEAs. They are expensive to conduct if you use it across the entire system. What makes more sense is to apply it selectively, once a catastrophic or significant hazard and its causes are identified through a safety analysis (e.g., hazard analysis or HAZOP), and if one of them is component failure, use the FMEA to further drill down to all the causal factors of the component failure that could lead to the hazard.
- Do not forget to try to assign probability and severity to the failure effects. This is very important because it gives the risk ranking of that component failure on the overall system and it helps to best determine if you really need to fix it or mitigate the failure consequences.

detecting the failure mode. This can be very useful if the failure mode is very severe. Different methods for detecting failure modes are instrumentation, feedback loops, inspections, testing, etc.

8.1.2 FAILURE MODES, EFFECTS, AND CRITICALITY ANALYSIS

FMECA is virtually the same as FMEA except that it also identifies the criticality of the component under study. The engineer emphasizes the probability of failure much more than in FMEA. The criticality is divided into its constituent parts:

PRACTICAL TIPS AND BEST PRACTICE

Use FMEAs to identify single-point failures in your new designs. Then ask the engineers to either design them out or support why they should be left in. If you go through that process (even a cursory review) on all designs before they are built or modified, you will develop products of significantly higher quality at a much lower cost. At the very least, it is a good sanity check.

- *Failure effect probability* (β): The β values are the conditional probabilities that the failure effect will result in the identified criticality classification, given that the failure mode occurs.
- *Failure mode ratio* (α): The failure mode ratio is the probability expressed as a decimal fraction that the part or item would fail in the identified mode.
- *Part failure rate* (λ_p): The part failure rate is the failure rate of individual piece, part, or component.
- *Operating time* (t): Operating time is the amount of time in hours or the number of operating cycles of the item per mission.

All of these are combined to give the *failure mode criticality number* (C_m)

$$C_m = \beta\alpha\lambda_p t$$

or, more specifically, an *item criticality number* (C_r). This is the sum of the failure mode criticality numbers under the severity classification and is written as

$$C_r = \sum_{n=1}^{j}(\beta\alpha\lambda_p t)n \quad n = 1,2,3\ldots$$

This information is then compiled in a criticality matrix, and the analysis can rank the items based on which is the most critical failure to the system.

8.2 HUMAN FACTORS SAFETY ANALYSIS

NASA's Magellan spacecraft eventually bound for Venus suffered a fire during ground testing. A technician incorrectly mated high- and low-voltage lines. He reached around into the back of the spacecraft—a blind mate, the connectors were of the same size and keyed the same—and connected the cables. Unable to visually verify the miss-mate, he powered on the test equipment and created a fire to the $400M spacecraft.

A factory used a large manifold that joined various production lines with packaging lines. The idea was to quickly be able to reconfigure the production and packaging lines based on the product going to market. The flexible piping from Production Unit A was incorrectly hooked to Packaging Line B, and liquid shampoo was dumped on to the packaging line floor for hours before anyone noticed.

And something that has happened to all of us, we put money in the coffee machine, the cup falls into place, but at an angle, the coffee flows past the cup into the drain. But we cannot open the little window to right the cup until the coffee has finished dispensing (presumably so that we do not burn our hands on the hot coffee). All of these are human factors mistakes. Interestingly enough, each of those human errors could have been avoided if the design had better considered how people would use the equipment.

Billings and Reynard (1981) say that 70%–90% of all system failures are due to human error. Table 8.2 (Reason, 2006) illustrates that this is a problem across all industries:

To try to mitigate this, engineers have used a potpourri of human factors controls. There really is no such thing as one human factor analysis technique—there are many. A handful of the more interesting (though not necessarily the most important)

TABLE 8.2
Estimates of Human Error

Industry	Estimates of Human Error (as percentage of all failures)
Jet transport	65–85
Air traffic control	90
Maritime vessels	80–85
Chemical industry	80–90
Nuclear power (United States)	70
Road transportation	85

names are task analysis, confusion matrix, expert estimation, THERP, HEART, SLIM-MAUD, human cognitive reliability model, operator action tree, and socio-technical assessment of human reliability. This section, however, will focus on more mundane definitions and techniques for figuring out how to deal with those pesky humans. Engineers, perhaps as an occupational hazard, tend to forget that people have to operate the machines and plants they design. As stated earlier in the book, because humans are so difficult to predict, especially under stressful conditions, engineers need to consider the human side of things in their designs, if they wish to make them safe. And, just because we are dealing with humans, this does not mean that there are not any quantitative human factors analyses available.

This section is really an extension of Section 5.6, which is one method of looking at how humans interact with the system. Operations and Support Hazard Analysis (O&SHA) will not be discussed here. However, other human factors tools will be described that can be used as inputs to O&SHA or the more hardware-oriented safety analysis techniques described in other chapters.

The field of study is called *human factors engineering*. Sometimes, engineers call it *ergonomics*. A complementary field is *human reliability engineering*. In system safety, we need to merge the three: understand how people act and react (human factors), design equipment that helps people do their jobs better not worse (ergonomics), and understand how to make the entire system more reliable in spite of the human element (human reliability). The military calls the practice human systems integration—the interaction between people (operators, maintainers, and support) and their systems. For practical purposes, it is useful to develop a global definition for any of these human factors or human reliability tools, none of which is solely dedicated to safety problems. An appropriate name for the merging of the three fields is *human factors safety analysis*, because it consists of the study of the *human activities associated with the interfaces among people, machines, and the operating environment.* Obviously, the operative words are *human activities*. All these tools focus on how well humans interact with their environment—something other safety analysis techniques tend to be somewhat weak in. As with all the other safety analysis tools described in this chapter, human factors safety analysis should not be used in lieu of other techniques, but in conjunction with them (e.g., hazard analysis or HAZOP). Human factors safety analysis fits very neatly with the other hardware-oriented system safety analyses. The goal is to fit the machine around the person, not fit the person to the machine. In particular, human factors look at

- Training—Equips the user with the knowledge, skills, and abilities required for the task
- Personnel (recruiting, retention)—Addresses all aspects of personnel requirements
- Habitability—Ensures all aspects of the living and working spaces (including environmental and operational) are designed for the operator in mind
- Engineering design features—Determines the design features of the equipment that minimize human error and reduce risk of accidents or injury
- Organizational—Identifies how the organization supports the success of the human operator through its infrastructure

Much research has been conducted in the fields of human factors and human reliability. As a result of the Three Mile Island nuclear nearmiss accident in the United States in the late 1970s, the U.S. Nuclear Regulatory Commission developed a standard for conducting human reliability analysis for commercial nuclear power plant operators. These quantitative human reliability analyses are plugged into the nuclear plant probabilistic risk analysis. The *Handbook of Human Reliability Analysis with Emphasis on Nuclear Power Plant Applications*, NUREG/CR-1278 (Swain and Guttman, 1983, updated May, 2011), is an excellent source of information that you are strongly urged to read. The document presents a methodology for identifying human errors and even predicting quantitative human error rates. There are numerous international standards in human factors and ergonomics developed, but the largest and best-known international human factors and ergonomics societies are Institute of Ergonomics and Human Factors (founded in the United Kingdom, it is the oldest professional body), Human Factors and Ergonomics Society, and the International Ergonomics Association (a federation of human factors and ergonomics societies from around the world). These societies have created many standards, guidelines, recommended practices, and certification programs that have also become ISO standards, American National Standards Institute standards, or standards in other countries. Ergonomic standards have been developed in many different industries and are especially created for industry-unique equipment such as computer workstations.

8.2.1 Performance and Human Error

To start, it is important to understand how people act, what influences their reactions, and what causes stresses on them. The end result is to better design our technological systems so that humans can use them without abusing them (or being abused by the system). All of us can cite hundreds of cases in our own plants where people have made stupid mistakes. Why did they do that? What was the result? Could it have killed someone?

People make mistakes as a result of a combination of causes. Some of those causes are within the individual (e.g., not understanding how to operate a particular piece of equipment), and some are external to the individual (e.g., poor lighting on the manufacturing floor causes the operator to operate a particular piece of equipment incorrectly).

Here is an interesting example. Itaipu Binacional is one of the world's largest hydro-electric power plants in the world. It sits on the border between Brazil and Paraguay and produces electricity for both countries. Brazilians speak Portuguese and Paraguayans speak Spanish, both languages are very similar but are separate languages. Imagine the challenges that their binational control room has to deal with from a human factors perspective. Both workforces speak their native languages, and carefully managing the linguistic and cultural differences becomes particularly important during an emergency.

Human performance is determined by certain factors that influence how people act. These factors are called *performance-shaping factors* (PSFs). PSFs are usually a complex confluence of items that affect the operator in a system and are divided into external PSFs, internal PSFs, and stressor PSFs. PSFs can greatly affect how safely a system is operated.

External PSFs are made up of all the conditions that an individual encounters—including the entire work environment, especially the equipment design and the written procedures or oral instructions. Three general conditions are included in external PSFs: situational conditions (things that influence the individual that are plant wide or company wide, such as plant shift schedules and holidays), task and equipment characteristics (factors that are related only to a specific task or piece of equipment), and job and task instructions (factors that influence how an operator is taught to do the task).

Internal PSFs are the factors related to the individual's previous training or experience in performing the task. Other PSFs are the individual's state of current practice or skill, personality and motivation, emotional state at the time of performing the task, and physical condition.

As workforces around the world become more and more internationalized, this becomes even more important. International corporations face these PSF challenges everyday; how do they manage a corporate identity and work procedures yet take into consideration local cultural norms (and what are those impacts on safety)? Review Chapter 4 on safety culture about how safety culture must be part of the SMS, no matter how large or global an operation.

NOTES FROM NICK'S FILE

Working with one of the world's leading global manufacturers was an instruction to me on how important corporate identity and local customs must reconcile the differences. This French company had operations all over the world, and they very much maintained a French work culture and ethic. Hiring locally, the two cultures had to work together. I realized the power of this when I suggested that the company uses a safety awards system to help motivate safe behavior and actions on the factory floor. One manager asked me in response, why are we rewarding people for doing their job? He pointed out to me that I was suggesting a very American approach to safety culture and safety behavior.

Stressor PSFs are much more difficult to understand and therefore are usually ignored. Unfortunately, these PSFs do influence how people can react in a hazardous situation. A stressor is a stress (which can be either positive or negative in terms of performance of the task) that is applied to the individual during the task. Psychological and physiological stressors are the two major groupings of stressors.

Some of the psychological stressors that directly affect mental stress are suddenness of onset; duration of stress; task speed; task load; high-jeopardy risk; threat of failure; boring, repetitive, or meaningless work; long, uneventful vigilance periods; and distractions (e.g., noise, glare, and flicker).

Some of the physical stressors that influence physical stress are duration of the stress, fatigue, pain or discomfort, hunger or thirst, temperature extremes, atmospheric pressure extremes, oxygen insufficiency, vibration, and disruption of circadian rhythm.

The interplay of all these PSFs can have a negative, positive, or mixed impact on how well the person performs the task. For example, if the work environment is too hot, the work is very repetitive, and the worker is overqualified for the job, the result is likely to be many, many errors. It may not be a problem if there are no safety concerns tied to these human errors. But what if the job is on the floor of a plant that manufactures razor blades? Inattention to detail can get a worker killed if he inappropriately handles the machine that cuts the razor blades from long metal sheets. Couple this with poor machine design and a disaster is just waiting to happen. Of course, the best solution would be to design the machine so that it is physically impossible for the operator to work on it while it is running.

Unfortunately, we see many of these types of accidents in the surface transport sector. Many train accidents have occurred over the years due to the driver operating a train from a cab that is not sufficiently cooled under hot operating conditions. The hot weather creates fatigue and inattention quickly slips in. This certainly is a key risk to be managed as many Middle Eastern countries build and connect their international rail lines over the coming years. Summer ambient temperatures easily reach over 120°F. This is without even considering how hot a diesel train cab can get during operations. It will be imperative to consider this during the cab design and operational testing.

Human error is an out-of-tolerance action within the human–machine system. It occurs when the human operator is mismatched with the task at hand. PSFs affect how well the operator and the task are matched. Obviously, a poorly designed system can set a person up to fail. There is no reason to blame or find fault with a person for doing a task incorrectly. The mismatching of the task and the person is the real reason why a person makes a mistake. Using another machine example, one company had a rash of injuries on its mechanical power press. Management blamed the inattention of the operators, spent a lot of money on training, and in the end got almost the same results. The machine was designed to be operated only with the strictest concentration (almost superhuman); the operator was set up to fail. Once a *presence-sensing device* was added to the press, creating a sensing field or area that signals the clutch/brake control to deactivate the clutch and activate the brake of the press when any part of the operator's body or a hand tool is within such the field or area, the injury rate dropped. The design change was significantly less than the training needed to operate the press safely and very much less expensive than the workman's compensation payments. Again, think of the Hazard Reduction Precedence discussed in Chapter 2.

Human error, if not the cause, is a significant contributor to many accidents. Better understanding how people act and react to the PSFs in their environment allows us to better design systems to be tolerant (or more robust) to human errors.

Popular assumptions about human error are that it is inevitable and there is little we can do and that people are careless, have a bad attitude, and do not pay attention. Many feel that the only solution is intensive training or some sort of negative reward (e.g., losing your job). The truth is that most people need to go through a trial-and-error period to learn. As our systems become more and more complex, an individual understands them less and less, and therefore, mistakes are likely to increase. So, to decrease the amount or significance of human errors that can lead to a hazard, you need to make the safe operation of your systems less dependent on how well people can operate them.

Again, the Three Mile Island commercial nuclear reactor nearmiss is a classic example of how human error exasperated an already dangerous situation. Some of the human errors during the crisis were

- Auxiliary feedwater failure not recognized for 8 min
- Stuck-open relief valve not recognized
- Loss of coolant not recognized
- High-pressure safety injection turned off
- Incorrectly opened drain line valve
- Steam in primary coolant system not recognized

According to Swain and Guttman (1983, 2011), if we imagine that human errors are really incorrect outputs from the human element, then we can see that human errors fall into two major categories, errors of omission and errors of commission. *Errors of omission* include such errors as leaving out an entire task or a step in a task. Errors of commission include

- Selection error: Selecting the wrong control, miss-positioning a control (miss-mating pieces, reversing pieces, etc.), and issuing a wrong command or wrong information
- Error of sequence
- Time error: Too early or too late
- Qualitative error: Too little or too much

Engineers love to quantify whenever possible, and we have done so even in the amorphous area of human behavior. We even attempt to predict quantitatively how people will commit errors. This value is called *human error probability* (HEP). Mechanical and electrical failure probability predictions are fairly controversial; human error probabilities are even more so. There is no single internationally accepted method of quantifying human behavior prediction. Human error prediction, however, is already being used in a variety of fields:

- The military is probably one of the biggest consumers of human error prediction, especially as related to crew resource management during combat. It is also important for weapon platforms (e.g., fighter attack aircraft).
- NASA and European Space Agency and other national space agencies.

- The commercial nuclear power industry uses it extensively in all probabilistic risk assessments (called probabilistic safety assessment in Europe).
- SNEF, the French engineering company, used it in determining panel layouts in their bullet trains. It is now common in high-speed rail design around the world.
- Private companies are investigating its use for intelligent highway systems.
- Aircraft associations are applying it to cockpit design.
- Off-shore oil production operations are starting to look at it more seriously.
- Interestingly, sports medicine is doing a lot of new work.
- It is used in marine systems.
- The chemical industry is starting to look at its use in plant operations centers.

HEP is measured by observation. It is the relation of the number of the observed errors to the total number of chances for errors:

$$\text{HEP} = \frac{\text{Number of errors that occurred}}{\text{Number of opportunities for errors to occur}} = \frac{n}{N}$$

This simplistic calculation really entails much more work. In developing HEPs, it is important to consider data probability distributions, data dependence, and uncertainty limits. Again, refer to the *Handbook of Human Reliability Analysis* (Swain and Guttman, 1983, 2011) for a step-by-step approach.

8.2.2 CONDUCTING HUMAN FACTORS SAFETY ANALYSIS

The objective of human factors safety analysis is to identify and correct human error situations that could lead to significant hazards. The analysis can be either qualitative or quantitative, depending on the level of detail desired and what the consequences are of a person making a mistake. The steps of a human factors safety analysis are as follows:

Step 1: Describe the system goals and functions. Then define the system hazards of interest. These are system functions that may be influenced by human errors. Use one of the other safety tools (e.g., HAZOP) described earlier.

Step 2: List and analyze the related human operations. A qualitative tool—task analysis—will be employed. It analyzes how the task is performed and what types of aids are needed to support performance.

Step 3: Analyze the human errors. Look at how the task can fail, what errors can occur, and how the system can recover from them.

Step 4: Screen the identified errors and decide which ones are worth quantifying.

Step 5: Quantify the errors and estimate how they will affect the rest of the system. Assess which errors would have the largest impact on the safety of the system.

Step 6: Recommend changes to the system that will reduce the impact of human errors. This can be done in two ways: reduce the likelihood that the error will occur or reduce the severity of the effect on the system (make the system more error tolerant or robust).

The purpose of step 1 is to understand where people fit into the system goals and functions. What are some of the assumptions people have toward the system? Is the system highly computer controlled? Do operators need to read dials and operate valves? Focus on the hazards identified in the other system safety analyses. Identify which of those are human induced or can be mitigated by human actions.

Step 2 requires that you take the information generated in the previous step and list the actual human operations (tasks) involved in those situations that create hazards. A *task* is any action necessary to conduct an operation or operate a machine or process. It can be long, involved, and complex, or it can be very short and direct.

After identifying the tasks, conduct a task analysis. *Task analysis* studies the relevant human elements in tasks and their potential for human error. The primary purpose is to analyze operator goals and the correct sequence of action. The engineer describes what the job performer does and then analyzes what can go wrong and why. There is grave danger in skipping this step. If you assume that human errors can be discovered without a formal task analysis, then the methodical approach to system safety is lost. Task analysis forces the analyst to consider systematically all aspects of the task. Figure 8.2 shows the procedure for conducting a task analysis.

The results of a task analysis are typically documented on a tabular form, such as those used for hazard analysis. The kind of information Swain and Guttman (1983, 2011) suggest are

- Brief description of task or step
- Instruments or controls that display task information
- Activity completed
- Cues used for initiation or completion of activity
- Comments or remarks
- Scanning, perceptual, or anticipatory requirements
- Recall required (i.e., long- or short-term memory)
- Interpretation requirements
- Manipulative problems
- Likely human errors
- Safety concerns resulting from the human error (added by author)
- Comments if confirmed

After the task analysis forms have been filled out, you may wish to develop an *operational sequence diagram*. This is a graphical depiction of human–machine interactions and decision-making flow in the performance of a task over time. Because this graphical system representation is so labor intensive, it should be used only for very complex systems, such as nuclear power plant operations.

One of the strong points of task analysis is that the human–machine interface is studied comprehensively. Because task analysis breaks down the tasks to their

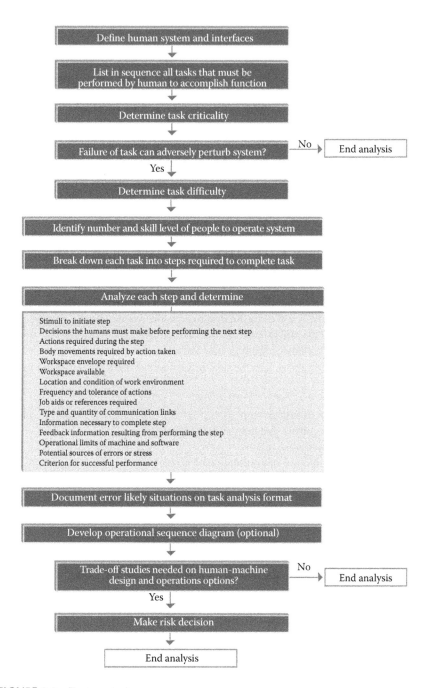

FIGURE 8.2 Tank analysis process.

individual steps and then attempts to analyze each step and the action's relation to the human operator, it is a powerful tool to help understand the system and pinpoint problems. Task analysis is very efficacious because it can be modified to analyze tasks that are complex or tasks that are very simple. This is one reason that it is so popular with human factors specialists. As can be seen by studying the task analysis process, it can be performed by nonhuman factors specialists and thus is useful for all engineers who wish to try to minimize problems caused by a human operator.

After conducting the task analysis, steps 3 and 4 require you to analyze the human errors that are important to the safety of the system. Pick out just the human errors and tasks that have significant consequences. Those are the errors that you want to quantify (if you want to use numbers). See in the Further Reading section at the end of this chapter for various sources for HEP tables.

In step 5 you merge that data into another system safety analysis and see how much impact the human errors have on the total system. In the nuclear power industry, it is common to add human error probabilities to the system fault tree.

And of course, the bottom line (step 6) is as follows: find ways to make the system safer. You can reduce the likelihood of human errors (reduce their probability of occurrence through engineering controls). Or it may be easier or cheaper to mitigate the effects of the human error.

PRACTICAL TIPS AND BEST PRACTICE

- Do not assume that people follow what is written down in the procedures. Many times, if not almost the majority of the time, operators do not follow exactly what the procedure says—even when quality assurance (QA) has to sign off. A number of years ago, a 90-foot, bright-yellow-colored scaffolding was left inside the Space Shuttle payload bay when the doors were closed. No one noticed until the orbiter was lifted vertically to be mated with the rest of the Shuttle, and everyone heard load crashes and bangs. Both operators and QA had signed the logbook stating that the scaffold had been removed.
- *Observe* the task you wish to study; do not rely on what is supposed to be done.
- Personally *observe* third-shift operations—lots of things happen at that time.

8.2.3   BRIEF EXAMPLE OF HUMAN FACTORS SAFETY ANALYSIS: MANUAL SWITCHOVER TO AUXILIARY FEEDWATER SYSTEM

This example is taken from Swain and Guttman (1983, 2011) and summarized later. Their results derived from a plant visit, review of procedures, interviews with the operators, and observations of the tasks performed.

Plans were underway to change from a manual main feedwater system in a pressurized water reactor at a nuclear power plant to an automatic switchover. The concern was whether the manual switchover was a safe procedure to follow during the transition to the automatic system. In different plants, this switchover could take from 5 to 60 min to perform. If the action was not performed in time, then the steam generator might run dry and cause a safety hazard.

A second operator, whose sole function was to maintain sufficient water inventory in the event of a transient (nuclear power industry term for the initiating process to emergency shutdown of a reactor), was assigned to the control room. This was in addition to the primary operator, who monitored the rest of the control room activities. The second operator was relegated to a small, confined space to perform his or her task. The plant viewed this job as training for becoming a primary operator, and everyone knew that the position was temporary.

Further, the plant had adopted a procedure to eliminate the need for decision making to initiate the auxiliary feedwater system. Whenever the plant was operating at more than 15% power, a reactor trip initiated, and the second operator would perform his or her duty. Many switchovers were performed at the plant, in both real and simulated cases. The second operator knew the task steps very well, and it was felt that there was little chance for human error in the performance of the task. The larger concern was the failure to begin the switchover procedures. Table 8.3 illustrates Swain and Guttman's analysis results.

The first step in the analysis was to consider the implications of only the primary operator performing the task along with his or her other duties. A HEP of 0.05 for the first 5 min was taken from handbook tables for annunciated displays. If the need to switch over does occur, there could be 40 or more other annunciators sounding, and the primary operator has to sort through all of this information simultaneously.

TABLE 8.3
Human Error Probabilities

At the End of x min	Primary Operator	Shift Supervisor	Joint HEP
Situation without Second Operator			
5	0.05	—	0.05
15	0.01	0.5	0.005
30	0.005	0.25	0.001
60	No change	No change	No change
Situation with Second Operator			
5	0.002	—	0.002
15	0.001	0.5	0.0005
30	0.0005	0.25	0.0001
60	No change	No change	No change

If time constraints were loosened from 5 to 15 min, it would result in a HEP of 0.01 (a reduction by a factor of 5). Stretching out the constraints to 30 min lowered the HEP to 0.005.

The shift supervisor was a natural backup to the primary operator, but he would not be available for the first 5 min because of other duties. Between 5 and 15 min into the emergency, he or she would be only *coming up to speed* and would not be fully cognizant of what precisely was going on. The conditional probability of the shift supervisor's failure to compensate for the primary operator's failure was 0.5 (equivalent to a high level of dependence). In 30 min, it changes to 0.25.

Swain and Guttman did not consider any estimates at 60 min because if the system had not switched over by the end of 30 min, the operators in the control room would be very much occupied with other tasks. Switchover performance would not improve until the other problems were under control.

Looking at the second operator, emphasis was placed on the passing of information through oral instructions. A HEP of 0.001 was determined, based on a 15 min response time. The estimate was doubled for the 5 min response because it was felt that the second operator would be somewhat unsure if this was really happening. He also would be inundated with the other alarms sounding. For the 30 min period, the HEP reduced to 0.0005.

The success probability, or human reliability probability, is

$$\text{Human reliability probability} = 1 - \text{HEP}$$

You can see that the success rate or reliability of the primary operator working alone is 95%—not very good. That means the operator will fail to perform the action in time 5 out of 100 times. Just increasing the reaction time to 30 min gives

$$1 - 0.001 = 0.999 \text{ or } 99.9\%$$

This is much better, but not as good as using the second operator. The answer to the question is that the second operator has a sufficiently low human error rate (even in the 5 min operation), which probably is safe enough as a temporary measure. Extending the time limit to 15 min increases the chances for success significantly. Of course, the engineer would have to look at the cost implications to those two alternatives.

Notice that the shift supervisor really does not add anything to help the situation, contrary to intuition. In fact, the shift supervisor will make a mistake 50% of the time in the first 15 min. He or she will make a mistake one-quarter of the time in 30 min.

Swain and Guttman used the data in this example from generic human error tables and then modified to what seemed to make sense. It is important to remember that these numbers are highly questionable. This does not mean, however, that they are completely useless. If you do not apply the numbers in absolute terms—they really mean that the operator will fail in 5 out of 100 instances—then you can use them for what they are most valuable for. You should compare which situation is best (based on the numbers): the single operator or the use of dual operators. With these estimates, you can get a rough order of magnitude, and that is very useful.

8.3   SOFTWARE SAFETY

There have been numerous accidents that were either induced by software errors or faults or software was a heavy contributor. Just a few examples from IEEE's Reliability Society 2009 Annual Technology Report (Wong et al., 2009) are as follows:

- Shutdown of Hartsfield–Jackson Atlanta International Airport due to a software-induced false alarm that indicated a suspicious device was found.
- Loss of communications between FAA Air Traffic Control Center and in-flight aircraft from a bug in a Microsoft system combined with human error.
- Tragically, the 2009 Air France Flight 447, killing all on board, had software contributing to the crash with discrepancies in the indicated airspeed readings.
- Emergency shutdown of Hatch Nuclear Power Plant after a software update was installed.
- Miscalculated radiation dosages at the National Oncology Institute in Panama due to the operator trying to *trick the software* to allow a modified procedure.
- Patriot Missile failed to track and intercept a Scud missile attack in Saudi Arabia and hit army barracks killing 28 due to a software bug.
- Power outage across the Northeastern United States and Southeastern Canada in 2003 that shut down nuclear power plants and slowed air traffic and impacted over 50 million people and cost $13 billion because a maintenance worker forgot to turn a control trigger back on after finishing maintenance work that caused various systems to shut down without warning and crash backup servers hosting a queue of alarms. Though this was not purely a software fault, software heavily contributed. (Obviously, human factors errors were also involved.)

Probably, one of the biggest nonevents (thankfully) was the year 2000 (Y2K) software scare that resulted in nothing. For those who may not remember, it was the fear that once we entered the Y2K, our industrial control systems would fail. The concern was that those industrial systems controlled by software or microprocessors that used formal dates as part of their operations would fail due to the hardware industrial system failing from the year changeover in 2000. Think about the industrial systems that we are dependent on: electricity (not just to give us power but also to power our computer and software systems), clean water and water distribution, healthcare, ATM (cash dispense machines), air traffic control, pumping gas into your car, and millions of other everyday examples. Y2K did create a lot of fear and expense, which demonstrated how little we understood of our own industrial control systems. It also emphasized the importance of software safety and that software systems are used to control industrial systems, and we need to look at both to be sure that the industrial system operates safely. Today, the significant uptick in cyber attacks on industrial control systems further makes the point.

And of course, accidents are much worse if created on purpose. We are now much more at risk of any type of accidents resulting from software-controlled systems because of our trend toward open, interconnected, and networked systems. Throw into that cloud computing and mobile technology (mobile devices are now very common in troubleshooting industrial systems), and we can see that software is much more important to safety today than just a few years ago. Probably, one of the best examples is the new trend in *smart cities*, putting an entire city's control under software and cyber systems.

Software safety is the newest member of the system safety field. With the incredible proliferation of computers and microprocessors to all countries of the world, their safety control becomes both paramount and difficult. With the advent of software-controlled systems, software safety has reached much higher visibility. And who could deny the concern we all have about these *invisible* systems controlling much of our lives and health? Software control is everywhere, managing our bank accounts, launching nuclear warheads, monitoring chemical dispensing in foods, flying commercial airliners, and controlling highly complex processes in chemical plants—the list is almost endless. Obviously, we need to integrate software safety into our safety programs; the question is how. On top of all this, we see how cyber security has a significant impact on the safe operations of industrial systems. Though this section is focused on software safety, it is important that even in the security field, *safe* software systems are important in industrial control systems. If the industrial control system is designed to be safe during normal operations, consideration should be given in that they are still safe even if there are cyber attacks. The hazard reduction precedence can help you in this regard. Software safety really is a field unto itself. It is highly recommended that you discuss with information technology (IT) specialists the problem of software and hazard control. This section briefly describes what software safety is. You should consult the Further Reading section at the end of this chapter for more detailed information.

There are numerous software safety tools on the market, some quite good. And you can even take some of our current tools and use them for analyzing software systems. The most common ones are software hazard analysis, software fault tree analysis, and software FMECA. These are good starts, but insufficient to do the job completely. However, before you can attack the problem of software safety, a few facts should be stated first:

- Software itself is not a hazard. Software cannot be safe or unsafe. Like hardware, software can either support or mitigate a hazardous situation. Software and its systems are enablers to safe or unsafe conditions.
- Software does not fail. Software codes do not break, like a motor. However, software operations can get stuck (in things like endless loops), like a valve poppet. This is similar to human factors: people and software do not fail; they just do not complete the mission that we wish them to complete.
- Many people feel that computers fail safely, but when computers do fail electronically, they can produce unforeseen voltage levels and thus flip bits.

- Treating software like black boxes that can either always fail or never fail (because computers do not fail) is incorrect, dangerous, and very costly.
- Reviewing every single line of code will produce safe software. That would be a Herculean task, however, and hyper-expensive. We should spend the majority of our efforts reviewing lines of code in software-critical systems.
- Health checking (a common term in software parlance) is not the same as assuring that a system is safe. Most health monitoring is to assure that the system performs as intended (which is not always the safest way).
- Many people confuse fault tolerance and safety. Making systems (including software systems) fault tolerant is good if the resulting fault would create a hazardous situation. If it does not, then it does not add to the safety effort.
- If a dangerous situation does arise, shutting down the computer-controlled system is not always the best practice. Many complex operations require a very complex back-out procedure to ensure that the system is not left in a hazardous state when power is shut off.

PRACTICAL TIPS AND BEST PRACTICE

Treat software like anything else in the system. The bottom-line question is always: What is the hazard? Answer that first. If software is somehow involved in creating the hazardous situation, then you need to deal with it.

8.3.1 SOFTWARE SAFETY ANALYSIS

The safety management system discussed in Chapter 4 presents the need to include all aspects of system operations in the safety process. Software use and control is no exception. A software safety program should be an integral part of the system safety program. In fact, it would be dangerous to segregate software safety from the rest of the safety process.

The European Union (EU) has a large-scale collaboration project called OPENCOSS, which is part of the EU's Seventh Framework Program. It uses the certification management process as a framework to ensure safety assurance of embedded systems in automotive, railway, and aerospace industries. This is very important because the cost of ensuring software safety effectiveness is nontrivial. The project is trying to find methods to reuse certification assets to help control costs without reducing safety. The project is a great example of the importance of sharing lessons learned across various industries; this is particularly useful in software safety.

One of the most important software safety standards is the International Electrotechnical Commission IEC 61508, *Functional Safety of Electrical/Electronic/Programmable Electronic Safety-Related Systems.* It applies to all industries that use safety-critical systems that are electronically or software controlled. It uses safety processes described in Chapter 4 and in other safety analysis techniques chapters (e.g., Chapter 5). What is central is their use of safety integrity levels (SILs). The SIL is a relative level of risk reduction that a hazard control provides. The American National Standards Institute (ANSI)/ISA S84 standard for the functional safety in the process industry and the European standard for railway applications EN 50128 and EN 50129, software for railway control and protection and safety-related electronic systems for signaling, respectively, also use SIL and software safety assurance tools.

Software engineers should work closely with other system and hardware engineers while developing the software. Software safety requirements should be developed at the same time that other software requirements are written. Those requirements then flow to the software development and test organizations. MIL-STD-882 defines the flow process from the general to the specific as software requirements development, top-level systems hazard analysis, detailed design hazard analysis, code hazard analysis, software safety testing, software user interface analysis, and software change analysis (configuration control).

Software requirements development: As with all requirements development, software requirements development is performed early in the system's requirement phase. Requirements should be written that force the software developer to examine the system and identify any software commands that could cause the system to become unsafe.

Top-level systems hazards analysis: A system-level software safety analysis is conducted at this level. Safety-critical software is identified. Each software functional module is evaluated for hazards.

Detailed design hazard analysis: The software safety analysis is performed at the level of databases, files, and other algorithms.

Code hazard analysis: Those safety-critical subsystem modules that have been identified are further scrutinized at the code level. This analysis reviews things such as fault tolerance, operations sequencing, timing, error detection, and recovery operations.

Software safety testing: At this stage, identified software is tested to verify its correct operation. Remember to test for both normal and abnormal operating conditions.

Software user interface analysis: This ties the human (and other hardware systems) together with the software. Does the software control what it should control? Can the human operator understand what the software is telling him?

Software change analysis: Software changes are constant in the development and deployment of software systems. Configuration control is the only way to ensure that you know how the software has been modified. In this stage of the software safety process, you must also verify that software changes do not invalidate hazard controls or create a hazardous condition. And, this all must be adequately documented, not just the changes but also the implications to the changes.

NOTES FROM NICK'S FILE

I was once working on a satellite that would be deployed on orbit from the Space Shuttle orbiter payload bay. To be safe, we had to ensure that the solar arrays (which would power the satellite) and its communications antenna would not deploy until the satellite had cleared the payload bay and would be at a certain safe distance from the Space Shuttle orbiter. To be safe, when the Shuttle launched, the satellite in the payload bay was not powered on. So our challenge was once on orbit: to deploy the satellite, reach a safe distance from the Space Shuttle orbiter, then unfurl the solar arrays, then power on the communications antenna (we also did not want it to radiate while in the Space Shuttle orbiter before deployment), and then start the mission. Because some of the hazards were considered catastrophic, their controls had to be two-fault tolerant (after two failures, the system would still remain safe). We could not use pure software controls. So we carefully designed the control system with watchdog timers to kick in after a certain time. We were holding our breath during the satellite deployment; thankfully, it worked beautifully, and the satellite exceeded its operational expectations and design life. One important issue was not to be dependent on software controls to prevent a catastrophic hazard.

In the system safety analysis process, you will come across IT-driven or microprocessor-based systems. While performing any of the system safety analyses, numerous hazardous situations will be discovered. The first step is to decide whether there are any software controls in those particular subsystems. If there are, then it can be considered a safety-critical subsystem. More formally, a *safety-critical subsystem is one in which the operations must work properly or a hazardous situation will result. Safety-critical software is a software within a control system that contains one or more hazardous or safety-critical functions.*

A software safety analysis must be conducted whenever software is used to

- *Identify a hazard*: If software is part of the monitoring system for hazards, then its failure could allow an undetected hazardous situation to arise
- *Control a hazard*: If the software does not operate as planned and a hazard is no longer controlled, then you must look closely at the software
- *Verify a control is in place*: This means that if the function of the software is to monitor and indicate whether the control is still viable, then it must be studied. This leads into the next example
- *Provide safety-critical information or safety-related system status to other systems or operators*: Also, if it is used to detect a safety-critical system fault, or failure, it should be analyzed
- *Recovery from a hazardous condition*: Any time software is employed to assist the system to recover from a hazardous situation, it should be looked at closely

Numerous software safety analysis tools are available. When you start to look at the large variety, you will quickly become overwhelmed. Remember that software safety analysis techniques are either top-down (from the systems level to the code level) or bottom-up (from the code level back up to the level of consequences on the system). Also, the various safety analysis techniques are more useful and applicable at particular levels of software development.

PRACTICAL TIPS AND BEST PRACTICE

- Look very closely at remote or embedded real-time software that controls safety-critical subsystems or hazardous operations.
- Be very careful with safety-critical and non-safety-critical software sharing a CPU. Something in the noncritical software could affect the CPU (such as locking it up) just when you need it.

The easiest way to keep things straight is to divide the safety tools into various categories: software safety requirements analysis, architectural design analysis (top-level systems), detailed hazard analysis, code analysis, test analysis, and user interface analysis and configuration control.

Software safety requirements analysis: The two primary tools for software safety requirements analysis are flow-down analysis and criticality analysis. *Flow-down analysis* does precisely verify that the proper safety requirements have been communicated to all appropriate parties and that they are correct, consistent, and complete. Checklists and cross-references are frequently used. Europe uses a mathematical modeling tool called *formal methods*. It is used to specify and model the behavior of a system so that system specifications can be developed.

Requirements *criticality analysis* is used to identify program requirements that affect safety. This is where safety-critical subsystems are often first determined. All items listed as safety-critical are then tracked through the entire software development process.

Of course, more standard specification requirements analysis techniques are also employed. The principal one is *specification analysis*. This follows the traditional methodology of evaluating the requirements as they stand alone and as they interface with the rest of the requirements set.

Architectural design analysis: Once the requirements phase has been completed, the software team passes on to the top-level systems design. As the design is laid out, the criticality analysis tracking system is updated with the new, more detailed information. This is performed primarily through software hazard analysis. Another tool is software FMEA.

The principal activity at this stage is to analyze the software architectural module design and identify what hazards are in the system. It is important that the hazard analysis verify that the program software safety requirements have been implemented in the design. This is very similar to a hardware safety analysis, which also documents which safety requirements are applicable to the hazard control; in fact, the format is virtually the same.

At this stage, system-level hazards should be identified. You should try to identify the potential causes and effects of those hazards. Software dependencies should also be noted. This is the moment when possible hazard controls should be recommended. The fact that the hazard is caused by a software glitch does not mean that the control must be via software. That is why it is very important that the hardware hazard analysis be conducted in concert with the software hazard analysis. Software user interface analysis is also conducted at this stage, verifying that the people, hardware, and software all work together.

Detailed design analysis: This is where much of the meat of the hazard analysis is performed. There now is sufficient software to check logic, interfaces, and constraints to the software modules. Revisit the hazard descriptions, causes, effects, and controls noted in the previous analysis level and update as needed. At this level, coding approaches to control hazards are recommended. Two of the more common detailed design analysis techniques are

- Software fault tree analysis (also called soft tree analysis)
- Petri-Net

Soft tree analysis is just like its cousin, fault tree analysis. It is particularly useful in software safety because programmers tend to use forward inference in their design logic. Software development is predicated on assumptions that the system is at a certain state before transitioning to another state. The soft tree analysis does just the opposite; as you know from fault tree logic, it starts with a top-level fault and works down, identifying the events that make that top event occur. This is especially useful in finding software faults because it is so vastly different from the way programmers perform their work. This is particularly important for safety-critical systems.

Soft tree analysis is a tool that engineers and software programmers can both understand. If the engineer is comfortable with fault trees, then it will not take long to feel comfortable with soft tree analysis. The logic gate symbols are the same. It can be used with any software language. NASA uses this method extensively on spacecraft control systems.

Petri-Net is a mathematical model that describes the system in graphical symbols. It is very useful for analyzing properties such as reachability, recoverability, deadlock, and fault tolerance. The biggest advantage of Petri-Nets, however, is that they can link hardware, software, and human elements in the system. The technique is also good for understanding software-timing issues in real-time systems. They can also be quantified with probabilities.

A Petri-Net models the system and its dynamic changes in state. As certain conditions occur, transitions from one state to another occur. A Petri-Net maps this dynamic process. Places represent conditions, and transitions are used to represent events. The change in state caused by *firing* a transition is defined by the next-state function. You can determine from a Petri-Net if a high-risk state is reachable in the system process. For example, we can design the system to assure that a particular hazard does not occur by assuring that transition 1 occurs before transition 2. Obviously, Petri-Nets are very useful in applications where there is a lot of complicated transitioning from one state to another. The chemical process industry is one example.

Timed Petri-Nets are models that consider timing issues in the sequencing. Untimed Petri-Nets do not consider timing. The most significant problem with Petri-Nets is that they are fairly expensive to perform and should really be used only on safety-critical software.

It has become quite popular to integrate timed Petri-Nets with software fault tree analysis. You can use the Petri-Net to describe the system architecture and then switch to software fault trees to describe the hazards in the system and the events that lead to that top event and keep switching back and forth to analyze the software safety of the system.

Code analysis: Code analysis goes into the very heart of the software system and verifies that the coded program actually accomplishes what it is designed to do. Software fault trees and Petri-Nets are used, as well as various other techniques. Some of the other tools used are code logic analysis, code data analysis, code interface analysis, measurement of complexity, code constraint analysis, safe subsets of programming languages, and formal methods and safety-critical considerations.

8.3.2 SOFTWARE TESTING AND **IV&V**

The purpose of software testing is to ensure that the software meets all its specifications. The purpose of software safety testing is to verify that all safety aspects of the software have been identified and appropriately taken to disposition. Because there are so many lines of code in most programs, it is impossible to test every line of code. You might wish to use Monte Carlo simulations with your testing program to help identify worst-case scenarios.

Because software goes through so many changes, tight configuration control is very important. After every change in the software, a change analysis should be conducted to verify that no new hazards have been introduced into the system. Also, you need to ensure that the software changes will not invalidate other hazard controls already in place.

Nothing has been said so far about the independent verification and validation (IV&V) organization. Many feel that by virtue of having an IV&V, the software systems will be safe. But as you can see, if the various levels of software safety analysis are not conducted, then the IV&V activity will be meaningless.

REFERENCES

Billings, C. E. and Reynard, W. D. 1981. Dimensions of the information transfer problem. In Billings, C. E. and Cheney, E. S. (eds.), *Information Transfer Problems in the Aviation System*. NASA-TP-1875. Moffett Field, CA: NASA Ames Research Center.

Reason, J. 2006. Human factors a personal perspective. *Human Factors Seminar*, Helsinki, February 13, 2006: http://www.vtt.fi/liitetiedostot/muut/HFS06Reason.pdf slide 6, downloaded February 1, 2014.

Swain, A. D. and Guttman, H. E. 1983, 2011. *Handbook of Human Reliability Analysis with Emphasis on Nuclear Power Plant Applications.* NUREG/CR-1278. Washington, DC: U.S. Nuclear Regulatory Commission. Updated and released, May 4, 2011.

Wong, E. W., Debroy, V., and Restrepo, A. 2009. The role of software in recent catastrophic accidents. This article is part of the IEEE Reliability Society 2009 Annual Technology Report. http://paris.utdallas.edu/IEEE-RS-ATR/document/2009/2009-17.pdf downloaded on February 7, 2014.

FURTHER READING

Boeing Aircraft Company. 1984. *System Safety Engineering in Software Development.* D180-28554-1. November 9, 1984. Seattle, WA: Boeing Aircraft Company.

British Standards Institute. 1991. *Reliability of Systems, Equipment and Components Part 5: Guide to Failure Modes, Effects and Criticality Analysis (FMEA and FMECA).* BS 5760-5. London, U.K.: British Standards Institute.

British Standards Institute. 2003. *Railway Applications. Communication, Signaling and Processing Systems. Safety Related Electronic Systems for Signaling.* BS EN 50129:2003. London, U.K.: British Standards Institute.

Calixto, E. 2012. *Gas and Oil Reliability Engineering.* Houston, TX: Gulf Professional Publishing.

Henley, E. J. and Kumamoto, H. 2000. *Probabilistic Risk Assessment and Management for Engineers and Scientists*, 2nd edn. New York: Wiley-IEEE Press.

Hoover, D. N., Guaspari, D., and Humenn, P. 1996. *Applications of Formal Methods to Specification and Safety of Avionics Software.* Hampton, VA: National Aeronautics and Space Administration, Langley Research Center.

IEEE Standards Association. 1994 (Reaffirmed 2010). *IEEE Standard for Software Safety Plans.* Piscataway, NJ: IEEE Computer Society.

International Electrotechnical Commission. 2001. *Railway Applications—Communications, Signaling and Processing Systems—Software for Railway Control and Protection Systems.* IEC 62279 (EN 50128). Geneva, Switzerland: International Electrotechnical Commission.

International Electrotechnical Commission. 2006. *Analysis Techniques for System Reliability— Procedure for Failure Mode and Effects Analysis.* IEC 60812. Geneva, Switzerland: International Electrotechnical Commission.

International Electrotechnical Commission. 2010. *Functional Safety of Electrical/Electronic/ Programmable Electronic Safety-Related Systems.* IEC 61508. Geneva, Switzerland: International Electrotechnical Commission.

Karwowski, W. 2006. *International Encyclopedia of Ergonomics and Human Factors*, 2nd edn., 3 Volume Set. Boca Raton, FL: CRC Press.

Keene, S. J. 1992. Assuring software safety. *Proceedings Annual Reliability and Maintainability Symposium*, New York.

Kirwan, B. 1994. *A Guide to Practical Human Reliability Assessment.* Boca Raton, FL: CRC Press.

Kirwan, B. and Ainsworth, L. (Eds.). 1992. *A Guide to Task Analysis.* London, U.K.: Taylor & Francis.

Kletz, T. A. 2001. *An Engineer's View of Human Error*, 3rd edn. Boca Raton, FL: CRC Press.

Larrucea, X., Combelles, A., and Favaro, J. 2013. *Safety-Critical Software.* May/June 2013 (Vol. 30. No. 3). IEEE Computer Society, pp. 25–27.

Leveson, N. 1995. *Safeware System Safety and Computers.* Boston, MA: Addison-Wesley Professional.

Leveson, N. 2012. *Engineering a Safer World.* Cambridge, MA: MIT Press. Also can download a draft e-copy at: http://mitpress.mit.edu/books/engineering-safer-world.

Misumi, I., Miller, R., and Wilpert, B. 2003. *Nuclear Safety: A Human Factors Perspective.* Boca Raton, FL: CRC Press.

National Aeronautics and Space Administration. 2004. *NASA Safety Software Guidebook.* NASA-GB-8719.13. Washington, DC: National Aeronautics and Space Administration.

Open Platform for Evolutionary Certification of Safety Critical Systems (OPENCOSS), European Commission, Seventh Framework Programme, www.opencoss-project.eu.

Radio Technical Commission for Aeronautics, Inc. 2011. *Software Considerations in Airborne Systems and Equipment Certification* DO-178C. Washington, DC: Radio Technical Commission for Aeronautics.

Rasmussen, J. 1981. *Human Errors. A Taxonomy for Describing Human Malfunction in Industrial Installations.* Risø-M-2304. Roskilde, Denmark: Risø National Laboratory.

Rasmussen, J. and Taylor, J. R. 1976. *Notes on Human Factors Problems in Process Plant Reliability and Safety Prediction.* Risø-M-1894. Roskilde, Denmark: Risø National Laboratory.

Reason, J. 1990. *Human Error.* Cambridge: U.K.: Cambridge University Press.

Reason, J. 2008. *The Human Contribution.* Boston, MA: Arena Publishers.

Reason, J. and Hobbs, A. 2003. *Managing Maintenance Error.* Farnham, U.K.: Ashgate Publishing Ltd.

Rierson, L. 2013. *Developing Safety-Critical Software: A Practical Guide for Aviation Software and DO-178C Compliance.* Boca Raton, FL: CRC Press.

Sanders, M. S. and McCormick, E. J. 1993. *Human Factors in Engineering and Design,* 7th edn. New York: McGraw-Hill.

Shimeall, T. J., McGraw, R. J., and Gill, J. A. 1991. Software safety analysis in heterogeneous multiprocessor control systems. *Proceedings Annual Reliability and Maintainability Symposium,* New York.

Smith, D. and Simpson, K. 2010. *Safety Critical Systems Handbook – A Straightforward Guide to Functional Safety, IEC 61508 (2010 Edition) and Related Standards,* 3rd edn. Oxford: U.K.: Butterworth-Heinemann.

U.S. Department of Defense. 2012. *Human Engineering Design Criteria for Military Systems, Equipment, and Facilities.* Mil-Std-1472G. Washington, DC: U.S. Department of Defense.

U.S. Department of Defense. 2012. *Military Standard, System Safety Program Requirements.* Mil-Std-882E. Washington, DC: U.S. Department of Defense.

9 Other Techniques

I am the master of my fate:
I am the captain of my soul.

Invictus, Book of Verses, 1888
William Ernest Henley

Fear is the foundation of safety.

Tertullian of Carthage, 160-225 BCE

Real knowledge is to know the extent of one's ignorance.

Confucius, 551-479 BCE

There are myriad ways to solve safety problems and almost as many analysis techniques with which to do so. This chapter describes other commonly used and accepted safety analysis tools. Of course, there are literally scores and scores more. In fact, the System Safety Society (Stephans and Talso, 1993) has documented 101 safety methodologies and techniques. A few of these safety techniques require at least a cursory explanation.

9.1 MORT

Management Oversight and Risk Tree (MORT) is a strictly qualitative safety tool that has fallen into disuse and is not as common as it once was, during its heyday in the 1970s. MORT was one of the engineering profession's first attempts to merge safety management and safety engineering. During the early 1970s, the U.S. Department of Energy (then called the Energy Research and Development Administration) decided that a more systematic approach to safety was needed in handling its nuclear weapons program.

Since then, MORT has been used considerably in accident investigation (one of its last high-profile uses, though perfunctorily, was the Space Shuttle *Challenger* accident). Its purpose is to analyze a system methodically and identify the interrelationships among the plant operations and management organizations. A predefined graphical tree, much like a fault tree (with similar symbols), analyzes management policy in relation to risk assessment and the hazard analysis process.

You work through the predefined tree comparing your management and operations structure to the ideal system safety tree structure. Like the fault tree, you work from the top event down to determine what oversights and omissions were in place that caused the accident or created an unsafe situation. The tree also forces you to look at the risk assumed by the management organization and whether it makes sense or not.

MORT is highly labor intensive, though very easy to learn. It takes about a day to learn how to *read* the tree. Its major drawback is that the tree is so large and unwieldy (it has 98 generic problems and over 1500 basic events) that it is very easy to get lost in the process. Another significant problem is that it assumes that there is an *ideal* safety system and that it can be predefined. Also, it does not lend itself very well to *tailoring* the tree to a smaller problem.

9.2 ENERGY TRACE BARRIER ANALYSIS

The qualitative tool called energy trace barrier analysis (ETBA) was initially developed as part of MORT. Since the demise of MORT, ETBA has been used extensively in hazard analysis. The purpose of ETBA is to identify hazards by tracing energy flow into, through, and out of a system. A hazard is defined as an energy source that adversely affects an unprotected or vulnerable target.

The typical types of energy sources found in a system are electrical, mechanical, chemical, and radiation. Appendix A, "Typical Energy Sources," can be used as a checklist. By following the energy path, you can determine if adequate controls are in place to ensure that undesired energy release to a vulnerable target does not occur.

ETBA identifies not only the energy sources but also the barriers to the undesired release of energy. Barriers can be anything from pressure container walls to gloves for handling cryogens. A barrier is what keeps the energy from being released in an undesired fashion or coming in contact with a vulnerable target.

The energy flow is traced through the operation. As the energy is traced through the system, each energy transfer point must be identified. Also, each physical and procedural barrier is considered to determine whether the energy still can cause undue harm. The procedure to conduct an ETBA is as follows:

- Examine the system and identify all energy sources.
- Trace each energy source's travel through the system, from beginning to end.
- Identify all vulnerable targets to the energy source along its travel path.
- Identify all barriers in the path of the energy.
- Determine if controls are adequate.

One particularly good point about ETBA is that it can be performed at any time in the system life cycle. It also is relatively inexpensive and rapid.

9.3 SNEAK CIRCUIT ANALYSIS

Sneak circuit analysis was standardized by Boeing in 1967 and is a formal analysis conducted on every possible combination of paths that a process (most typically electrical circuits, though it could also apply to process flows) could take. The intent is to identify all the paths in the circuit that are designed in and not created due to failures. In other words, the analyst tries to find *sneak* paths, timing, or procedures that could yield an undesired effect. These sneak or latent paths are found in systems

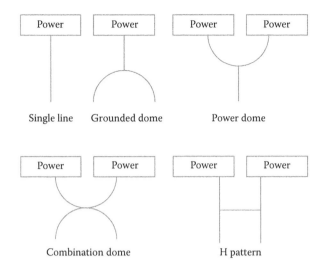

FIGURE 9.1   Sneak circuit topological patterns.

that are not operated frequently and thus are not easily identified or tested when a system is put into place. Though it is more common in reliability engineering, it is also a system safety tool.

The procedure is to review the system engineering drawings and translate the system drawings into a series of topological patterns that describe the entire circuit. The heart of sneak circuit analysis is predicated on the concept that there are five basic topological patterns that are vulnerable to sneak paths. The five electrical topological patterns are shown in Figure 9.1.

You review the drawings and identify all the places in the circuitry where one or more of these patterns or combinations of patterns are found. The patterns assume that some sort of switching operation is found between the power and ground points in each pattern. After the patterns have been identified, a set of questions or clues is applied to each node, attempting to find all the sneak paths.

Sneak circuit analysis is usually performed with complex computer codes and is very expensive. It only becomes cost-effective on subsystems that are safety critical, such as an aircraft control system. Obviously, sneak circuit analysis should be teamed with the software safety analysis tools discussed in Chapter 8. This is a very powerful combination, but not cheap, certainly, very important for the most safety-critical circuits of very high-risk systems.

9.4   CAUSE–CONSEQUENCE ANALYSIS

Cause–consequence analysis uses symbolic logic trees similar to fault trees. You start with an accident or failure scenario that challenges or adversely impacts the system and then develop a bottom-up analysis. Failure probabilities are calculated and incorporated into each step of the analysis, and thus quantify the tree.

A fault tree can be used to arrive at the event that challenges the system. The probability of occurrence is included here. From that *top* event, the consequences are identified. Each consequence could have a variety of potential outcomes, representing incremental levels of success or failure. Each of these levels also has probabilities associated with it. Then the severity of each consequence is determined.

9.5   ROOT CAUSE ANALYSIS

Root cause analysis is a tool that focuses on identifying the root causes of an accident, fault, hazardous condition, or other undesirable event. It is particularly useful during accident investigations; the National Transportation Safety Board uses the technique for its investigations. It is now used in patient safety analysis, common in many countries and in particular with the UK National Health Service. Though there is no single method, it should consider a few key points. It should focus on people, process, and technology and how they contribute to the actual event being investigated.

One such example is the approach that the U.S. National Institute of Standards and Technology (NIST, 2009) used to investigate an accidental plutonium spill at a Boulder, Colorado, facility. In summary, the steps the team took were as follows:

- Gather prior accident documentation and develop plan of action and approach methodology.
- Collect data (e.g., operating procedures, process flow maps, and generate timeline).
- Conduct detailed document review and analysis; document gaps in the process flow maps.
- Create *initial or preliminary list* of potential causes.
- Use fault tree analysis to further investigate causes.
- Develop cause-and-effect chart (fishbone diagram) to illustrate cause–effect relationships.
- Generate root cause analysis report.

NOTES FROM NICK'S FILE

I was part of the NIST team evaluating the investigation of the plutonium spill. One of the most important discussions we had was to figure out which safety tools to use in our root cause analysis. Ultimately, we picked a combination of event and causal factor analysis (which identified time sequencing and surrounding conditions), cause-and-effect analysis (illustrated through fishbone diagrams), and fault tree analysis (which helped determine the faults or errors and the corresponding events), and then the results were summarized in a gap analysis. What is important to remember, do not be beholden to one method or another; use what makes sense for the problem at hand. And do not be afraid to combine tools as required. Remember, construction workers have many tools in their tool box. You should too.

9.6 BOW TIE ANALYSIS

Bow tie analysis is a tool that has become very popular in the last few years, especially because of the ease in which it can display cause–consequence of a particular hazardous condition. It is a qualitative tool that combines the fault tree to determine the causes and how the fault could occur, with the event tree, which documents the consequence of the hazardous condition. It became much better known in the mid-1990s when Royal Dutch/Shell used it to better understand the Piper Alpha disaster. The process industry uses it not only to assess the hazards and risks but also as a very effective communication tool to illustrate the cause–consequence–control and how it can impact a hazardous condition. In reality, it really isn't a new analytical tool, but rather, a very good visualization tool.

At the center of the diagram is the hazardous condition or undesired event that is to be studied. It is very similar to the top event of a fault tree. On the left-hand side, the hazardous condition is broken down to determine all the causes (or threats to the system that would cause the hazardous condition) that could lead to that event (imagine a fault tree rotated 90°) and the barriers that could mitigate that threat. On the right-hand side are the controls (that could mitigate the hazardous condition) and the corresponding consequences if the controls are in place. Figure 9.2 illustrates a bow tie example.

9.7 DISPERSION MODELING

Chemical emission and dispersion modeling is a quantitative tool that straddles environmental engineering and system safety engineering. Dispersions can happen through the atmosphere, soil, or water. In planning for emergency response for process plants, tanker truck crashes, or rail car accidental chemical dispersions, one of the steps to determine how serious a chemical release would be is to perform a dispersion model of the accident. Many models are currently in use, but an internationally recognized

FIGURE 9.2 **(See color insert.)** Bow tie example.

model (actually a suite of various models) developed by the U.S. National Oceanic and Atmospheric Administration and the U.S. Environmental Protection Agency is Computer-Aided Management of Emergency Operations (CAMEO). CAMEO has been designed to be in concert with the Awareness and Preparedness for Emergencies at Local Level of the United Nations Environment Program. It was created especially to help emergency first responders to better plan how to respond to chemical releases.

CAMEO is organized into eight general information areas. The categories were selected for their universal applicability. Maps and calculations are part of the system, so a total dispersion map and hazard analysis can be generated. Some of the features of the system are the diverse types of databases used:

- Hazardous chemical
- Facility information
- Chemical inventory
- Hazard analysis functions
- Incident reporting
- Population information
- Transportation
- Shipper information

CAMEO can interface with other detailed air, water, and soil dispersion models. For example, Mexico uses its own dispersion model—Sistema de Información Rápida de Impacto Ambiental (SIRIA)—in conjunction with CAMEO. Radian Corporation's Complex Hazardous Air Release Model (CHARM) is another model that is frequently used.

One of the biggest advantages of dispersion modeling is that you do not have to look at unlikely worst-case scenarios but can investigate the gamut of scenarios. Dispersion modeling also fits neatly into other safety analyses and risk assessments.

9.8   TEST SAFETY

There really is no test safety analysis technique; however, it is worthwhile to define test safety. Simply put, *test safety is the process to assure that there is a safe operating environment during all systems and prototype testing.* The author has detailed elsewhere (Bahr, 1988) the management and engineering interfaces needed to ensure a successful and safe test. Obviously, this program was developed for organizations that are primarily research and development oriented, where considerable testing is performed.

The aim of a good test safety program is to integrate the system safety process into the test process. The test environment can be thought of in three layers: test facility, test bed, and test article. A safety analysis, or some sort of safety input, is typically needed at each level.

The test facility safety analysis can be conducted with a facility hazard analysis. The facility hazard analysis is performed when the facility is brought online and periodically when facility modifications take place.

A hazard analysis can be performed on the test bed and thus certify that the test bed is safe to be used to conduct a variety of tests. Shorter hazard analyses are performed on the test article and how the three layers interface together.

Many research and development organizations hold some sort of *test readiness review*, especially if the unit to be tested is of very high dollar value. Test engineers and test conductors present the status of the test preparation and at that time can also address the safety analyses conducted.

NOTES FROM NICK'S FILE

I was involved in ground testing of the redesign of the solid rocket booster after the *Challenger* accident. Our test safety program used a combination of safety tools with the hazard analysis as the central tool to identify test hazards and track controls to closure.

9.9 COMPARING THE VARIOUS TECHNIQUES

A plethora of safety analysis techniques are available to the practicing engineer, some of them are very complex and involved, and some are quite simple. They are applied at different phases of the system life cycle. Some of the safety tools are quantitative and some are not. All of the safety analysis techniques do have some sort of cost attached to them; some are very expensive and others are not as costly. After reading about the different safety techniques described here, you are probably somewhat overwhelmed with trying to decide which one to use. This section will help you to decide.

One very important note is that all of the techniques presented in this book are recognized and accepted in their particular industry. Many of them are also being applied in numerous other industries. All of them have proven track records and are accepted by national and international organizations such as the U.S. Occupational Safety and Health Administration, the International Standards Organization, and national safety organizations in most countries of the world (where such standards exist). For example, the Mexican government suggests that companies use hazard and operability analyses (HAZOPs), hazard analyses, or fault tree analyses in their safety and environmental risk assessments. All are good, thorough, methodological, formal, and analytical tools that do serve the purpose of identifying, controlling, and mitigating the effects of hazards and enhancing the safety of a system or process.

PRACTICAL TIPS AND BEST PRACTICE

Remember that exercising one safety tool does not preclude using other ones at the same time. The real *art* in system safety is combining these tools and using only what is most appropriate and to the right level of detail.

9.9.1 ADVANTAGES AND DISADVANTAGES

If you pick a safety tool that is not quite appropriate to the application, you still have not erred significantly. All of the techniques are comprehensive analyses.

The real problem is that the results may not give you sufficient data to solve your safety problems efficiently. Performing hazard analyses on a power plant operation without looking at how the human operators interact is one such example. You also may spend more time and money than you really needed to spend.

To decide which tool or tools to use, first ask yourself a few questions and then refer to Table 9.1.

- Why are you doing this? Did an accident occur? Is there regulatory pressure? Are there labor disputes?
- What kind of results are you looking for? Is it just for compliance? Is it to solve safety and process inefficiencies? Who is your audience? Is it to be used as input to additional safety analyses? Is it to investigate an accident and make sure it does not occur again? Is it to update past analyses? Is it, or should it be, quantitative?
- How does it fit into your company or plant system safety program? How does it fit into my safety management system? How do I want to use the data?
- Identify what information (line drawings, calculations, test reports, operations procedures, other safety analyses, accident investigation reports, etc.) is needed to produce the desired results? How easy is the information to obtain?
- Think about your system or process. What kind of process is it (chemical, drilling, packaging, etc.)? How large and complex is it? Is it highly automated? How involved are people and software? How critical are the operations to the overall mission of the plant or company?
- What kind of hazards and risks appear to be involved? Will large populations be affected? Will workers be exposed to toxic levels or other hazardous situations regularly? Do accidents result in explosions, fire, toxic spills, etc.?
- What kinds of resources are available to do the work? How skilled are the engineers performing the safety analyses? How much do the engineers understand the process they are analyzing? How many people do you have available to do the work?
- How much time do you have available to complete the analysis?
- How much are you willing to spend? Can savings from the analysis be used to help finance the safety work?

PRACTICAL TIPS AND BEST PRACTICE

- Do not hesitate to combine some of these tools. For example, when doing a hazard analysis or HAZOP, add a human factors analysis (as a subset of the overall hazard analysis) if human operators play a significant role. Or, if the HAZOP has identified particularly dangerous deviations of the process resulting from a failure in the system, do an FMEA or fault tree analysis of just that critical subsystem.
- Think of the safety analysis techniques as individual tools in a toolbox. Mix and match as needed.

TABLE 9.1
Comparison of Safety Analysis Techniques

	Hazard Analysis	Facilities Hazard Analysis	Operations and Support Hazard Analysis	HAZOP	What-if/Safety Checklists	Fault Tree Analysis	Failure Modes and Effects Analysis	Human Factors Analysis	Software Safety Analysis
Primary industry	All industries	Manufacturing; facilities in all industries	Manufacturing; aerospace	Chemical/process petroleum, food processing	Chemical/process, manufacturing	Aerospace, nuclear	Aerospace, nuclear	All industries	All industries with safety-critical software
Applications	Identifies systems/subsystems, hazards, controls, consequences; hazard tracking; accident investigation	Identifies and assesses facility hazards; accident investigation	Identifies hazards, controls, consequences in human-controlled operations	Identifies process hazards; assesses hazard controls	Systems review; systems modes; identifies possible accident scenarios; contingency planning; compliance verification	Systems review to identify events that lead to a hazard; accident investigation	Primarily reliability analysis; used to determine how equipment fails; accident investigation	Analyzes systems where people play an important role; accident investigation.	Models system failures and hazards that are controlled by software; especially useful for sequencing and timing
Life-cycle phases	All phases	All phases	Operational, maintenance, disposal phases	Early in design, operations, and modifications phases	All phases	Any phase after requirements definition	After design is finalized	All phases but especially during operations	All phases

(Continued)

TABLE 9.1 (*Continued*)
Comparison of Safety Analysis Techniques

	Hazard Analysis	Facilities Hazard Analysis	Operations and Support Hazard Analysis	HAZOP	What-if/Safety Checklists	Fault Tree Analysis	Failure Modes and Effects Analysis	Human Factors Analysis	Software Safety Analysis
Type of results	Identifies hazards, controls, and consequences for total system; hazard tracking	Evaluation of total facility hazards; identifies and assesses hazards and controls; hazards tracking	Evaluation of overall system operations and procedures; input to procedures and manuals	Comprehensive listing of process-related hazards; identification of consequences of process deviations and suggested controls	List of hazards or safety concerns	Identifies faults that lead to hazard; predicts probability of top event; list of cut sets; graphical representation of system	List of all component failures, causes, and consequences; identifies single-point failures	Evaluation of human error, hazards created, consequences, and possible controls	Identifies software that creates system hazards; identifies controls to software-generated hazards
Quantitative/ qualitative	Qualitative	Qualitative	Qualitative	Qualitative	Qualitative	Qualitative and quantitative	Qualitative and quantitative	Qualitative and quantitative	Qualitative and quantitative
Time required,[a] simple/small systems	1–3 days	3–5 days	1–2 days	6–12 days	12–25 h	9–18 days	2–8 days	6–13 days	6–12 days
Time required,[b] complex/ large systems	3–12 weeks	2–4 weeks	5–10 days	3–12 weeks	2–4 weeks	6–13 weeks	3–8 weeks	6–10 weeks	6–13 weeks

No. of people needed	1–2	1–2	1–2	4–8	1–2	1–2	1–2	1–2	2–3
Skill level needed	Minor training; understands system well	Minor training; understands system moderately well	Minor training; moderate understanding of system	Moderate training; very good understanding of process	No special training; cursory understanding of system	Minor training; good understanding of system	Moderate training; very good understanding of system	Significant training; moderate understanding of system	Significant training; significant understanding of software system
Cost	Moderate	Moderate	Minor– moderate	Moderate– expensive	Inexpensive	Moderate– expensive	Moderate– expensive	Moderate– expensive	Expensive
Strengths	Assesses total system risk very thoroughly; looks at all aspects of a system; good for identifying hazards and controls; can be easily integrated w/ other safety analyses; easy to learn	Good for any kind of facility; fits very well into facility acquisition cycle; maximizes facility operational readiness; very useful for building models; can be easily integrated w/ environmental system	Integrates people and hardware hazards; very good at evaluating safety of procedures and checking maintenance activities; identifies training requirements	Very thorough; identifies ultimate consequences and evaluates existing safeguards; good for complex system; finds process inefficiencies also	Very inexpensive; do not need a lot of data; very user friendly; can be used quickly and easily	Very thorough; very good for complex system; easy to learn; good for accident investigation; quantifies top event; provides visual model of safety system; provides logical causes for top event; can be quantified; models system functions well;	Very thorough; good for identifying failure causes; identifies single-point failures; good for complex system	Identifies human error and ways to mitigate mistakes; very thoroughly identifies ways to *people-proof* the system; very good at evaluating safety of procedures; qualitative and quantitative	Very thorough; involved in all life-cycle phases; quantifiable

(Continued)

TABLE 9.1 (*Continued*)
Comparison of Safety Analysis Techniques

	Hazard Analysis	Facilities Hazard Analysis	Operations and Support Hazard Analysis	HAZOP	What-if/Safety Checklists	Fault Tree Analysis	Failure Modes and Effects Analysis	Human Factors Analysis	Software Safety Analysis
						good for identifying redundancies and fault tolerance			
Weakness	Not quantifiable; can be time consuming	Useful only for facility hazards	Difficult to imagine all possible situations	Can be expensive, time consuming; needs a well-defined system; needs a well-disciplined team	Not as complete as other analyses; not good for complex system; not good for identifying interdependency	May be costly. Does not model all faults, only those that lead to predefined top event; time consuming	Addresses failures not safety issues; must be performed fairly late in design; human error not addressed	Quantification can be misleading; very difficult to model human behavior; data gathering can be difficult; can be expensive	Expensive; labor intensive; some tools are difficult to use; needs special training

[a] The time required to evaluate simple/small systems is assuming a system equivalent to a single pump station.

[b] The time required to evaluate complex/large systems is assuming a system equivalent to one complete process unit in a processing plant.

Table 9.1 is a matrix of the major safety analysis techniques described in previous chapters. Use it as an easy reference to get a feel for the different types of analyses. However, it is very limited and should be used only to get a notion of how the tools stack up with each other. For example, the costs of these techniques are highly dependent on the complexity of the system. Also, the time required to complete the analysis is also contingent on the size and complexity of the process. You should be very careful not to accept the time to perform the analysis too literally. It is necessary to give some sort of idea, but it is only that—an idea—not an absolute time frame. It is highly dependent on the process, how mature the system is, the team assembled, how experienced the crew is, and how many problems there may actually be.

REFERENCES

Bahr, N. J. 1988. The Johnson Space Center Test Safety Program. 88-WA/SAF-l. *American Society of Mechanical Engineers Winter Annual Meeting*, Chicago, IL.

National Institute of Standards and Technology. 2009. *Root Cause Analysis Report of Plutonium Spill at Boulder Laboratory*. http://www.nist.gov/public_affairs/releases/upload/root_cause_plutonium_010709.pdf downloaded May 17, 2014.

Stephans, R. A. and Talso, W. W. 1997. *System Safety Analysis Handbook*, 2nd edn. Albuquerque, NM: System Safety Society.

FURTHER READING

Bahr, N. 1998. Development of a comparative evaluation method for safety analysis tools. Thesis (M.S.). University of Maryland, College Park, MD.

Buratto, D. L. and Goody, S. G. June 1982. *Sneak Analysis Application Guidelines*. RADC-TR-82-179. Rome Air Station, NY: U.S. Air Force.

Center for Chemical Process Safety. 2008. *Guidelines for Hazard Evaluation Procedures*, 3rd edn. Hoboken, NJ: Wiley. 2008.

Clardy, R. C. 1977. Sneak analysis: An integrated approach. *International System Safety Conference Proceedings, 1977*. New York: System Safety Society.

Clemens, P. L. 1982. A compendium of hazard identification and evaluation techniques for system safety application. *Hazard Prevention*, March/April 1982.

Goldberg, B. E., Everhar, K., Stevens, R., Babbitt, N., III, Clemens, P., and Stout, L. 1994. *System Engineering "Toolbox" for Design-Oriented Engineers*. NASA Reference Publication 1358. Washington, DC: National Aeronautics and Space Administration.

Kletz, T. A. 2010. *Process Plants: A Handbook for Inherently Safer Designs*, 2nd edn. Boca Raton, FL: CRC Press.

Knox, N. W. and Eicher, R. W. 1977. *MORT User's Manual, for Use with the Management Oversight and Risk Tree Analytical Logic Diagram*. ERDA-76/45-4, SSDC-4 (Rev. 1). Washington, DC: U.S. Research and Development Administration.

Mannan, S. 2012. *Lee's Loss Prevention in the Process Industries: Hazard Identification, Assessment and Control*, 4th edn. 3 Volumes. Boston, MA: Butterworth-Heinemann.

Ministry of Defense, United Kingdom. 2012. *Applied R&M Manual for Defence Systems*. GR-77. U.K.: Ministry of Defence.

National Institute of Water and Atmospheric Research. 2004. *Good Practice Guide for Atmospheric Dispersion Modelling*. Hamilton, New Zealand: National Institute of Water and Atmospheric Research, Ministry for the Environment.

Okes, D. 2009. *Root Cause Analysis: The Core of Problem Solving and Corrective Action*. Milwaukee, WI: ASQ Quality Press.

Rahej, D. 1986. Testing for safety. In Bass, L. (ed.), *Products Liability: Design and Manufacturing Defects*. New York: *Shepard's*/McGraw-Hill.

Roland, H. E. and Moriarty, B. 1990. *System Safety Engineering and Management*, 2nd edn. New York: Wiley-Interscience.

Trost, W. A. and Nertney, R. J. 1985. *Barrier Analysis*. DOE-76-45/29, SSDC-29. Washington, DC: U.S. Department of Energy.

10 Data Sources and Training

Knowledge is of two kinds.
We know a subject ourselves, or we know
where we can find information upon it.

Samuel Johnson, 1775

A little learning is a dangerous thing.

An Essay on Criticism, 1711
Alexander Pope

Rewards and punishment are the lowest form of education.

Zhuangzi, c 286 BCE

Human history becomes more and more a race between education and catastrophe.

The Outline of History, Chapter 41, 1920
H.G. Wells

As the famous saying goes, "The devil is in the details." The system safety analyses we perform can be only as good as the data we obtain. We must clearly understand the data limits and assumptions. Otherwise, the safety analyses will be suspect. Luckily, one good thing about living in an information age is that there are multiple sources of data, including system safety data. But information alone is not knowledge, and knowledge alone is not necessarily wisdom. What we do with the information to create wisdom is what needs to drive system safety.

Because system safety is really a field that crosses all parts of engineering and society, you should not limit your thinking to the traditional data sources. Obviously, the first place to start is with your company, looking at the historical data on similar systems, past accidents (and near misses—we can learn a lot from what we barely averted), trend analyses, engineering reports, and analyses.

The next step is to obtain copies of any pertinent government, industry, or international engineering standards. Chapter 2 provides a number of places to look.

Before looking at various ways to obtain data, remember to be careful about misunderstanding the data you gather and the data you present to others. Quantifiable data are not always the best. During the 1960s, when NASA was working on going to the moon, numerous calculations were crunched that *proved* it was too risky a project. Those calculations *proved* that we could not send a man to the moon and bring him back safely within the decade. Luckily, NASA threw away the numbers and set out to solve the problem another way. A more mundane but no less poignant example is assuming that the electronic part failure data published in Mil-Std-217, *Reliability Prediction of Electronic Equipment*, have taken into account all possible failure modes and have been able to quantify them. The fact that something is stated to fail 1 in 10,000 cycles does not mean that it will fail on the 10,001st cycle or the 9,999th. Keep in mind the context of the data you use. When using data, be sure to always cite the source and especially any assumptions used.

If you use quantitative data, be sure to include confidence limits. If you are unsure of how good the data are, say so. The fact that the numbers are questionable does not mean that they must be disregarded. Even if you use failure data that seem very unreasonable, you can still include upper and lower limits that would help bound the problem. It is possible to have fairly good confidence that the number lies somewhere between lower and upper bounds, even if you do not know the exact number. Also remember to include confidence intervals on the answer, not just the input data. Consult a reliability engineering book before manipulating failure data.

It is usually very difficult to obtain data showing exactly what you are looking for. Data on the failure rate of needle valves using hydraulic fluids on drilling rigs in the Amazon jungle probably do not exist or, at the very least, are proprietary. Of course, data do exist on the reliability of needle valves. The problem is to use that data appropriately.

When you start searching for data, you will immediately find that most of it is generic. In other words, you may find information about the failure rate of a resistor or a valve, but not much more than that. Some sources do attempt to provide some data on how the failure rates change in different environments. The environment in which the equipment is operated greatly affects the failure rate, and it would be a Herculean task to list all the failure data of components in all operating conditions. The failure rate of a valve pumping a highly viscous fluid such as sludge in the winter is different than the failure rate of a valve pumping water in the summer.

If you cannot find the failure rate for a large item, break it into its constituent parts. There are data on pump failures but not on pump station failures. Divide the pump station into the pump, motor, valves, piping, instrumentation, etc. Failure rates do exist for all those components. However, you will have to take into consideration the operating characteristics and conditions.

Another way to generate more useful information from generic sources is to use expert judgment. Start by locating failure rate data and then show the data to experts in your company or others in your industry. Using expert opinions and Bayesian updating or other Delphi techniques, you can convert the data and massage into more realistic values. Consult a good statistics or reliability engineering book.

As said many times before in this book, failures do not necessarily mean hazards, and we can have hazards without failures. When looking for safety data, specifically hazard information, it is important to treat it with the same trepidation that you treat taking failure data from one context to another. A system operating in one condition will not necessarily have the same hazards operating in a different one.

NOTES FROM NICK'S FILE

I was working on a signaling system for a railway in the Arabian Desert in the Middle East. The conditions were very harsh: blinding and biting sandstorms, extreme heat (over 120°F), constantly moving mountains of sand onto the railway track and switches, random camels and camel herds crossing, and extremely remote operations. We had much discussion on how well various signaling systems, designed for more benign weather conditions, would operate here. We debated quite a bit on what additional hazards that we had to contend with that the designers and maintainers would not have thought of.

10.1 GOVERNMENT DATA BANKS

One of the largest collections of equipment and human error data is the U.S. Government Printing Office and now the U.S. government Internet sites. Numerous data books are published (and updated periodically and many times published online) from the various executive branch and independent agencies such as the Department of Defense, NASA, the Environmental Protection Agency (EPA), the Nuclear Regulatory Commission, and Occupational Safety and Health Administration (OSHA). Many of these documents are currently under review and are being updated as either joint government–industry documents or are totally managed by industry. And, best of all, most of these are on the Internet.

One such example is the Government–Industry Data Exchange Program (GIDEP). This is a cooperative activity between the U.S. and Canadian governments and industry to gather and disseminate useful failure and safety information to all concerned parties. Some of the participating organizations are the U.S. Army, Navy, and Air Force, the Department of Labor, the Federal Aviation Administration, the Department of Energy, the National Institute of Standards and Technology, and the Canadian Department of Defense. Literally hundreds of private companies and industrial organizations are also part of the network. None of the information is classified or proprietary.

Some of the kinds of data available are engineering, metrology, reliability and maintainability data, failure experience, and safety hazards and experiences.

A special part of the program is the ALERT program. GIDEP immediately notifies all participants of actual or potential problems with parts, components, materials, manufacturing processes, test equipment, or safety conditions. The list seems almost infinite and very long per industry, but here are just a very small handful of other useful U.S. government sources (many countries have their own information sources):

- General data that can be useful—https://www.data.gov/
- Chemical Hazards and Emergency Medical Management—http://chemm.nlm.nih.gov/
- NIOSH Pocket Guide To Chemical Hazards—http://www.cdc.gov/niosh/npg/default.html
- Material safety data sheets (MSDSs)—http://www.msdssearch.com/

- Environmental safety data sources—http://www.epa.gov/enviro/html/data_source.html
- Aviation Safety Reporting System—http://asrs.arc.nasa.gov/
- Food safety—http://www.foodsafety.gov/

There are various database groups in Europe and pretty much every country around the world. One in the United Kingdom is the Atomic Energy Authority at Warrington, United Kingdom, which maintains a system reliability data bank. Other great safety data sources are

- Safety statistics—http://www.hse.gov.uk/statistics/
- General safety information—http://www.hse.gov.uk/legislation/services.htm
- Safety Data Sheets in Australia—http://www.safeworkaustralia.gov.au/sites/swa/whs-information/hazardous-chemicals/sds/pages/sds
- International traffic safety data—http://internationaltransportforum.org/irtadpublic/index.html
- International Chemical Safety Cards (accessible in many languages besides English)—http://www.cdc.gov/niosh/ipcs/icstart.html
- International Air Transport Association safety data—https://www.iata.org/whatwedo/safety/pages/safety-data.aspx
- Institute of Safety in Technology and Research—http://www.istr.org.uk/othinf.shtml
- European health and safety statistics and sources—http://epp.eurostat.ec.europa.eu/statistics_explained/index.php/Health_and_safety_at_work_statistics
- Health and safety sources in Canada—http://www.ccohs.ca/resources/

Henley and Kumamoto (2000) provide a list of American and European government and industry data sources.

10.2 INDUSTRY DATA BANKS

Large corporations that build hardware usually have internal records or databases of failure and safety information. Unfortunately, the data typically are available only to company employees. Often, however, calling the vendor is sufficient for them to send you the information you may need. Insurance companies are another source of information.

Search individual safety associations and safety societies from around the world. Most of them have listings of safety data sources that are searchable, and their lists are very useful. The Electronics Industries Associations maintains a large database of electronic safety problems. The Consumer Product Safety Commission is another good source. The System Safety Society (Stephans and Talso, 1997) offers a long list of data sources that are found in general industry.

And, of course, do not forget about MSDSs. MSDSs are great sources of safety information and should be read, not just filed away.

10.3 CREATING YOUR OWN SAFETY KNOWLEDGE MANAGEMENT SYSTEM: SOME SUGGESTIONS

One of the problems with data banks is that they primarily hold failure data. Failure data are very important to understand how and when a piece of equipment can fail, but they do not necessarily tell us whether the equipment is safe or not. System safety analysis cannot be totally computerized anymore than designing a bridge can. We can put a lot of structural analysis tools (e.g., finite-element analysis) online, but the pure creativity of designing a bridge still has not been fully automated.

What we really need is information that indicates safe or unsafe scenarios. A *scenario is an event that is composed of all the equipment, people, environment, and operations of a system.* By being able to gather, store, and disseminate safety scenario information, we can refer to it to help us solve new safety problems.

PRACTICAL TIPS AND BEST PRACTICE

- *Do not forget the Internet.* Use it. This is probably one of the easiest, and certainly the fastest, way to find organizations that have useful data. It seems like an obvious suggestion, but many have not really used it well.
- But on the other side, be very careful of the Internet too. Just because it is on the Internet does not mean that it is correct or true. Double check sources and use only reliable and reputable sources.
- Use Appendix D as a starting point to find more safety data.

We would be foolish not to take advantage of the cheap price of personal computers and the proliferation of local and wide area networks and the cloud. Placing safety information on these networks and in the cloud, easily accessible to all engineers, is very advantageous. Collaboration sites, such as SharePoint, are very useful for sharing safety data.

It is worthwhile to develop an interactive knowledge-based system that will be your safety knowledge management system and that has the plant or system:

- Design
- Assessments
- Hazard identification and evaluation
- Compliance verification
- Safety and failure history

Keeping this kind of information in one place, easily accessible (e.g., online), also makes it vastly easier to produce documentation if an OSHA inspector comes through. Law already requires some of this information. More important, it puts everything in one place so that you can actually *see* the entire system safety process, analyze the data, and develop better, more efficient, and less expensive safety for your plant or system. This should all be part of your safety management system.

Think of the safety knowledge management system as having three parts: system design and operations (and maintenance) information, safety information, and the safety management system information. Some kinds of information to put into your safety knowledge management system include

- A list of hazardous materials used in the system (quantities, how, where, and when used)
- MSDSs for all hazardous materials
- Systems or process design information (basic system descriptions, layout drawings, engineering drawings, etc.)
- A list of safety-critical systems and subsystems
- Best design practices
- Best operational and maintenance practices
- Test history
- Failure history
- The information stored in the system safety portion of the safety knowledge management system should include
    - Prior and current safety analyses
    - Accident and near-miss histories
    - Safety standards
    - Identified hazards
    - Known causes of identified hazards
    - Proven hazard controls
    - List of hazard consequences
    - Hazard logs and risk registers
    - Your hazard tracking system

PRACTICAL TIPS AND BEST PRACTICE

Best practice companies pretty much have all of this online: the entire safety management system and the data described earlier. This meets ISO certifications to ensure the most recent and approved procedures, data, and information are used.

All of this may seem like an avalanche of information, hardly worth the effort, but most of it, at least the systems engineering data, is already stored electronically. If it is not, then you can scan the information in.

The safety knowledge management system should also use a keyword search engine. The database programs should be designed so that you can search on any of this information. For example, if you wish to know the failure and safety history of your pumps, that should be easy to pull up. Then you will want to list all the systems that have fire hazards; the program should list all components and subsystems that have fire in the hazard description. Likewise, listing all of the controls for fire hazards will be very useful, including who has been safety trained and when they were certified.

Of course, this system, like any filing system, is good only if it is kept up to date. Every time you conduct a safety analysis, its hazard description, causes, controls, and consequences should be entered into the safety knowledge management system. The next time you have to analyze a similar system, you will have very good, hard data on the safety of the system. One system that a private company developed has over 1500 hazards listed, which makes it much easier to do future safety analyses if they already have a strong database of information. The results of the safety analysis can be input, and any open hazards can be tracked in this system.

With all of this information, you can also trend accident results, open hazards, or other data. This will give you something substantial to use to develop a set of metrics, key performance indicators, or key safety or risk indicators, or to assess whether the hazard controls are really worth the cost (or the opposite, if the lack of hazard controls is worth the price).

Even though using training to control hazards is the least desirable technique, training still must be carried out. Training is essential because it provides employees with the necessary information to do their jobs safely; decreases the likelihood of accidents, and increases productivity and efficiency in system operations. It is vital that managers, employees, and others on company property know what to do in the event of an emergency. More important, system safety training and awareness is a very valuable tool to help everyone to be cognizant of a dangerous situation and to help solve it. Having everyone conscious of safety and able to identify hazardous situations before they get out of control is significantly less expensive than cleaning up afterward. If a company has all employees helping, then it is much less likely that an accident will occur.

10.4   SAFETY TRAINING

The UK Health and Safety Executive (2013) states that the cost of accidents in the United Kingdom in 2011 was £13.8 billion. Ray and Bishop (1995) point out that in 1992, the United States lost over $115.9 billion and 8500 deaths occurred due to workplace accidents. They point out that there still are significant accidents because employees are performing unsafe work practices and they are not very motivated to change. Of course, designing out the hazards is usually the best solution. But if you are a rail organization that transports chlorine, you cannot get rid of the chlorine, and training is one piece of the solution to ensuring safe passage.

Chapter 4 illustrated that OSHA requires employers to train their employees and other on-site subcontractors about the hazards that exist in the workplace. This training should be twofold: employee training (including subcontractors) related to the everyday hazards and the training needed to respond appropriately to an emergency situation. And, of course, safety training must be targeted. Senior executives are responsible for the safety, health, and risks to their employees and the public, and they require a very different kind of safety training than frontline staff who works in hazardous conditions such as on an assembly line.

There are three types of safety training: initial training, refresher training, and new training for changes in system architecture or operations. All organizations should include the three kinds of training in their safety training program. Of course, the safety training program must be part of the safety management system discussed in Chapter 4.

10.4.1 Employee Safety Training

Before actually developing employee safety training, there are a few basic steps that a company should go through. First, a training that needs assessment should be conducted. Why are you doing the training? Did you just go through a recent flurry of accidents? Did OSHA cite you? Did the system safety analysis identify problem areas that could be resolved or mitigated by training? Who is the training for? Is it for only production-line operators, managers, supervisors, or subcontractors?

After determining what the safety training needs are, it is very important to decide what purpose you wish the training to serve. Is it to lower accident rates in the maintenance group? Is it to give overall safety information and awareness to employees about the hazards around them? There should be a direct correlation between the training needs and the overall and specific purpose of the training.

The next step is to assess what people are doing now on the job. What are the specifics of the job that are hazardous? What specific tasks are hazardous? What have employees done in the past in this kind of situation? At this stage, you are gathering the raw material that will be incorporated into training plans and course materials.

Then you must determine what budget you have to develop, implement, and maintain the training. It is critical that all safety training records be kept. There have been many instances where companies presented in court safety training records that showed that the company was adequately training their employees even though there was an accident. And there have been many cases where companies could not produce the records and therefore had no substantiating evidence that indicated they were not at fault.

Some of the topics and subject matter should be taken from the system safety analyses. The analyses identified hazards, their consequences, their controls, and the verification of the controls. This is very important information to pass on to the individuals who work close to these systems. The safety training must be a cogent part of the safety management system. Some of the kinds of information that should be included in a safety training program and safety training courses (whether corporate developed or contracted) are

- Essential features and requirements of OSHA (or other relevant occupational safety and health codes—both local and national)
- Individual employee rights and responsibilities
- How the safety training program fits into the safety management system
- Corporate policy and procedures about safety
- The company system for reporting injuries, illnesses, and hazardous conditions
- Actual hazards in the workplace and how they should be controlled
- Safe work practices
- What to do in the event of an emergency
- First aid procedures
- Use and care of protective equipment

10.4.2 EMERGENCY PREPAREDNESS AND RESPONSE TRAINING

It is never pleasant to imagine what would happen if an accident occurs. Unfortunately, we must do it and even prepare for one. During the preparation of an emergency preparedness drill for the Washington, DC, subway system, one firefighter was complaining that he saw no reason for him to participate in a subway disaster since it did not fall into his jurisdiction. A year later, an airliner taking off from the Washington National Airport struck the 14th Street Bridge. At about the same time, there was a subway train emergency. All of the emergency response units were working on pulling aircraft passengers from the icy Potomac River. The firefighter was the first to respond to the subway incident.

Chapter 4 discusses emergency preparedness programs and illustrates a typical plan. Everyone who could be affected by the emergency should go through regular emergency training. Some items that should be included in emergency drills are

- Evacuation procedures
- Shutdown of equipment during an emergency
- Firefighting and first aid
- Crowd control and panic prevention

Emergency drills do not always have to be full dress rehearsals. Full dress rehearsals can be very expensive, especially for large, complex operations. One solution is the use of *tabletop exercises*. A tabletop exercise is one in which the personnel work through the emergency in a group using paper and pencil. For example, the response of the crew at a hydroelectric plant could be practiced as a tabletop exercise. This is an excellent method by which to discuss what emergency response techniques are best. Obviously, after doing a tabletop exercise, it is important to do at least one dress rehearsal to verify that the team will truly respond as intended. It is best to try to do one full-scale emergency response training exercise or drill every few years.

PRACTICAL TIPS AND BEST PRACTICE

- Think carefully about the makeup of the course participants. Should they be only managers? Should they be only engineers or production-line employees? Should the course be for line supervisors?
- There are advantages to mixing up the group a little. It gets together groups that traditionally do not talk to each other, and perhaps they can learn something from each other.
- The downside is that the type of safety information needed for the different groups could be vastly different.

10.4.3 PERSONNEL CERTIFICATION FOR HAZARDOUS OPERATIONS

If your plant or facility has hazardous operations (as identified in the system safety analysis), then you should set up a personnel certification program for conducting hazardous operations. Anyone who performs or controls hazardous operations or who uses or transports hazardous material must understand the dangers involved. Typical hazardous jobs that should be certified are high-voltage electricians, welders, power-tool operators, heavy-equipment operators and riggers, aerial basket and truck platform operators, and boiler plant operators.

To ensure that individuals truly understand how to do the hazardous operations safely, you should have a good certification program. Permanent records should be kept, and scheduled refresher courses must be given. Some things to include in a certification program are

- Certification examination
- Physical examination
- The necessary classroom and hands-on training
- Written (and demonstration) test of safe working practices
- Recertification schedule

NOTES FROM NICK'S FILE

Years ago, one of my duties as a program manager was to run a large training program. It was a very hands-on set of courses for technicians building high-reliability systems. About 6 weeks after the completion of every course, we contacted the participant's supervisor and asked them to rate how well the employee was actually using the knowledge gained from the training. This helped us determine the effectiveness and practicality of our training programs. It also was very motivating for the employee and supervisor to reflect on the importance of the training.

10.4.4 SAMPLE SAFETY TRAINING COURSE OUTLINE FOR A MICROPROCESSOR PRODUCTION PLANT

Table 10.1 is an outline for a safety training course outline for a microprocessor manufacturing plant. There are many organizations, companies, government entities, safety associations, industry associations, and many others that conduct very good safety training programs. If a safety training program is outsourced, it is important that the company brought in to conduct the training configure the training to the reality to what hazards and risks that employees will face. High-hazard industries should have their own in-house training programs so that the unique safety considerations can be adequately discussed and employees trained on it.

10.5 SAFETY AWARENESS TRAINING

There are numerous ways to make people think about safety. Putting up posters, writing articles for the company newsletter, hosting safety workshops, and having a safety awareness day are all good. Positive incentives (not negative incentives for

TABLE 10.1
Sample Safety Training Course Outline for a
Microprocessor Manufacturing Plant

I.   Introduction
       Management's responsibilities
       Right-to-know laws
       OSHA and EPA
       Employee rights and responsibilities
       Company safety management system
       Corporate safety policy and procedures
       Corporate and plant system safety program
       Company system for reporting injuries, illnesses, and hazardous conditions
II.   Hazard classification of materials
       Health hazards
       Flammables
       Reactive and corrosive chemicals
       Compressed gases
       Cryogenic liquids
III.   Chemical labeling
IV.   Plant mechanical hazards
V.   Hazard recognition and control
       Plant operations
       Procedures for identification of engineering and administrative controls
VI.   Personal protective equipment
VII.   Hazardous materials storage, dispensing, and handling
       Incompatible materials
       Storage classifications
       Handling hazardous materials
VIII. Waste disposal
IX.   Industrial hygiene
       Health standards
       Health assessment programs
       Air quality monitoring program
X.   Emergency procedures
       Notification and reporting
       Hazardous spill procedures
       First aid procedures
       Fire and ambulance response

accidents—that only leads to the hiding of accidents) for having the least number of lost-time injuries or mishaps are good. Chapter 4 addresses safety culture and ways to improve it. Safety awareness training is one way to improve safety culture and is also worthy.

Of course, having a safety management system in place is the first and most important step. And it has to happen. As part of the safety management system, you may wish to have a safety representative in each plant functional area. For example, if you run a plant that manufactures personal hygiene products, you could have a safety rep for each major process point: razor blades, shampoo, toiletries, warehouse and shipping, labs, and administrative offices.

The safety reps should be regular employees from each area, not necessarily safety engineers. The group of safety reps can meet quarterly and discuss current safety problems: open hazards identified in the safety analysis, safety training opportunities, safety awareness programs, and any other important safety issues.

The safety reps are the eyes and ears of the system safety program and the safety management system. They can help significantly in ensuring that the system safety program is adequate and is being implemented.

The amount of special safety training needed for the safety reps is minimal. They need to know the hazards and safe work practices in their part of the plant. The safety reps should also be trained in conducting workplace safety inspections and monitoring the safety program's effectiveness. The safety reps will prove to be very useful sources of information when it comes time to update the system safety analysis or make a change in the process.

Safety awareness training can use a combination of delivery mechanisms. eLearning is very cost-effective and easy to repeat on a regular basis. It also makes it easy to track and maintain safety training records. Of course the best learning is by doing, and that is also true for safety and safety awareness training. It is very useful and important to have employees participate in safety demonstrations to best understand the concepts (and consequences) of safety awareness.

REFERENCES

Henley, E. J. and Kumamoto, H. 2000. *Probabilistic Risk Assessment and Management for Engineers and Scientists*, 2nd edn. New York: Wiley-IEEE Press.

Ray, P. S. and Bishop, P. A. 1995. Can training alone ensure a safe workplace. *Professional Safety* 40(4): 56–59.

Stephans, R. A. and Talso, W. W. 1997. *System Safety Analysis Handbook*, 2nd edn. Albuquerque, NM: System Safety Society.

United Kingdom Health and Safety Executive. 2013. Costs to Britain of Workplace Fatalities and self-reported injuries and ill health 2011/2013. Health and Safety Executive, U.K. http://www.hse.gov.uk/statistics/pdf/cost-to-britain.pdf, downloaded February 14, 2014.

FURTHER READING

There are a number of safety-related college courses and undergraduate and graduate study programs available in every country around the world. Many private companies give short courses in general and specific safety subjects. Engineering societies such as the American Society of Safety Engineers and the American Society of Mechanical Engineers also give short courses. Probably the most elaborate system of safety courses (though only for the chemical industry) is the Center for Chemical Process Safety of the American Institute of Chemical Engineers. The International System Safety Society and the American Society of Safety Engineers have many safety courses and links to safety information.

Center for Chemical Process Safety. 1990. *Guidelines for Process Equipment Reliability Data*. New York: American Institute of Chemical Engineers.

Fucigna, J. T., Cleven, A., and Pepler, R. D. 1973. *Basic Training Program for Emergency Medical Technicians—Ambulance, Concepts and Recommendations*. Washington, DC: Dunlap and Associates, U.S. Government Printing Office.

Haddow, G. and Bullock, J. 2013. *Introduction to Emergency Management*, 5th edn. Oxford, UK: Butterworth-Heinemann Publishers.

Kapur, K. C. and Lamberson, L. R. 1977. *Reliability in Engineering Design*. New York: John Wiley.

Kirkpatrick, D. and Kirkpatrick, J. 2006. *Evaluating Training Programs: The Four Levels*, 3rd edn. San Francisco, CA: Berrett-Koehler Publishers.

National Aeronautics and Space Administration. 2011. *NASA System Safety Handbook*. NASA/SP-2010-50. Washington, DC: National Aeronautics and Space Administration.

Poe, R. J. and Skewis, W. H. 1990. *Handbook of Reliability Predication Procedures for Mechanical Equipment*. DTRC-90/010. David Taylor Research Center. Washington, DC: U.S. Department of Defense.

Powell, A. E. 1993. An analysis of system safety training requirements for facility acquisition managers, planners, and engineers. Masters thesis. University of Maryland, College Park, MD.

Smith, E. R. 1981. The correlation between the predicted and the observed reliabilities of components, equipment, and systems. Report No. NCSR R18. Warrington, U.K.: UK Atomic Energy Authority.

Swain, A. D. and Guttman, H. E. 2011. *Handbook of Human Reliability Analysis with Emphasis on Nuclear Power Plant Applications*. NUREG/CR-1278. Washington, DC: U.S. Nuclear Regulatory Commission.

U.S. Department of Defense. 1991. *Reliability Prediction of Electronic Equipment*. Mil-Hdbk-217F. Washington, DC: U.S. Department of Defense.

U.S. Department of Transportation. 1995. *Recommended Emergency Preparedness Guidelines for Urban, Rural, and Specialized Transit Systems*. UMTA-MA-06-0196-91-1. Washington, DC: U.S. Department of Transportation.

11 Accident Reporting, Investigation, Documentation, and Communication

… so unlucky that he runs into accidents which started out to happen to somebody else.

Archy's Life of Mehitabel, Archy Says, 1933
Donald Robert Marquis

Men may second their fortune, but cannot oppose it; that they may act in accordance with, but cannot infringe, its ordinances.

I Discorsi, II, 29, 1517
Nicolo Machiavelli

Study the past if you would divine the future.

Analects, fourth century BCE
Confucius

When you assume an attitude of suspicion, you overlook no clue.

Foucault's Pendulum, 1988
Umberto Eco

In 1984, Charles Perrow wrote in the first edition of his book, *Normal Accidents, Living with High-Risk Technologies*: "If interactive complexity and tight coupling—system characteristics—inevitably will produce an accident, I believe we are justified in calling it a *normal accident* [italics his], or a *system accident*.… It suggests, for example, that the probability of a nuclear plant meltdown with dispersion of radioactive materials to the atmosphere is not one chance in a million a year, but more like one chance in the next decade." Perrow was much more accurate than he imagined—Chernobyl occurred in 1986 and Fukushima was in 2011. But in between the two, there were others—thankfully with minimal radiation release and few deaths—here are some of the accidents with radiation release (there were other accidents with no release):

- Hamm-Uentrop, Germany (1986)
- Sosnovy Bor, Leningrad Oblast, Russia (1992)
- Tokaimura, Ibaraki Prefecture, Japan, two killed (1999)
- Mihama, Fukui Prefecture, Japan, four killed (2004)

The statistic of one accidental release per decade seems to be holding up. None of us wish accidents to occur, but unfortunately they do. What is important is to learn what mistakes were made and to ensure that corrections have been implemented. Also, it is important to remember that we should not wait until an accident occurs to investigate how well a safety system works.

Near misses, or almost-accidents, can be even more revealing than an actual accident. The reason that near misses are so interesting to improving safety is that when something does not go wrong, we tend not to fix it. Many people mistakenly believe that if no accident occurred, that demonstrated how well the safety systems work. The only problem is that something happened that we did not predict. So, yes, it does demonstrate how well the system works, but also how well it does not work.

Too many engineers involved with accident investigation do not take full advantage of the gold mine of information that an investigation can produce. Most accident investigation reports address only the narrow cause of a particular accident; too many times they leave out the systemic reasons why the accident occurred. What should be remembered is that the purpose of an accident investigation is not only to find the causes of the accident but also to find ways to prevent another accident in the future. The accident investigation system is thus part of the safety management system (SMS) (see Chapter 4 for more discussion of SMSs).

We all are very well aware of high-profile accident investigations. Yet, because of these investigations—and the subsequent corrective actions, improvements, and other changes—we do have much safer systems. One great example is the transportation industry. It is quite remarkable how safe aviation is, and that is because of the mandatory accident investigations that are conducted. Probably the most famous accident investigation body in the world is the U.S. National Transportation Safety Board (NTSB). It is an independent federal agency that is charged and financed by the U.S. Congress to investigate every civil aviation accident and significant accidents in other modes of transportation (e.g., rail, roads, maritime, pipelines). Neither the president nor congress can turn off an NTSB investigation once it starts. Though the results are nonbinding, most transport systems do implement the vast majority of their recommendations. Many other countries have created their own independent investigation boards. What is critical about the NTSB and its counterparts around the world is the nonpunitive nature of investigations. Chapter 12 discusses regulatory regimes and safety and illustrates how to correctly balance punitive and nonpunitive accident investigations. Chapter 11 focuses on how to actually conduct an accident investigation.

11.1   ANATOMY OF AN ACCIDENT

First, it is important to think about how an accident can occur. Chapter 2 briefly discusses the makeup of an accident. It is worthwhile to reflect in more detail here. Government and industry are facing many fluid changes and additional challenges and pressures to their operations; Figure 11.1 illustrates some of these.

When considering how an accident can occur, or its anatomy, it is important to realize that it is very rare, if ever, that a single event will lead to an accident. Chapter 2 discusses preliminary, initiating, and intermediate events that can lead up

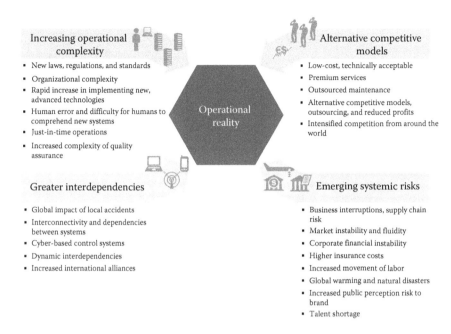

Increasing operational complexity

- New laws, regulations, and standards
- Organizational complexity
- Rapid increase in implementing new, advanced technologies
- Human error and difficulty for humans to comprehend new systems
- Just-in-time operations
- Increased complexity of quality assurance

Alternative competitive models

- Low-cost, technically acceptable
- Premium services
- Outsourced maintenance
- Alternative competitive models, outsourcing, and reduced profits
- Intensified competition from around the world

Operational reality

Greater interdependencies

- Global impact of local accidents
- Interconnectivity and dependencies between systems
- Cyber-based control systems
- Dynamic interdependencies
- Increased international alliances

Emerging systemic risks

- Business interruptions, supply chain risk
- Market instability and fluidity
- Corporate financial instability
- Higher insurance costs
- Increased movement of labor
- Global warming and natural disasters
- Increased public perception risk to brand
- Talent shortage

FIGURE 11.1 **(See color insert.)** Changing operating environment and its challenges.

to an accident. As Figure 11.1 illustrates, there are many pressures on your systems that can influence how an accident can occur. But this also means that you can use the concept of the many events that must be realized for an accident to occur to your benefit. Probably the most famous person to do this is James Reason with his so-called Swiss cheese model (Reason, 2006). Figure 11.2 demonstrates how it works.

The figure shows that if you can prevent one of these conditions, then you can prevent the accident. Of course, it may not be so easy to ensure that the one that you choose actually holds up. The figure implies that the human operator committed an unsafe act (an active failure or action), which caused the accident. In reality, there were a number of other latent conditions that could have also contributed to a particular accident. These latent conditions may have existed for months or even years. When you investigate an accident, it is important to go down to the root causes, and not just the symptoms or what appears obvious, to ensure that all the accident mechanisms have been identified. Again, refer to the hazard reduction precedence in Chapter 2 on the hierarchy of hazard controls.

There are many reasons why you should investigate an accident or near miss: first, to find out what went wrong. It is imperative to understand the cause of the accident so that it can be prevented in the future. Depending on the severity of the accident, it may be a legal requirement to investigate. And it is important for you and the regulators to know if there were compliance issues related to safety regulations, rules, or the company SMS. Investigating the accident is necessary to determine workers' compensation claims. And in the end, a company will want to understand the total cost that resulted from the accident, from material damage, lawsuits, lost time, reputation, and other factors.

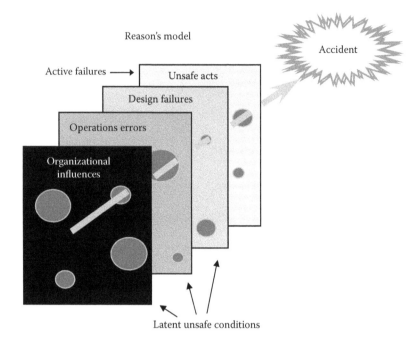

FIGURE 11.2   (**See color insert.**) Sequence of events leading to an accident ("Reason's model").

11.2   ACCIDENT INVESTIGATION BOARD

An accident should be investigated by a group of individuals who are as free from bias and independent from the accident participants as reasonable. It should ensure that the investigation is as free from external influences as possible. Creating an accident investigation board can do this. The board's purpose is to

- Serve as an independent body to determine the root causes of the accident
- Develop recommendations of how to prevent reoccurrence or similar accidents
- Improve safety of the system investigated
- Focus on causes to the accident rather than apportioning blame to the participants

The last item is very important. Too many times people will want to assign blame to individuals and discipline them. The problem with this is that experience has shown that people involved will hide evidence to avoid blame. Unfortunately, it makes it very difficult to determine root causes when this occurs. As Chapter 8 discusses, most human factor errors are really due to people being set up to fail. You have to design systems to fit around people, not the other way around. Of course legal liabilities complicate any accident investigation.

The board must have the power to actually conduct the investigation, or it will be worthless. That means it must have sufficient funding and time to reach a conclusion, have the power to require persons to answer questions, and produce documents or other materials relevant to the investigation. It must also have the power and ability to enter the premises where the accident occurred to visually inspect.

PRACTICAL TIPS AND BEST PRACTICE

- Provide an honest, unbiased investigation.
- Determine the detailed root causes and conditions that led to the accident.
- Treat all parties with dignity and fairness.
- Present detailed facts to support each investigatory point and conclusion.
- Provide recommendations and corrective actions to improve safety.

Before an accident occurs, the board or the safety office should prepare and publish accident reporting and investigation procedures. The reporting procedures should detail employee responsibilities for notification and reporting. The elements that should be detailed in an accident investigation report should also be determined beforehand. The accident investigation procedure should be clearly tied to the SMS and not conflict with it in any way. The investigation report results must be fed back into the SMS; otherwise, there will be no lessons learned and most likely the accident will occur again.

11.3   REPORTING THE ACCIDENT

The first step in developing an accident investigation system is to set up a system that allows employees to report accidents (without retribution). If employees fear for their jobs, not only will they not report accidents, they will also hide information that could possibly prevent accidents. And of course companies are obligated by law to post any reportable injury. A *reportable injury is an occupational death, injury, or illness that must be recorded on OSHA Form 200, Log of Occupational Injuries and Illnesses.* Other countries have comparable reporting systems.

During new-employee orientation, employees should be told that they are not only empowered to report any accident they witness, it is also their duty. It should also be emphasized that the purpose of an investigation is to discover cause-and-effect relationships in the accident and to develop corrective actions. The purpose is not to assign blame but to determine how to improve system operation and reduce the chance for further errors.

Employees also will see this as management involvement and interest in their concerns. Overall, morale can increase—not decrease—after an accident if it is handled in a professional manner. Every company will define the kinds of accident events that must be reported. Of course, local and federal laws will also determine what is reportable.

11.3.1 SETTING UP A CLOSED-LOOP REPORTING SYSTEM

If there is an immediate threat to human life, employees should immediately call emergency responders (911 or an equivalent number) if the plant has trained emergency response personnel. An accident report form must be filled out within 24 h of the incident. It is critical to capture as much information as possible about events leading up to the accident. Prompt and factual reporting of accidents facilitates accident investigation and acquisition of necessary information that will eventually go into the system safety analyses and be incorporated into the company SMS. If the accident is serious, an investigation board should be set up immediately. All accidents should be investigated.

The accident report should be put online (in the safety knowledge management system, see Chapter 10) so that all accident statistics can be tracked. This will allow you to trend useful information to determine if the hazard controls are adequate or if there is some underlying problem that has not been resolved.

After the immediate threat to human life has been resolved, it is important to contact the safety representative (see Chapter 10) for that particular area. The safety rep should track all the incidents that occur in his or her area. If the rep is not informed of what happened, he or she will not be able to prevent its recurrence.

The corporate chief engineer should manage the accident reporting system for the entire company. If accident investigation does not get the priority and visibility needed, it will be ineffectual, and there will be no way to ensure that the accident (or one like it) will not happen again.

11.3.2 EXAMPLE OF AN AUTOMATED SYSTEM

An example accident report is shown in Table 11.1. If this accident report is input or scanned into a database program of your safety knowledge management system, you can sort and trend the data.

11.4 FORMING AN INVESTIGATION BOARD

The company should first decide how accidents are to be classified—from minor to catastrophic—and then use this classification system for deciding how the investigation board will be formed. Using the Hazard Risk Index discussed in Chapter 5 is another good method. NASA (No date) uses a very good technique for accident classification. A slightly modified version is shown in Table 11.2 for your immediate use.

11.4.1 SELECTING THE INVESTIGATION BOARD

The investigation board should be composed of individuals from various backgrounds, such as design, operations, and management. The actual makeup of the board depends on the accident. Be careful not to limit yourself by not including people who may prove helpful simply because you thought the accident had no causal factors in their areas of expertise. The board has the freedom to add or subtract technical advisors, as needed, though the core board members should not change.

TABLE 11.1
Sample Initial Accident Report

Acorn Company   **First Report of Accident or Near Miss**

Accidents that result in personnel injuries or equipment damage must be reported to the safety office (x2222) immediately. This form must be submitted to safety within 24 h of the reported incident. *Attach additional sheets if more space is required.*

Section 1 (To Be Completed by Reporting Employee)   **Date of Report:**

Name: Office Phone:

Title: Mail Code:

Home Address: Home Phones:

Date of Accident: Time of Accident: Location of Accident:

Area Supervisor: Supervisor Notified? Yes ☐ No ☐

Description of Accident:

Facilities, Tooling, or Other Equipment Used at Time of Accident:

List of Work Orders or Procedures In Use at Time of Accident:

Name of Witnesses: Office Phone:

I Hereby Certify That To the Best of My   Employee Signature: Date:
    Knowledge and Belief All Above
    Statements Are True and Correct

Section 2 (To Be Completed by Investigating Official)

Was Medical Treatment Given?   Yes ☐ No ☐ Location: Date:

Describe Medical Treatment Given:

Employee: Returned to Work ☐ Sent Home ☐ In Hospital Care ☐

Corrective Action Initiated:

Area Safety Rep. Notified?   Yes ☐ No ☐

Investigation Board Initiated?   Yes ☐ No ☐

Investigating Official Signature: Date:

TABLE 11.2
Accident Classification Scheme

Type A accident: An accident that causes death and/or damage to equipment or property in excess of $1,000,000. Type A accidents will be investigated by a board appointed by the company CEO.

Type B accident: An accident that causes permanent disability to one or more persons, hospitalization of three or more persons, and/or damage to equipment or property equal to between $1,000,000 and $250,000. The company vice president responsible for this site will appoint the investigation board.

Type C accident: An accident that results in occupational injury or illness, which results in a lost-work day case or a restricted-duty case or damage to equipment or property of $25,000 to less than $250,000. The responsible division chief will appoint the investigation board.

Incident: Any accident of less than type C severity of injury to personnel or equipment or property damage less than $25,000. The responsible plant manager will appoint the investigation board.

Near miss: Any unsafe act that did not result in injury and little or no property damage, nor interruption to productive work, but that possesses high potential for any of the foregoing. The responsible line supervisor will lead the investigation, with a copy of the report to the plant manager.

The board should have broad powers to investigate what caused the accident and to develop recommendations to prevent further occurrences. An impartial individual should chair the board. The chairperson should be picked in accordance with Table 11.2.

The basic responsibility of the investigation board is to investigate the accident and write a report that documents the investigation, conclusions, and recommendations. The investigation should focus on the following:

- Determining the sequence of events that led up to the accident. It should also investigate the consequences of the accident.
- It should determine the accident initiating events. Was it a failure? Did the accident occur as part of normal operations? What kinds of factors led to the initiating events? Were people involved? What was the state of the equipment involved in the accident?
- Determine the causal factors (or probable causes) that led to the initiating events. Look at all system factors (human operators, operating environment, equipment design, equipment operations, maintenance, etc.).
- Developing recommendations of how to prevent a recurrence of the accident. Recommendations should address not only the specific accident but also methods to control future accidents. If other safety concerns are discovered during the investigation, include recommendations of how to control or mitigate these hazards.

11.4.2 CONDUCTING THE INVESTIGATION

There are five phases in accident investigation: preparing for the investigation, gathering the evidence and information, analyzing the data, discussion of the analysis

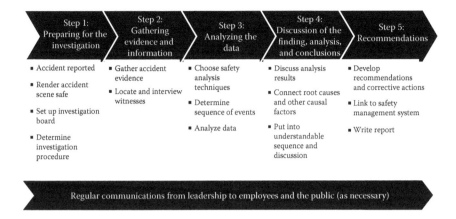

Step 1: Preparing for the investigation	Step 2: Gathering evidence and information	Step 3: Analyzing the data	Step 4: Discussion of the finding, analysis, and conclusions	Step 5: Recommendations
▪ Accident reported ▪ Render accident scene safe ▪ Set up investigation board ▪ Determine investigation procedure	▪ Gather accident evidence ▪ Locate and interview witnesses	▪ Choose safety analysis techniques ▪ Determine sequence of events ▪ Analyze data	▪ Discuss analysis results ▪ Connect root causes and other causal factors ▪ Put into understandable sequence and discussion	▪ Develop recommendations and corrective actions ▪ Link to safety management system ▪ Write report

Regular communications from leadership to employees and the public (as necessary)

FIGURE 11.3   Conducting the accident investigation.

and conclusions, and recommendations to prevent further recurrence. These five phases clearly mirror the objectives of an accident investigation outlined earlier. Figure 11.3 illustrates the process.

> *Step* 1: *Preparing for the investigation.* Once an accident has been reported and all personnel are safe from harm, the first report of accident (within 24 h of the accident) is submitted. The proper authorities classify the accident, an investigation board is convened, and the investigation starts. Obviously, depending on local and national laws and regulations, it may be necessary to report the accident to regulators.

Of course the first thing to do is to render the accident scene safe (after the first responders have arrived). The accident scene must be rendered safe for investigators to sift through the wreckage. A security perimeter should be set up immediately and access controlled.

The initial work a board performs largely determines the outcome of the investigation. If the board is sloppy, the results could be inconclusive, incorrect, or even misleading. Crucial accident evidence and information could be lost forever, and it might become impossible to truly determine what happened even if a new board is formed.

The board chairperson selects the other members and calls an initial meeting to discuss the investigation protocols. Individual assignments are given, and requests for specialized technical support groups (fire protection engineers, lab analysis, operations review of procedures, etc.) to support the board are made. The specialized technical support groups can be formed and dissolved during the course of the investigation, depending on their needs and applicability to that particular phase of the investigation.

Regular board meetings should be held to coordinate investigation activities. It is critical that the technical support groups also report on their progress. The board may have to redirect the technical support groups as the investigation unfolds. The board chairperson or other board members may also wish to meet regularly with individual technical support groups to review their results.

It is crucial that board members meet on a regular basis to discuss the investigation and its progress. These meetings should be closed door; since the board must investigate all possibilities, rumors of one theory or another may cause severe concern among the employees. However, the company should schedule regular briefings for the employees (and the press, if necessary), updating the status of the investigation. Board meetings should have limited access, but the investigation should not be secret.

Before the investigation can actually start, it is critical to determine the investigation procedure. How will the investigation unfold? What are the protocols? Hopefully, the company will already have an authorized accident investigation procedure developed as part of the company SMS.

While one group is locating all witnesses or participants in the accident, another should start developing interview questions. Tools and records that witnesses used during the accident should be impounded for close scrutiny.

> *Step* 2: *Gathering evidence and information.* Studying accident evidence and interviewing witnesses is the primary mode of gathering information. If evidence is not properly preserved, you may lose an important source of information or even a chance to determine what happened.

The accident scene should be thoroughly photographed by still camera or video camera, as appropriate, *before removing any evidence*. Exotic cameras may need to be used to help in the investigation. Ultraviolet and thermal cameras can provide evidence not available in the visible light spectrum. Thermal cameras are particularly good for locating heat sources that may have caused an explosion.

Make a grid map of the accident scene. Measurements and locations should be put on the grid map before the physical evidence is removed for analysis. Tag all recovered evidence. Do not allow any physical testing on the evidence until the board member responsible gives approval.

The board should appoint a technical support group to go out to the engineering and management offices to gather written information and documentation that may prove useful to the investigation. The kinds of information that are very helpful are engineering drawings, test reports, maintenance reports, operating procedures, quality control logs, lab reports, design specifications, reliability analyses, and system design analyses. Obviously, gather all previous safety information, such as system safety analyses, risk registers, hazard-tracking and corrective action–tracking reports, failure histories, and past investigation reports. A few items that many investigators do not always realize are very valuable are budget schedules and constraints, equipment installation procedures, equipment operating records, personnel and training records, engineer's notes on a system, and weather reports.

Witnesses should be located and interviewed as quickly as possible. They should be interviewed separately and in private. Questions should be in the form of an interview and not an interrogation. Be sensitive to people's concern for assuming that you are blaming them by virtue of the fact that they have been invited to present information. Also be sensitive to the reverse: Some people may immediately assume they did something wrong and could have prevented the loss of life. Sometimes a witness may not truly remember.

In one case, an investigator was interviewing the driver of a subway train and asked him if he saw the victim fall in front of the train. The driver repeatedly stated that he did not kill the hapless victim, in fact there was no death, in spite of the fact that the driver himself helped pull the body from the tracks. Obviously, the emotional effect of having killed someone was too strong for the driver even to recall it.

And, of course, do not use the interview to blame anyone, especially who you are interviewing. As discussed earlier, even though an individual may have caused an initiating event due to his or her inappropriate action, blaming the individual only forces them to hide evidence. If evidence is hidden, you may not get to the root cause. Creating a *blame-free* investigation atmosphere is critical to success. However, that does not mean that individuals should not be held accountable. Use this step to gather information; assigning accountability is the last action, after all the facts are in.

NOTES FROM NICK'S FILE

During the Sydney, New South Wales, Australia, Waterfall rail accident investigation, I was the lead to audit the regulator's effectiveness and the railway's response and SMS. The lawyers (especially via the press) were very defensive of their clients' culpability for the accident. Among various findings, the investigation found that there was no single individual responsible but that the SMS was woefully inadequate and that many managers did not have the appropriate safety competencies to do their job. This, coupled with a poor SMS, left managers set up to fail.

Step 3: *Analyzing the data.* Once the information has been gathered (and even before all the information becomes available), data analysis begins. Whenever possible, it is best to analyze the actual system measurements (e.g., motor rpm) rather than pseudomeasurements. The purpose of Step 3 is to understand the accident based on the available data.

Developing the actual (as opposed to what people believe occurred) sequence of events is crucial. Various safety analysis techniques discussed in Chapters 5 through 9 are excellent methods to use in attempting to reconstruct a timeline of activities. These tools are also very useful for linking accident sequences to sequence causes. The analysis should clearly indicate primary, contributory, and other causal factors that led to the accident.

The fault tree has been used in many accident investigations and has proven to be very useful in determining how the top event (the accident in this case) came about. Failure mode and effect analysis is particularly useful for understanding how a particular component (identified through the fault tree) failed. There also are cases where *what-if* analysis has helped determine accident sequences. Many use cause–consequence analysis to understand the accident path. The bow-tie method is also useful in understanding the link between hazard causes and controls.

NOTES FROM NICK'S FILE

My favorite accident analysis tool is fault tree analysis. I've used it for every investigation that I've been involved in. It is great to visually show others (especially nontechnical people) what went wrong (faults). It can show engineering failures, but more importantly, it shows management faults (e.g., lack of communications between departments, inadequate staged-gate reviews of designs).

One very useful tool in accident investigation analysis is to simulate the accident again. The simulation can give tremendous information. It is important to try to replicate as much as possible of the accident conditions and environment. A successful simulation is strong corroborating evidence of why and how an accident occurred.

Step 4: *Discussion of the findings, analysis, and conclusions.* After analyzing the sequence of events and the reasons or causes to these events, the board must put all of this into context. This is probably the most difficult part of the investigation. It is very easy to jump to conclusions. It is crucial that the board use a thorough and disciplined approach to discussing the analysis results, as they would in their regular engineering jobs. Conclusions should not be reached in haste. Bad conclusions will come back and haunt the board and the company.

Whenever possible, substantiating evidence that supports the conclusions should be presented and documented. Never suppress evidence or analysis that could contradict your conclusions. It is infinitely better to admit to a company error and rectify the problem so that it never occurs again than it is to hide it. There are hundreds of cases where companies attempted to hide the facts and results of the accident investigation. Apart from the immorality, the legal ramifications are tremendous, with not only large fines but also jail time possible. And in some cases, it may be impossible to find the true causes of the accident. This is still important if the information is proprietary, highly confidential (e.g., trade secrets), or classified. Of course, in these circumstances, senior leadership will have to be deeply involved (and government agencies if pertinent information is classified).

Step 5: *Recommendations.* This is the most important part of the investigation. This phase is composed of listing the investigation findings and giving specific recommendations (and sometimes the recommendation is to conduct further testing, analysis, or investigation) to preventing the accident from recurring.

It is imperative to link the accident investigation findings, corrective actions, and recommendations to the SMS. If it does not plug back into the SMS, then the risk of reoccurrence (or worse) is high. All of this will be documented in the accident investigation report.

TABLE 11.3
Accident Investigation Employee Responsibilities

- All personnel and departments must assist in an accident investigation as required.
- Departments and units involved in the accident should support development and implementation of corrective actions.
- The Safety Office should disseminate the Accident Investigation Report to the appropriate parties per the safety management system.
- The Safety Office should ensure that appropriate changes in training have occurred, as required by the investigation.
- Department managers should communicate investigation findings, conclusions, and recommendations to employees and contractors under their supervision.
- Department managers and the Safety Office (as defined by the safety management system) should maintain the documentation of the investigation findings, conclusions, and recommendations and track the status of corrective actions.
- The company should make available the investigation report to the regulators or other appropriate authorities.

It is worth noting that all parties in the company have various responsibilities during the accident investigation. Table 11.3 lists some of those responsibilities.

Regular communications from leadership to employees and the public (as necessary). Communicating regularly and in a transparent way is imperative to maintain company brand reputation and credibility. The lack of transparency during the Fukushima nuclear accident investigation had a significant impact on government investigators, plant operator, and others. Section 11.6 addresses communications in more detail.

11.4.3 INVESTIGATION REPORT

The written report should be submitted to the appropriate company authorities. The kinds of information that should go into the investigation report are as follows:

- *Executive summary*—Briefly summarizes the accident findings, conclusions, corrective actions, and how the investigation was managed.
- *Summary of report findings and recommendations*—Goes into more detail and explanation of the accident findings and what recommendations should be put in place (and why).
- *Investigation procedure*—Is usually in an appendix. It should be a detailed description of how the accident was investigated, setting up the board, using technical support groups, conducting interviews, performing analyses and tests, etc. This is important because it will demonstrate the robustness of the investigation and the credibility of the investigators.
- *Background and introduction to accident*—Describes the operational environment occurring at the time of the accident, the precursor events that may have set up the situation to occur, and other relevant data that can help the reader understand the conditions at the time of the accident.

- *Sequence of events that led to accident*—Is a detailed timeline that shows all the steps, events, and activities that occurred before, during, and after the accident (including how the company and others immediately responded to the accident). It should include not just what happened during the accident but also the time first responders arrived, what they did, when regulatory agencies were notified, etc.
- *Analysis methodologies and techniques*—Are sometimes put in the appendix. It describes the kind of safety analysis tools that were used in the course of the investigation (e.g., fault tree analysis, FMECA) and how and why they were used.
- *Analysis results and discussion*—Is the information gathered and analyzed. But it should not just include raw analysis; it is important to also discuss what the analysis results mean and how the various results tie together to give a picture or mosaic of what occurred and why it occurred. In this section, probable accident causes, precursor events, contributing factors, and other details are discussed.
- *Conclusions*—Are the bottom-line results, what it all means, determined from the investigation. This is derived from the prior section.
- *Detailed findings and recommendations (include how recommendations fit into the SMS)*—Focus on tying the findings to concrete, actionable recommendations. It is helpful to use a side-by-side table to show how specific recommendations will correct particular findings.
- *Minority reports (if necessary)*—May be included if the board is not in 100% agreement to what caused the accident. This does happen, and for very large and severe accidents, it may be necessary to have a section to allow minority or contrarian reviews to be put forth. This is very important to show impartiality, transparency, and differing views. But in the end, some sort of majority conclusion and recommendations must be made so that the company can move forward.
- *Appendices*—Include physical characteristics of the scene, interview reports, details on probable causes and contributing factors, and other relevant information used to analyze the data (i.e., lab reports, photographs, detailed analyses, interview transcriptions, maintenance and operating records, detailed safety analyses).

11.5 DOCUMENTING THE ACCIDENT

The investigation report is the primary method for documenting the accident and the corrective action. The governing organization needs to decide what the distribution of the accident report should be. If there is a significant amount of proprietary information, then obviously it cannot be widely disseminated. However, proprietary or other sensitive information can be segregated into a separate report so that the findings and recommendations can be disseminated.

It is very important that any corrective action be implemented as soon as feasible. Letting the accident report languish in someone's files can only lead to future problems.

Any new safety information should be immediately incorporated into the system safety analyses and SMS. The investigation results (or summary of results) must be made available for future safety analysis.

PRACTICAL TIPS AND BEST PRACTICE

The deftness with which a company responds to press inquiries and public concerns is an indication to the public of what kind of corporate citizen the company is. Remember that safety sells. Be as forthcoming as possible and use your response to the accident to show your concern with *always* trying to provide the safest work environment and product.

11.5.1 RETENTION OF RECORDS

Any information used in the accident investigation should be retained. If the paper takes up too much space, then scan the information electronically. If a court case results from the accident, you will be glad that you kept all the information.

11.5.2 PUBLIC RELEASE OF INFORMATION

The investigation board should not organize press briefings. This should be left up to the company public affairs office. However, the board chairperson should be present at the press briefings. The chairperson should offer as much information as possible. Obviously, the press will ask what caused the accident and what will be done about it. It is up to the board and the company to decide how much information they wish to release. However, the perception (which has a very strong influence on how a news story is reported) of stonewalling or hiding something is much more damaging to the corporate image than releasing some facts. Remember the company still must operate in the community long after the accident has passed. It is not worthwhile to appear callous to the public.

11.5.3 ACCIDENT INVESTIGATION LESSONS LEARNED

Table 11.4 lists some lessons learned and things to consider during an accident investigation.

11.6 COMMUNICATING THE ACCIDENT TO THE PUBLIC

We have all seen on TV the results of not being forthcoming to the public when an accident occurs. Withholding information is—unfortunately—all too common. It does not benefit the company or government agency. In the end, it will eventually come to light, and those who purposely hid information will be held liable and probably be subject to criminal prosecution.

All companies should have a crisis management plan, a plan that guides them on what to do during a crisis, such as an accident. This crisis management plan should

TABLE 11.4

Accident Investigation Lessons Learned

- Do not jump to conclusions too early in the investigation. It is okay and normal to use a hypothesis-driven investigation, with each analysis step supporting or refuting the hypothesis. Start the investigation from the beginning, not from the middle or end.
- Poor analysis means poor results—spend the necessary time to get the analysis right—go to the ROOT CAUSE and the SUPPORTING CONTRIBUTING FACTORS that led to the accident.
- Make sure to use a systems approach, as described in Figure 5.1, and look at all subsystems that make up a system.
- Too many engineers put in lots of data but no analysis. Raw data does not mean anything! It is the context of the data that is important. For the investigation report to have impact and relevance, the data must be described in its correct contextual environment.
- Unfortunately, it is not uncommon that recommendations are not implemented. Consider setting up an implementation committee to oversee and ensure that recommendations and corrective actions are validated to be appropriate and verified to be in place.
- Make recommendations concrete, pragmatic, actionable, and realistic. If they are too general or generic it will be impossible to implement.
- Make sure that lessons learned and corrective actions and recommendations are plugged back into the safety management system, this also includes new training, new procedures, and maybe new equipment or instrumentation.
- Spend time selecting your team. Make sure that they are competent and that they can stand up to scrutiny. Include their bios in an appendix of your accident report.
- Teach the team how to do the investigation (especially interview protocols) before they go out and gather evidence and interview witnesses. There is a correct way to interview witnesses showing respect, not use leading questions or making accusations, etc.
- Think of the Swiss Cheese approach, what events, conditions, etc., could have occurred that would have prevented the accident or mitigated its consequences.
- Share lessons learned and accident results to all employees so that they know how it was resolved and will be prevented in the future.

---

also include a crisis communications plan—how to communicate to the public and to employees during a crisis, such as an accident investigation.

11.6.1 Developing a Crisis Communication Plan

During an accident investigation, especially a very-high-profile one, everyone is under tremendous stress and pressure, and it is very easy to make mistakes, especially on how to communicate information to those outside the investigation. It is a good idea to use a crisis communication plan. It can be used for accident investigations (which is usually a crisis) as well as any other company crisis. Though large companies will probably have this well defined (in spite of the fact that we have repeatedly seen large organizations fail to communicate well during an accident), here are some of the key factors to include in a plan:

- *Define the crisis communication team and crisis communication plan before the event*: This should be predetermined and where possible staffed with professional communication specialists. But the crisis communication team will also need a representative from the accident investigation board. That board member must ensure that information communicated publicly is accurate and not misleading. Most communication specialists are not technical, and they will struggle on how to communicate complex technical information. This is one reason why companies many times look worse in front of the public.

- *Predetermine likely crisis communication scenarios*: If you already have a hazard analysis of your most likely hazards and accident types, how would you communicate to the public and employees if those scenarios became real? It would also be helpful to have some generic press releases prepared so that you do not have to invent on the spot. Of course, press releases and communications have to be specific to the situation.
- *When an accident occurs, the team should (with senior leadership) define the key messages that will be given*: Many may see this as subterfuge, but it is not. It is important to understand that how you say things has significant impact. You would not go into a big presentation with your leadership without having practiced and thought out the presentation beforehand, and you should not during an accident investigation.
- *Train whoever will be the company spokesperson and make them media savvy as well as well informed on the status of the investigation*: Interviews on the nightly news are usually given in sound bite answers (a reality, whether we like it or not), which is very different from long-form interviews in a more technical setting. Obviously, how this spokesperson delivers the key messages is very important. Be transparent as much as possible.
- *As soon as an accident occurs, develop a media response plan*: Assume that the accident news will go viral immediately. This response plan will determine who will talk for the company; how to address the various media outlets, especially social media, key themes, or messages to be communicated; and what kinds of media challenges that could exist for the company.
- *Have a plan for how senior leadership should interact with the media*: They will be confronted at public meetings, speeches, and anywhere they go and should be prepared to respond to a pursuing press.
- *Understand that the accident will be reported through different media outlets*: We all understand mainstream media including the local news broadcasters and national news programs and newspapers, but there are also specialized news reporters. It is not uncommon that high-hazard, high-visibility industries have confrontational media that exist solely to challenge large corporations and their activities. You cannot ignore them. Many times they break a story in an obscure media outlet and echo it into mainstream media. You want to address what they say so that you can be prepared for later reporting and the echo chamber effect. Remember the Internet is 24/7 and many if not most people nowadays are getting news from the Internet. Internet reporting and blogs will have tremendous impact on your image, not just broadcast and print media.

- *Your public relations staff (or whoever would do this for you) should track what is going on in the media*: It is critical to know what others are saying on blogs, other Internet sources, and in the general and specialized media *before* you speak publicly. If you are giving daily briefings, know what others are saying about the company and its response before you go public. If there are inaccuracies in the media (and there most likely will be), correct them immediately.
- *Make sure to have a plan of how to communicate information to employees*: Communicate to them that they are not authorized to speak to the media, only company spokespersons can. But also keep them up-to-date on the investigation. Whether you like it or not, they will be discussing it with family and friends.
- *Practice, practice, practice*: The crisis communication management team and spokesperson should practice how to communicate with the accident investigators, respond to hostile press, respond to a hostile public, and always, always be as open and transparent as possible.

11.6.2 Common Mistakes: What Not to Say and Do

When we are under pressure, we make mistakes. Here are a few key thoughts when speaking publicly about an accident investigation:

- *No lying to the media or anyone else*—Being truthful, transparent, and not misleading has a much more positive impact than ever lying, plus you could go to jail.
- *Do not ever blame the media*—There was an accident, and they did not create it. Blaming the media only makes you look culpable and trying to change the conversation. Always be polite and try to give them as much information as possible.
- *Blame employees*—This is a great way to get them to shut up during the investigation. There may very well be employee negligence, let the investigation determine that. Get all the facts and put it in perspective and then release.
- *Protect or defend tainted leadership or managers*—Again, give only facts and focus on being transparent. Let the investigation determine what happened, put it in context, and then release.
- *Minimize the accident*—This is particularly terrible if lives were lost or people's property was destroyed. Be honest on what transpired and show sincere sympathy for those impacted, including employees and the public.
- *Afraid to correct mistakes*—If you gave inaccurate information early on, correct it immediately. You can always explain that it was during the heat of the accident that conflicting reports came out. Do not spend too much time on the error—correct it quickly and move on.
- *Avoid saying "no comment"*—Makes you look guilty and trying to hide something.

REFERENCES

National Aeronautics and Space Administration. No date. NASA Procedures and Guidelines For Mishap Reporting, Investigating, and Recordkeeping. *NASA Procedures and Guidelines*. NPG: 8621, Draft 1. http://www.hq.nasa.gov/office/codeq/8621d1ax.pdf, downloaded February 14, 2014.

Perrow, C. 1984. *Normal Accidents, Living with High-Risk Technologies*. New York: Basic Books, pp. 4–5.

Reason, J., 2006. Human factors: A personal perspective. *Human Factors Seminar*, Helsinki, Finland, February 13, 2006. http://www.vtt.fi/liitetiedostot/muut/HFS06Reason.pdf, downloaded February 14, 2014.

FURTHER READING

Coombs, W. T. 2011. *Ongoing Crisis Communication: Planning, Managing, and Responding*, 3rd edn. Los Angeles, CA: SAGE Publications.

Ferry, T. S. 1988. *Modern Accident Investigation and Analysis*. New York: John Wiley.

Grant, S. and Powell, D. 1999. Crisis Response & Communication Planning MANUAL, Ontario Ministry of Agriculture, Food and Rural Affairs, Guelph, Ontario, Canada. http://www.foodsafety.ksu.edu/articles/313/crisis_communication_planning_manual. pdf, downloaded February 21, 2014.

Hathaway, W. T., Knapton, D. A., and Rudich, R. A. 1993. New York Metropolitan Transportation Authority Safety Investigation. DOT-VNTSC-FTA-93-4. Cambridge, MA: U.S. Department of Transportation, John A. Volpe National Transportation Systems Center.

Hendrick, K. and Benner, L. 1987. *Investigating Accidents with STEP*. New York: Marcel Dekker.

Kletz, T. A. 2009. *What Went Wrong? Fifth Edition: Case Histories of Process Plant Disasters and How They Could Have Been Avoided*. 5th edn. Oxford, UK: Gulf Publishing.

National Aeronautics and Space Administration. 1983. *NASA Safety Manual*, Vol. 2, *Guidelines for Mishap Investigation*. NHB 1700.1 (V2). Washington, DC: National Aeronautics and Space Administration.

National Transportation Safety Board. http://www.ntsb.gov/.

National Transportation Safety Board. 2002. *National Transportation Safety Board Aviation Investigation Manual Major Team Investigations*. National Transportation Safety Board. https://www.ntsb.gov/doclib/manuals/MajorInvestigationsManual.pdf,   downloaded May 20, 2014.

Oakley, J. 2012. *Accident Investigation Techniques*, 2nd edn. Des Plaines, IL: American Society of Safety Engineers.

Perrow, C. 1999. *Normal Accidents, Living with High-Risk Technologies*, Updated edn. Princeton, NJ: Princeton University Press.

U.S. Chemical Safety Board. http://www.csb.gov/, downloaded May 20, 2014.

U.S. Nuclear Regulatory Commission. 1975. *Reactor Safety Study: Assessment of Accident Risks in U.S. Commercial Nuclear Power Plants*. WASH 1400 (NUREG-75/014). Washington, DC: U.S. Nuclear Regulatory Commission.

12 Government Regulations and Safety Oversight

All government, indeed every human benefit and enjoyment, every virtue, and every prudent act, is founded on compromise and barter.

Speech on Conciliation with America, 1775
Edmund Burke

The care of human life and happiness, and not their destruction, is the first and only legitimate object of good government.

To the Republican Citizens of Washington County, Maryland, 1809
Thomas Jefferson

I don't try to describe the future. I try to prevent it.

Ray Bradbury, 1992

There is a famous American saying that "you can't live with or without your spouse," which is also the way that most people feel about government regulations. We all know how important and critical they are yet we also know what a pain it is to comply with, and sometimes we do not understand why we should—a necessary evil some would say. Yet, we clearly understand when we do not have enough regulations *after* a disaster has occurred. Our challenge today is to change the paradigm and develop appropriate, balanced regulations *before* we have to suffer through a disaster.

An explosion and resulting fire killed 29 mine workers in 2010 (the worst mine disaster in the United States in 40 years) at the Upper Big Branch Mine South in Montcoal, West Virginia, United States. Though the fault of the accident was clearly the mine operator, an independent report (National Institute for Occupational Safety and Health, 2012) and the final investigation report laid much blame on the Mine Safety and Health Administration (MSHA), which failed to heed warning signs or implement their own regulations. The independent report states, "Therefore the IP's (Independent Panel's) overall analysis suggests that if MSHA, had engaged in timely enforcement of the Mine Act and applicable standards and regulations, it would have lessened the chances and possibly could have prevented the UBB (Upper Big Branch) explosion." Interestingly, the independent report emphasized that enforcing current regulations, not writing new ones, might have prevented the accident. Clearly, a lesson learned from this accident is writing regulations and standards are not enough; enforcement is what makes regulations and standards reality.

The January 31, 2003, rail accident in Sydney, New South Wales, Australia, killing seven on board, illustrated a different regulatory problem. At the time of the accident, a coregulatory process, a method in which the government shares regulatory authority with industry partners, was applied. On face value, coregulation would suggest an optimal way to have industry and government work together to regulate industry. In New South Wales, this was done through jointly endorsed codes of practice and giving some regulatory responsibility to an industry body. Essentially, the rail operator would need to demonstrate an appropriately integrated safety management system (SMS). The Special Commission of Inquiry into the Waterfall accident report illustrates some of those challenges (McInerney, 2005). Among many findings, it did document that the Independent Transport Safety and Reliability Regulator (ITSRR) failed in two primary areas: "...Issues relating to a lack of perceived independence and proper allocation of resources between compliance, accreditation and policy functions...and...Inadequate approach to the safety accreditation of RailCorp (rail operator)."

Interestingly, ITSRR became operational in 2004 to give independence from the Ministry of Transport to better regulate safety. This independence means that the Minister of Transport or the New South Wales Premier cannot impede an ITSRR investigation or prosecution. So, legally ITSRR has the means to regulate and enforce safety. The Honorable Mr. McInerney, the Commissioner of the Waterfall investigation, further chastises the Ministry of Transport in the same report cited earlier stating that key individuals within the regulatory body lacked essential qualifications, no processes were in place to measure the effectiveness of the regulatory process, insufficient resources were applied to carry out responsibilities, and there was no overarching policy to frame regulations under the coregulatory model.

The study of history is so important. The Honorable Mr. McInerney also led the Glenbrook rail accident investigation, in which two trains collided also killing seven, in December 1999. It is important to remember that a series of rail accidents had occurred immediately before Glenbrook and Sydney was due to host the 2000 Summer Olympic Games. One of the key findings of the Glenbrook accident (McInerney, 2001) was the need to create an independent rail inspectorate. Various iterations of a rail safety act were promulgated and ITSRR was born right at the time of Waterfall. One can only imagine how upset the commissioner was to see his recommendations not adequately implemented after the Glenbrook accident. As a result, he set up an oversight board to report quarterly to the public on the progress of both the government (ITSRR and others) and the rail operator to implement all of the Waterfall recommendations.

A key lesson learned from the Australian examples is the importance for true independence of oversight, even though industry should be heavily participating in the regulatory process. Another lesson is just because an investigation board recommends something and the government implements the recommendations, it does not mean that they will be effective. So the question is what is the right balance of all these factors?

Just imagine the same discussion on food safety. "So, we operate in a world where consumer confidence is a key pillar of the global food system and consumer expectations for food safety are high. Most consumers understand that food is not risk free. They are not asking for the impossible. But they do expect that everyone involved in producing, processing, transporting, and marketing food is doing everything they reasonably can to prevent problems and make food safe," said the deputy commissioner for Foods and Veterinary Medicine of the U.S. Food and Drug Administration

(Taylor, 2012). Deputy Commissioner Michael Taylor at the same China International Food Safety and Quality Conference and Expo in 2012 emphasized key actions that should be considered for any safety regulations.

> I want to make one important point about food safety and consumer confidence. And it's simply this: the things we need to do to improve food safety are the very same things we need to do to strengthen consumer confidence. These are not two separate efforts. Food safety and consumer confidence are the product of a common effort that includes five key themes:
>
> 1. Food industry commitment and responsibility for food safety
> 2. A comprehensive systems approach, from farm to table
> 3. Credible and effective government oversight
> 4. Genuine public–private collaboration and partnership
> 5. Transparency on the part of industry and government

These five themes should always be considered and implemented by government safety regulators. The purpose of regulations is not just to regulate but also to ensure public confidence. This of course should be balanced with public–private collaboration and partnership. There is much discussion in world capitals that industry is too involved in writing regulations. In reality, we do want strong industry involvement. They have to live the regulations and balance the impact on their bottom line and jobs. But, it is also not illegitimate the fear that industry can hijack the process and have it serve only its own ends. That is why point number 5 in the deputy commissioner's remarks is central to public and consumer confidence in the regulations.

PRACTICAL TIPS AND BEST PRACTICE

- Always consider the five themes listed earlier and dedicate government (and public–private) efforts to address each one in its own right, as well as, together as one program.
- Be transparent and communicate *VERY* frequently with industry and the public; though the battle may seem tough, in the end everyone will win.
- Do not be scared to allow industry lawyers into the conversation. Better get an early and informal understanding of where they are headed before actions start. Also, many times, they will have very useful input that can avoid costly confrontations, especially if they have a better understanding of the intent of the regulation. Of course, in the end, the government is responsible for the public good.
- To the immediate prior point—trust, but verify.
- Use collaboration and outreach for stakeholder engagement. Educating stakeholders is the best way to get buy-in. Even though you have government power, you want buy-in from stakeholders if you want to have a successful regulation.

12.1 SAFETY REGULATORY OVERSIGHT

The purpose of this chapter on safety regulations is not to describe how the regulatory process works but to focus on what are some of the key elements and how best to use in the area of system safety. Every country and government agencies within countries look at the regulatory process in a similar way. Figure 12.1 illustrates the relationship between government oversight and industry or the operator/owner. It is always important to remember that the operator/owner is the risk owner. Though the government is required to ensure safety to the public, it is always the responsibility of the operator/owner to be safe, whether the government sufficiently regulates them or not.

The strategy and policy, at the top of the pyramid, is the government executive branch or ministry that creates the legal framework that should be passed by the country's legislature. In many countries, this legal framework is set as a safety act or safety legislation. This gives the legal authority to regulate. At the next level, regulation/oversight, the responsible government agency sets up the regulatory process and enforces the safety law. At the third level, operations, the risk is owned by the industry that has been regulated. They are responsible for implementing the safety regulations and reporting safety metrics to government regulators. Now it is clear the importance and advantage of strong stakeholder engagement and collaboration throughout all levels. It is not to industry's advantage to have bad regulations nor is it to the government's advantage. Both need to closely work together to balance regulations.

Many countries and government safety oversight agencies are evolving on how they conduct safety oversight. Following are some tips to think about and avoid (see Figure 12.2).

12.1.1 KEY COMPONENTS OF A SAFETY REGULATORY REGIME

History has taught us that many safety oversight programs around the world were borne out of safety failures and legacy issues. Usually something catastrophic occurred, an investigation ensued, blame was laid somewhere, regulations were created to avoid repeat, and oversight and enforcement responsibility was housed in an agency that seemed logical at the time. When we look at lagging safety indicators, yes, that approach can work. But as discussed earlier, we want to prevent the future accidents from happening so that means that we have to also rely on leading safety indicators (see Chapter 2 for more). This is done through a risk-based approach. Businesses have reacted to the growing regulatory environment, increased business complexity (across the entire value chain), and the increased requirement on accountability by creating governance, risk management, and compliance mechanisms. Government regulators should also consider these three concepts as they design their oversight framework and programs (some businesses would say that would help government regulators to better understand the impact of their regulations):

1. *Governance*—Requires that government departments think carefully where safety regulations should sit, which division is responsible, how is policy separated from enforcement, and how best to ensure strong information sharing vertically and horizontally between government entities and the private sector. Governance is the formal structure that ensures that government

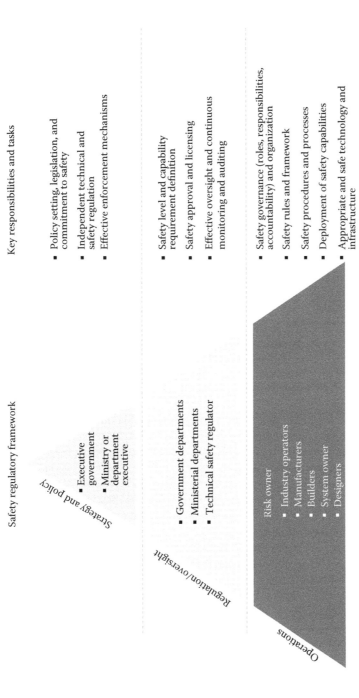

FIGURE 12.1 (**See color insert.**) Government–industry regulatory relationship.

Strategy and policy
- Missing or incomplete legal framework to regulate safety oversight and define safety responsibilities
- No safety regulator or formal mechanism to perform safety oversight
- Or, safety regulator in same government body as operations
- No comprehensive safety strategy and policy guidelines
- No independent accident investigation arm from the regulator (different from regulator enforcement investigation)

Regulation/ oversight
- Inadequate safety oversight program
- Lack of senior executive commitment to safety oversight
- No clear safety oversight objectives
- Inadequate safety enforcement regime in place
- Poor safety auditing system
- Insufficient resources and inadequate staff competencies
- Lack of safety certification or accreditation process

Operations
- Operations not separate body from regulator
- Lack of senior operations executive commitment to safety implementation
- Safety analysis is reactive (responding to accidents) rather than predictive or risk based
- Unclear safety responsibilities at front-line staff level
- Poor safety awareness, commitment, and communication
- Inadequate safety management system
- No rigorous safety and risk assessment process
- Unlinked accident investigation results to process improvements
- Unbalanced operations vs. safety (especially operational demands result in critical safety steps omitted)
- Blame culture present and overemphasis on misbehavior
- Employees do not have a deep understanding of how their jobs impact safety

FIGURE 12.2 Typical gaps to avoid in safety oversight.

plays an oversight role that has sufficient commitment, resources, and wherewithal to ensure industry builds and operates systems safely.

2. *Risk management*—Are the oversight processes that understand and evaluate the safety risks that industry poses and faces. It determines if controls are sufficient and monitors and tracks mitigation across industry. Its trending actions are very important to determine how best not to repeat past mistakes, but also it must be forward leaning and also determine where do the next unseen risks lay.

3. *Compliance*—Verifies that government safety rules are followed and safety issues reported appropriately and that enforcement actions are taken if they are not followed.

The three concepts outlined earlier can be translated into cogent government safety oversight program elements. Government safety regulators around the world, and in many industries, follow a successful approach that is similar to which the International Civil Aviation Organization (ICAO) employs that is well illustrated through the Irish Aviation Authority. The intent is to oversee the entire life cycle, from cradle to grave, of the system value chain. That means they will review how aviation products are designed, built, operated, deployed, and retired, including understanding the impact

FIGURE 12.3 Government safety regulatory elements. (Modified from Irish Aviation Authority, State Safety Program, https://www.iaa.ie/state-safety-programme, downloaded October 14, 2013.)

of the supply chain in the manufacturing process. Their safety oversight program is illustrated in Figure 12.3 (Irish Aviation Authority, 2013).

The purpose of the safety policy and objectives is to define the regulatory structure and responsibilities. The safety risk management function is to identify safety hazards and assess the safety risks across industry. It also defines how SMSs are defined and measures the industries' safety performance. Safety assurance is the regulatory mechanism such as requirements, operating regulations, and implementation oversight. This is where industry compliance data are collected and analyzed and enforcement documented. Safety promotion improves industry understanding and knowledge of government oversight regulations and compliance guidance.

Many countries are still grappling with what makes a safety act. A safety act, or safety law, should delineate the statutory safety policy, whereas regulations should be developed that detail how the safety act will be met and would include key regulatory objectives such as the following:

- Place a high value on effective identification, management, and control of risk.
- Work to ensure that system operations are constantly improving safety.
- Promote public confidence in the industry regulated.

As part of this, it is important that duties and the chain of responsibilities be clearly delineated. This means that operators have a duty to ensure the safety of operations. Product builders and maintainers also have a duty to ensure the safety of its product operation. And of course, workers have a duty to take reasonable care for their own safety and the safety of others. This chain of responsibilities must be clearly delineated in the law.

Many governments use a *reasonably practicable* test to ensure reasonable efforts expended for safety implementation. The intent is to first eliminate the risk to safety, if not reasonably practicable, then reduce the risk to a reasonably practicable level.

There is a (Safe Work Australia, 2013) great definition of elements of what constitutes a *reasonably practicable* level:

- Likelihood (or frequency) of the hazard or the risk concerned occurring
- Degree of harm that might result from the hazard or the risk
- What the person concerned knows, or ought reasonably to know, about the hazard or risk and ways of eliminating or minimizing the risk
- The availability and suitability of ways to eliminate or minimize the risk
- After assessing the extent of the risk and the available ways of eliminating or minimizing the risk, the cost associated with available ways of eliminating or minimizing the risk, including whether the cost is grossly disproportionate to the risk

Table 12.1 illustrates a notional rail safety act. It is a modified version of the Australian Rail Safety Act in 2006.

12.1.2 Description of Different Regulatory Oversight Models

Though there are many ways to look at safety regulatory oversight, there are really three kinds of safety regulatory models: prescriptive (sometimes called command and control), self-certification (sometimes called management based), and coregulatory (also sometimes called management based). There is a newer, fourth model that is frequently discussed in government circles: the performance-based model.

Prescriptive model—Is a set of detailed safety requirements that an operator or product developer must meet:

- Uses a government command and control approach to safety.
- Sets very prescriptive rules defined in a code of government-generated regulations that specifically define how products shall be engineered, tested, approved, and operated and maintained.
- Uses precise design, testing, operating, and maintenance requirements criteria that must be met.
- Compliance is measured by heavy and frequent inspection by government inspectors.
- Rule violations are addressed by issuing fines.
- All or part of operations can be shut down if violations are not corrected.

An example of this approach is the U.S. Federal Railroad Administration, Part 49 of the Code of Federal Regulations. This is not too surprising for an industry that is over 200 years old (though they currently are looking at how to incorporate an SMS approach into regulations). Probably an even more famous example is the U.S. Occupational Safety and Health Administration (OSHA) 29 CFR 1910, *OSHA Regulations for General Industry*. Though it is highly prescriptive, it too is migrating into a more management-based regulation such as the particular section 29 CFR 1910.119, *Process Safety Management of Highly Hazardous Chemicals*, and the 29 CFR 3133, *Process Safety Management Guidelines for Compliance*.

The problem is that the prescriptive model is very labor intensive and costly both to the regulator and the regulated industry and can be cumbersome for mature industries.

TABLE 12.1
Notional Rail Safety Act Outline

Part 1: Preliminary
- Purpose
- Objectives
- Rail safety act scope
- Definitions
- The concept of ensuring rail safety

Part 2: Occupational Health and Safety
- Summary of current regulations
- Applicability to rail workers

Part 3: Administration
- Rail safety regulator (RSR) description
- Functions
- Information to be included in annual reports
- Delegation
- Powers of RSR rail safety officers

Part 4: Rail Safety
General Duties
- Safety duties of rail transport operators
- Duties of designers, manufacturers, and suppliers of rail equipment and/or operations
- Duties of rail workers

Part 4: Rail Safety Accreditation
- Purpose of accreditation
- Application for accreditation
- Demonstrating accreditation
- Coordination with RSR
- Prescribed conditions and restrictions
- Penalty for breach of condition or restriction
- Surrender, revocation, or suspension of accreditation
- Immediate suspension of accreditation
- Application for variation of accreditation

Safety Management
- SMS
- Compliance with SMS
- Review of SMS
- Safety performance reports
- Interface coordination
- Emergency plan
- Health and fitness program
- Alcohol and drugs
- Fatigue management program
- Assessment of competence

Part 4: Rail Safety
Investigating and Reporting by Rail Operators
- Notification of notifiable occurrences
- Investigation of notifiable occurrences
- Audit and inspections by RSR
- Audit and inspection of railway operations of rail operators

Part 5: Enforcement
- Power to enter places
- Limitations of entry powers
- Notice of entry
- Securing a site

Part 6: Review of Decision
- Review by the RSR
- Appeal process

Part 7: General Liability and Evidentiary Provisions
- To be defined in subsequent phases

Part 8: General
- Development of compliance codes
- Approval and effect of compliance codes and guidelines
- Safety act implementation timelines

Source: Modified from Australian Rail Safety Bill Model Provisions, National Transport Commission, Rail Safety Bill Regulations, Sydney, New South Wales, Australia Government, April 2006.

The model does not easily foster innovation or safety improvements because the operator tends to focus strictly on compliance, *checking the box* and not necessarily improving. Along with this, it means that industry will primarily do what is mandated and tend to avoid voluntary improvements. The prescriptive model is punitive in nature and therefore has the unintended consequence to encourage industry not to report problems. Experience has shown that it also tends to foster a contentious relationship between the regulator and industry.

Though many regulations started this way and have since migrated to a different model, it does have useful and appropriate applications. It can be very useful for new technology or systems that have not been regulated before, and there is little experience or data to truly understand all of the hazards. This is particularly true for countries that are creating a regulatory regime for a particular industry for the first time. It certainly is possible, and oftentimes desirable, for developing countries that are not currently regulating a particular industry to start with a prescriptive approach. If the industry is not very mature and there is no previous regulatory regime, industry most likely will have a difficult time understanding how to apply a more elaborate regulatory model. And of course, a government can migrate from a prescriptive approach toward a more management-based approach over time.

NOTES FROM NICK'S FILE

I was working on a transformation project with a government agency in a developing country. The challenge was that the regulator and operator were under one roof. We separated the two and suggested using a prescriptive regulatory approach since the operator had never been regulated before.

Self-certification model (sometimes called management-based)—Could have some prescriptive elements but mostly requires that a system safety program is developed and certified to be safe; key elements include the following:

- Federal and/or the local government jurisdiction set safety guidelines.
- Many times, the federal government will delegate oversight to a local jurisdiction (or Notified Body), primarily for compliance testing and audits.
- A federal or local body can conduct compliance audits.
- The operator or product developer produces a safety certification package that describes how safety is met.
- The operator or product developer *self-certifies* that the system is safe.
- Risk-based process for regulations.
- Government–industry collaboration is strong.
- Rewards innovation.

It is also called a management-based model because it relies on an SMS as the core element that must be certified. A key differentiator with the prescriptive model is that it is more focused on identifying hazards, evaluating them, and then managing the risks. Probably one of the best-known examples is the CE certification given to European Economic Area countries for manufactured products. A manufacturer or an authorized representative or importer indicates (self-certifies) that a product, with a low safety risk, meets all applicable legal requirements. Those with higher safety risks are tested and test results used to help demonstrate self-certification through a third party or Notified Body certified by the government to act as an auditor. The manufacturer self-certification can be elaborate and typically includes documentation that describes the general system or product, hazard analysis and risk assessment to determine risks to the public, design and fabrication drawings and technical information, list of standards used in the design and build, test reports and outside certifications, and quality assurance program and results.

The two biggest challenges to self-certification are how strong the SMS is and how well the company follows it. Though the CE certification requires much documentation and this is great for highly repeatable processes and products, it is critical to ensure that a cogent SMS is in place to ensure safety is always considered. For many years, the United States used a self-certification process where state and local jurisdictions performed compliance functions on local transit systems. What they found out is that each of the 50 states of the United States audited in a different way and there was no uniformity to the oversight process. It also had a business impact on companies building transit systems because they would have to meet different standards in different states. Another disadvantage is that it is strongly dependent on mature safety programs in place in the industry. If industry safety programs are weak, then it becomes difficult to ensure that safety is effectively maintained in the manufacturing process. Also, this method does not always foster safety performance goal setting, if the SMS remains stagnate and is not regularly evaluated and improved.

Coregulatory model (also sometimes called management based)—Is very similar to the self-certification model in that it fosters strong government and industry collaboration. Its major elements are as follows:

- Government sets general safety guidelines.
- Industry develops a safety process and SMS that receives government accreditation.
- Many times, the safety accreditation is based on a safety case approach.
- Government reviews and accepts accreditation.
- Government conducts frequent audits to ensure that industry follows the accreditation.

The biggest challenge is how independent is the regulator and how well the regulator understands the industry (though one can ask that question of all regulatory models and regulators). At the heart of the coregulation model is the safety case. The Australian offshore petroleum industry also uses a safety case approach as part of the coregulation model. It is based on the concept that legislation sets the broad safety goals to be met and that the system designer and operator develop the appropriate methods to achieve these goals. As with all safety legislation worldwide, safety is the responsibility of the designer and operator. Therefore, the designer and operator should be responsible for figuring out how best to apply safety to their systems. Many industries, like the Australian offshore petroleum industry, were heavily regulated in a prescriptive fashion. But rapid changes in technology made it difficult for the regulator to keep up. Coregulation, with the safety case at the center, meant that both industry and government would determine what is considered safe.

Other industries are now looking at this concept as a possible new oversight model. For example, the U.S. Food and Drug Administration is considering for the infusion pump medical device using a safety case process. But the coregulation model and safety case approach is only as good as the oversight process itself. The Glenbrook and Waterfall rail accidents illustrated shortfalls in this method.

The coregulation model, if managed well, can be a more nimble oversight process and truly allow for more dynamic improvements to the system. The problem is that it may or may not be risk based, depending on how the regulator and industry define it. Another challenge is that the process can be overly bureaucratic. Many governments have fallen into the trap of conducting elaborate safety cases that include so much documentation that it sometimes can miss how well safety hazards are actually managed. Typically, a design safety case and an operational safety case are submitted as part of the accreditation process. But oftentimes they are two independent data files, developed by different teams that do not clearly connect design to operations.

It is worthwhile to spend some time on what exactly is a safety case. The *safety case is an evidence-based process that uses a structured argument to prove that the system or product is considered acceptably safe given a certain operating environment.* In other words, a safety argument is made stating that the system and operations are safe and evidence is given that demonstrates the validity of the argument. A *reasonability test* is used to determine how far to go in controlling hazards. The United Kingdom has coined the term *as low as reasonably practicable* (ALARP) as that reasonability test, and ALARP is now used in many countries around the world.

It all started with the 1988 Piper Alpha North Sea Oil Platform accident, killing 167 people and causing over \$3.6B in insurance claims. The subsequent Cullen Inquiry led to the development and promulgation in 1992 of the Offshore Installations (Safety Case) 1992, later updated in 2005 (United Kingdom, 2005). Now many industries (especially in the United Kingdom, Australia, New Zealand, and Europe), in addition to petroleum, use the safety case process including aviation, nuclear, rail, and military hardware. See Chapter 2 for more on ALARP.

The principal elements discussed in a safety case are

- Scope and details of the system and operations
- Management system that ensures safety (including governance and staff competencies)
- Applicable standards, codes, regulations, and requirements that the system and operating environment will meet
- Objective evidence that demonstrates that applicable standards are met
- A risk identification, evaluation, and management process that appropriately controls risks and evidence of compliance with the risk management process
- Evidence that residual risk is considered acceptable
- Independent assessment (independent safety assessor) that verifies and validates that the safety argument and evidence are acceptable

The safety case is oftentimes divided into an engineering safety case—which focuses on the design and deployment (including testing) of the system and the operational safety case—which focuses on the nominal and emergency operations of the system. Both should consider the entire life cycle from design, build, operations, maintenance, modifications, and disposal and should consider the build and operate/maintain system supply chain. Figure 12.4 illustrates how the system acquisition is considered in the safety case. The illustration taken from EN 50126, *Railway*

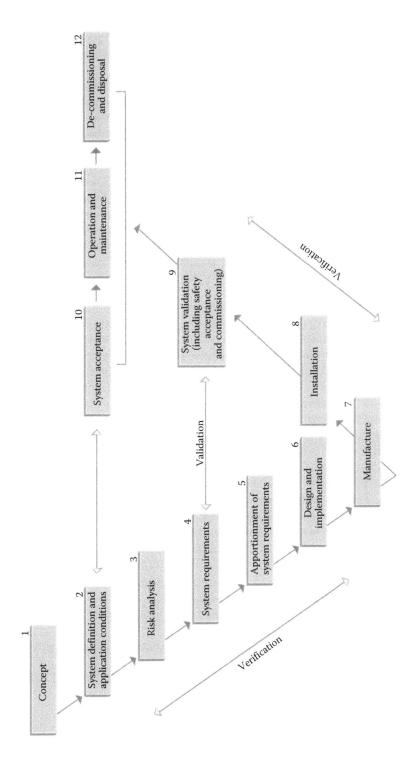

FIGURE 12.4 V-model for safety case.

applications, The specification and demonstration of reliability, availability, maintainability and safety (RAMS) Basic requirements and generic process, indicates the weak link between the engineering safety case and the operational safety case. A strong SMS must closely link the two together.

Central to the safety case is the hazard control and risk management process, ensuring that risks are well managed. ALARP is used to determine how far you should go to control the hazard. The safety case process is very thorough and highly labor intensive and, if you are not careful, could become overly bureaucratic. It is just a snapshot in time of the safety of the system, so that means that it must be maintained to still be relevant. This is where bureaucracy can take over if not careful.

The strengths of the safety case are that a cogent, reasonable argument is made as to why something is safe. It is based on demonstrated evidence (typically tests and engineering analysis), and it focuses on the operational environment and the corresponding hazards related to design and operations. The primary disadvantage is that it can be overly focused on the bureaucracy of the safety case and argument and difficult to determine what the appropriate level of safety that is acceptable. This is where ALARP is used.

Performance-based model (sometimes called outcome based)—Many times also called a market-based regulation that relies on specifying an outcome without dictating the specific requirements to reach the outcome. This is usually based on a *safety case* type of approach, mixed with a quantitative risk assessment that quantitatively determines ALARP. The quantitative risk assessment is compared with a cost–benefit to societal risks (e.g., loss of life, significant environmental degradation). Of course, the challenge always comes to what will society agree to as an acceptable number of deaths. There has been a significant amount of concern raised in some quarters that emphasize that industries that use a performance-based approach will lose the focus on identifying all the potential hazards because of its strong focus on only the highest risks that impact society. Much of those calculations are based on probabilistic risk assessments and therefore not necessarily based on all the potential hazards. This can make it very difficult to identify hazards that can lead to a Fukushima nuclear power–type disaster *black swan* event, because the risk of that accident scenario was considered too low to address.

Key features of the performance-based model are as follows:

- Mixture of a coregulatory and self-certification model.
- Federal regulator requires a safety case submittal.
- Notified Bodies, third-party bodies, which are required to audit safety cases via a contracted independent safety auditor.
- Process that relies on industry defining what society will tolerate as an acceptable level of residual risk.
- Annual quantitative safety tolerability levels are set by industry and approved by government.
- Complete audited safety case (including quantitative safety tolerability levels), conducted by a third-party-independent safety assessor, presented to government for final approval.

This approach is most common in very high-hazard operating environments such as the U.S. commercial nuclear power industry. Others, like the United Kingdom, Australia, and Norway, use it in offshore oil and gas. The UK Safety and Health Executive own the process in the United Kingdom and have also applied it extensively into rail transport. In fact, it has become ubiquitous across the EU, especially in rail transport.

12.1.3 CASE STUDY: SETTING UP A SAFETY OVERSIGHT BODY FROM SCRATCH

It is worthwhile to spend time discussing how one country created a safety oversight body from scratch. One good example is what the United Arab Emirates (UAE) is doing in Abu Dhabi and Dubai Emirates. Abu Dhabi is still nascent and under development, whereas Dubai already has a functioning program.

The UAE has seen phenomenal growth since their founding in 1971. Today, its gross domestic product (GDP) is 192 times larger than what it was 40 years ago. After Saudi Arabia, it is the second largest economy in the Middle East. Around 60% of its GDP is based on oil revenues and is ranked sixth in the world's known crude oil reserves. The UAE leadership understands that this is a finite resource and their dependence on its income is a significant risk to long-term economic sustainability. Across the UAE, they have embarked on an ambitious plan to reduce their dependence on oil revenues by diversifying the economy through increased tourism, manufacturing, migrating to a knowledge economy, improved education and health care, and a percentage of their energy needs met by renewable resources. A Louvre museum (the first outside of Paris) and the world's largest Guggenheim museum are slated to open soon. The Middle East's first United Nations body, the International Renewable Energy Agency (IRENA), is located in Abu Dhabi. Dubai, the more famous cousin, has one of the largest international airports in the world, is a leading reexport hub, and already is a top travel destination.

Central to Abu Dhabi's sustainable growth is an integrated public transport system. Abu Dhabi currently is building a light-rail and metro subway in the capital city. The Abu Dhabi Department of Transport (DoT) is responsible for safety oversight. But the DoT is a very young organization and how it is creating its surface transport safety oversight process is interesting to review. They have an integrated transport plan that guides them on all people and goods mobility in Abu Dhabi. In addition, they created a safety and security master plan. The safety and security master plan covers ferries and water taxis, buses, roads, intelligent transport systems, freight (truck and rail), commuter rail, light-rail, and metro.

As part of the Safety and Security Integrated Public Transport Master Plan, DoT conducted a benchmark study to look at how other countries regulated surface transport safety. In addition, to help advise the DoT, they set up an international advisory group. The DoT felt that it was very important to have a mix of advisors that could bring world-class best practices from leading transport systems but also include advisors from Middle Eastern countries with existing rail systems and who understood local customs and norms. One of Abu Dhabi's challenges was how to integrate these best practices in a culturally sensitive and sustainable way so that safety is always paramount but local customs are still honored and respected.

The DoT made many fact-finding trips to leading transit systems around the world and identified best practices in regulatory oversight. As part of the trips, they looked at not only how their regulatory regimes were set and how they managed safety risk but also how they operated their transit systems during nominal, emergency, and big event situations. Among the world-class transit systems they visited in London, Sydney, Singapore, and New York, they also visited both Vancouver and London immediately before the Winter and Summer Olympics to better understand not only how they operated safely but also how they prepared for one-off, large-scale events. One of the primary outcomes of the visits, and detailed analysis of the strengths and weaknesses of various regulatory regimes, was to model theirs on a self-regulatory model very similar to the New South Wales' Australia Independent Transport Safety Regulator. Central to operator accreditation is the safety case process (though currently no quantitative societal risk targets are being developed). The DoT requires that the transport operator hire an independent safety assessor to review their safety case package before submittal to DoT for final approval.

One of the key regulatory challenges (besides creating a regime from scratch) was how the DoT safety oversight process fits with the federal level promulgated by the National Transport Authority (NTA). The NTA does have a law that allows them to regulate transport safety across the UAE. A challenge is that the UAE is really a confederation of emirates with weak central authority but strong local emirate autonomy, very different from many western governments. Through workshops, interviews, meetings, and negotiations, much stakeholder engagement was conducted with NTA and other key stakeholders to determine federal and emirate jurisdictions. This was uncharted territory for Abu Dhabi and stakeholder participation and acceptance was and still is critical. The UAE uses a consensus-driven government approach to develop government entities and oversight programs.

Dubai (a separate emirate from Abu Dhabi) has a much more mature public transport system with an operating metro and light-rail system, albeit, only a few years old. The Roads and Transport Authority (RTA) is both a transport regulator and an operator. Dubai has followed a similar evolution of searching for world's best practices in the development of RTA's oversight process. They have implemented those best practices and their regulatory and oversight process has been in place for a number of years and has some very interesting aspects worth looking at.

The RTA corporate strategy and governance group is where the safety and risk regulation and planning office sits. Though the corporate safety and risk affairs are housed here, the day-to-day safety activities are located in each of the operating entities' quality, health, safety, and environment units. The corporate safety and risk department has six divisions that it oversees:

1. Certification and regulation (railway safety regulator)
2. Crisis and business continuity management
3. Policies and planning
4. Audit and reporting
5. Accident investigation
6. Enterprise risk management

The RTA has both BS OHSAS 18001 and ISO 14001 certifications. They are very active in promulgating safety among their contractors with many programs including a safety awards, including a safety personality of the year award.

What is unique about Dubai's safety risk management is that they have developed and implemented an enterprise risk management program across the entire RTA. Along with the certifications discussed earlier, they have also obtained ISO 31000, Risk Management System, and ISO 22399, Incident Preparedness and Operational Continuity Management System certifications. These actions have been central for their vision for providing a Safe and Smooth Transport for All program.

The enterprise risk management (ERM) program integrates all the risks across their operational and corporate sectors and ensures linkage to their crisis management and business continuity programs. Recently, they have launched an online ERM system that allows users to identify, analyze, report, and trend all risks across the RTA and its operations. It generates 62 different reports that are shared across and up and down RTA. Using ERM as a central hub is very logical toward linking the various different types of risks that RTA faces. It also helps ensure that safety risks are appropriately balanced with business and other corporate operational risks.

12.2 SAFETY OVERSIGHT FUNCTIONS AND GOVERNANCE

As we all know, safety oversight is important. But we also know that there can be too much of it. After creating the legal framework to regulate and determine the regulatory model to be used, now it is important to understand how to find the correct balance of oversight functions. If a government agency does not spend time on this, the number of divisions and offices can get out of control and significantly dilute effectiveness and confuse the regulated industry. For example, if risk management activities are mixed in with audit and compliance, it not only becomes overly cumbersome and difficult to manage but also confuses the office mission, potentially resulting in inadequate resources allocated to either function. Enforcement should not be part of the accident investigation activities because it prevents a thorough blame-free accident investigation and thus inhibits gathering all the data necessary to determine causes to the accident. Also, when looking at how to determine balance and calibrate resources, it is important to foster information sharing horizontally across the organization. Many times, organizations do not communicate horizontally across offices. As mentioned earlier, mixing regulatory oversight activities with operational activities is a very clear conflict of interest, though it is not uncommon. As a safety regulator, new or old, Table 12.2 lists some questions to ask as a first step.

12.2.1 MORE EFFECTIVE SAFETY SERVICE DELIVERY

Once the safety regulator has reflected on the safety oversight and governance questions and determined how the organization currently is constructed, it is important to think about how to deliver services more efficiently. Can functions and resources be bundled? Are they too fractured? It is important to determine how best to allocate safety resources within the regulator to maximize safety oversight capability with minimal government resources. Here are a few key questions to think about:

- Does the regulator provide safety approval and certification services to industry? If so, which services are provided, how much of it is actually safety related, and are redundant, superfluous, or peripheral nonsafety services included or bundled in the program?

TABLE 12.2

Oversight and Overnance Questions

Safety Oversight and Governance Questions

- Is there currently a high level of safety oversight?
  - Do core technical areas have strong engineering experience in design, manufacturing, and operations?
  - Are there strong technical relationships with industry?
  - Is there a historically strong industry safety record?
- Are safety and risk management activities fully and effectively integrated?
  - Are core safety and risk management activities spread throughout numerous sections and subgroups?
  - Is there a central or core safety and risk management group?
  - Are roles and responsibilities well known by staff?
  - Are there goals, objectives, and trending?
- Is there in-depth understanding of the industry risk profile?
  - Is there an integrated, systems-based risk assessment process in place?
  - Does the risk assessment cover the entire industry and appropriate parts of their supply chain?
  - Is it tracked and trended appropriately?
- Are long-term operational risks well known?
  - Is there sufficient insight into actual operational conditions and corresponding infrastructure?
  - Are regular operational risk assessments conducted and/or updated?
  - Are risks tracked and trended adequately?
- Are safety approval and certification/accreditation processes effective?
  - Are there detailed safety approval, certification, and regulation deviation processes in place and followed?
  - Are safety approvals and certification processes centralized?

- Are resources adequate and allocated appropriately today and into the near/medium term?
- What is the impact of safety to industry during the system life cycle (design, build, test, manufacture, operate, and retire)?
- How does this impact rulemaking and regulations?
- Are there various processes in the approval and certification program that can be adapted to increase the impact on safety while reducing resource requirements?
- How does your organization's resource allocation compare with other government agencies and countries?
- Are staff adequately trained, motivated, and have appropriate insight into industry safety problems?

It is not unusual to see government safety oversight resources distributed between three key functions:

1. *Preventative oversight*—Focuses on certifications and approval processes. These typically are activities that are conducted before a system can be approved for use or put into operations.

2. *Operational actions*—Emphasizes inspections, audits, and other compliance activities with industry. It can also include monitoring of industry operations and investigation of incidents and accidents.
3. *Safety rulemaking*—Concentrates on creation of laws and the development and promulgation of regulations.

There is no set rule on what the correct distribution of resources on the three functions should be; industries vary greatly in which functions are more resource intense based on the industry maturity, government safety oversight model, government–industry interaction (collaborative or contentious), complexity of industry operations, high-risk industry operating environment, and public exposure to risk. But certainly, there are economies of scale that can be considered when looking at detailed safety activities within each of the three functions. Many times, activities that do not require unique skills can be bundled.

PRACTICAL TIPS AND BEST PRACTICE

- Consider bundling generic services but be careful with highly technical or unique knowledge activities—they may not make sense to bundle.
- Spend extra time considering the unintended consequences of bundling, splitting activities, or redistributing resources. You do not want to make things worse.
- Investigate how technology can help you create efficiencies, especially with increasing service throughput and efficacy.
- Look for opportunities to increase throughput in services with minimal additional resources. Look at areas where bundling is possible and combine services where logical.
- First focus on bundled areas that will give the most impact.
- When making service changes, it is imperative that staff understand their new roles and have adequate time to migrate to the new approach. Developing a transition plan is imperative.
- Understand which services (e.g., back office activities like human resources, procurement) can go into a shared-services unit, while safety unique services remain separate.
- Be sure that the correct services are in the correct functional area. For example, regulation writing should be in safety rulemaking, not prevention oversight.
- Do look internationally, and at other government agencies, to see how they allocate resources. But, no two agencies will have the same distribution.
- Remember, just because you increase resources, if processes remain the same, you may not get any real improvement in safety.

NOTES FROM NICK'S FILE

Once, working with a European safety regulator, we were optimizing and rebalancing their safety oversight services. One of their big challenges was an aging workforce and lack of adequate technical skills in the government sector. We had to very carefully figure out how to appropriately bundle their resources but also consider their demographics and include improving technical skills as part of the rebalancing.

12.3   SAFETY OVERSIGHT ORGANIZATION OPTIONS

Now that generic and unique services have been identified and appropriately bundled or separated and an understanding has developed of what kind of resources are needed to run these services, it is important to think about the governance or organizational structure that best suits the oversight agency. As aforementioned, every agency and country distributes preventative, operational, and rulemaking activities in different ways; the next step is to determine which strategic organization's options best suit the agency. It seems like there is an infinite number of organizational design options and approaches. Figure 12.5 illustrates 10 key organizational criteria that focus on delivery of safety oversight services.

Each of these evaluation factors should be considered independently as well as jointly. The evaluation criteria questions help focus the oversight agency to be sure that as analyses are performed and international benchmarks are evaluated, the key organizational characteristics are identified and integrated. An agency can also take these evaluation criteria and modify them for its specific mission, intent, goals, and objectives. They can then be given a weighting factor and then compared with various organizational design options.

12.3.1   SAMPLE SAFETY OVERSIGHT ORGANIZATION

Every safety oversight organization will have some sort of mixture of preventative, operational, and rulemaking functional capabilities. The safety oversight organization should also include the regulatory elements illustrated in Figure 12.3, which include these elements:

- Safety policy and objectives
- Safety risk management
- Safety assurance
- Safety promotion

Figure 12.6 illustrates the corresponding organizational chart (it assumes that shared services, such as human resources and procurements, are located elsewhere).

The advantage of developing a similar organization chart as shown earlier is that it clearly delineates the critical organizational functions. It is important to note that the

	Evaluation factor	Evaluation criteria questions
1	Impact on safety (short and long term)	• What is the impact on safety? Can it be improved? • What effect can be expected over time?
2	Accountability	• Are decision rights clear, well stated, and designed for easy compliance? • Are responsibilities and accountabilities clear and in place?
3	Span of control	• How does the new organization impact daily work? • Is there a balance between independence and organizational integration?
4	Organizational hierarchy	• Does the organization comply with governance and legislation? • Does leadership have suitable *line of sight* into the organization?
5	Agency philosophy	• Does the organization fit agency philosophy, goals, and objectives? • How well does the organization fit into current agency structure?
6	Organizational structure	• Does the structure prevent silos, foster interagency/organizational collaboration? • Is knowledge management appropriate? Are work flows efficient, communications fluid and timely?
7	Resource requirements	• How costly, resource intensive is the structure? Is staff composition manageable? • Are there any special skills required in the organization?
8	Client expectations	• Are client (internal and external) needs and expectations met efficiently? • Can clients easily navigate the structure to get what they need when they need it?
9	Employee motivation	• Does the organization support employee motivation and impact? • Does it support sustainable career growth and career path?
10	Implementation feasibility	• How long would implementation take? Cost? What future developments impact implementation? • What is the expected *political* reaction? What political communications are necessary?

FIGURE 12.5 Key safety organizational suitability evaluation criteria.

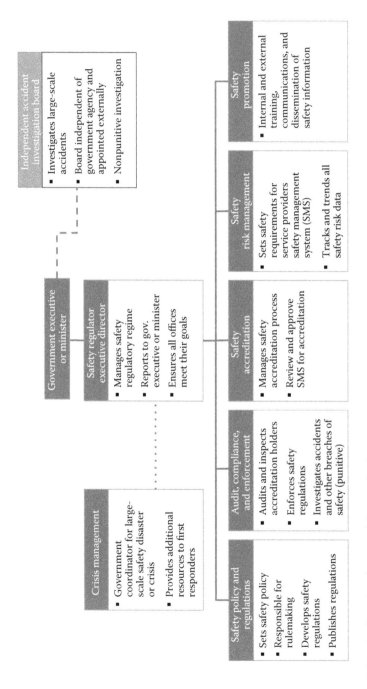

FIGURE 12.6   Sample safety oversight organization.

chart also shows an independent investigator and that the audit, compliance, and enforcement department will investigate accidents. Chapter 11 discusses in detail how accident investigations should be performed. The difference between the two groups is that the audit, compliance, and enforcement group has punitive responsibilities and will investigate the operator to determine culpability, liability, and noncompliance with regulations. These are the enforcement teeth of the regulator. Experience has proven time and again that enforcement investigators will not get full cooperation during investigations and at times can be very difficult to get all the information needed to prevent future accidents.

Because the independent accident investigation board is a *blame-free* investigation process, it cannot determine culpability nor levy punitive measures. The idea is that it can investigate all the way to the root causes of the accident. Because it is a nonenforcement investigation, the investigation can bring to light all relevant facts without fear of blame to the individuals and organizations involved. The United Kingdom and other Commonwealth countries have a similar process to the National Transportation Safety Board (NTSB) with their Royal Commissions.

12.3.1.1   Safety Regulator Executive Director Office

This is the executive-level office that is ultimately responsible for providing long-term, continuous safety oversight of industry. Typically, this office is fairly small in that its focus is to lead and manage the entire oversight organization. Each of the directors will report to the executive director. Its primary functions include

- Strategic coordination of oversight functions and safety regulations
- Final approval of matters related to the safe design, build, operation, maintenance, and disposal of industry systems (official signature for safety accreditation)
- Advice for the minister (or appropriate Cabinet-level official) about safety matters, especially safety regulations
- Issues appropriate enforcement, including punitive actions, fines, and compliance investigations, and dissemination of safety awareness information to government and industry

12.3.1.2   Crisis Management Office

This office is not a first responder to industry crises; rather, its purpose is to serve as a top-level management office to ensure that large-scale industry crises receive adequate and on-time emergency management resources. Many government agencies will have a full-time staffed skeleton crew of crisis managers working in a crisis management situation room. When a significant threshold has been passed, then the situation room (or crisis management center) is fully activated and representatives of relevant departments are seconded into the crisis center to help manage information flow and rapid deployment of resources (both human and material). It liaises directly with responders on the ground for real-time data, assesses the situation, and determines how best to support resources and keep senior government leadership closely informed of events on the ground.

It is important to remember that the emergency, accident, or unfolding disaster or crisis, many times, if not most, will not involve direct government activity. But that does not mean that the government agency can wait for the industry operator to respond. Almost any large-scale disaster illustrates the point. The Deep Water Horizon accident in the Gulf of Mexico in 2010 was so large that the U.S. Coast Guard was the incident responder. Though the state of Louisiana and other states each had activated their crisis management centers, the accident was so horrific and fluid that a crisis management center was also activated at both state and federal levels.

12.3.1.3   Safety Policy and Regulations Office

This office is responsible for developing, deploying, and amending safety laws (for legislature and executive approval), regulatory, and policy programs. The design of this office differs between the different regulatory models, depending on where the oversight agency resides between prescriptive and performance-based regulations. Though safety accreditation is managed elsewhere, this office ensures that the accreditation process is designed to meet safety regulations. Clearly, it will need to closely coordinate activities with the safety accreditation office.

12.3.1.4   Audit, Compliance, and Enforcement Office

The primary responsibility is to ensure industry compliance with safety regulations. This office conducts regular audits and inspections of industry through the entire system life cycle. It also has significant enforcement powers through a legal framework (such as a safety act). Trending industry performance and improvement is critical to ensure that appropriate corrective actions are put in place in a timely manner. The audit process works in concert with the safety accreditation office to ensure that accreditation holders meet their safety obligations and safety regulations. The principal focus of an industry audit should be to verify and validate the accreditation holder's SMS. Trending results are very important and should serve as feedback to the safety accreditation process. If industry safety trends are tending in the wrong direction, then the safety accreditation process (and potentially safety regulations) may need to be modified. This is especially important with the deployment of new or highly exotic technologies.

The teeth of a regulator sit within the enforcement function of this office. Enforcement escalation and penalties are typically defined in the safety act or other legal framework related to safety oversight. The executive director, as the government regulator, has legal authority to impose appropriate sanctions and penalties on the accreditation holder. Of course, through enforcement actions, the regulator can rescind accreditation or impose significant financial or other sanctions, up to and including system shutdown. Clearly, this should not be taken lightly or used too heavily either. Sanctions should be based on data gathered through audits, analyses, and other types of inspections. This office should closely follow an appropriate sanction escalation policy before reaching the final action of system shutdown.

The periodicity of auditing and inspections varies from industry to industry and should be based on the industry maturity, technology complexity, and high-hazard environment. Some people feel that audits must be unannounced to ensure full honesty and transparency. However, experience has borne out that it is very difficult to hide structural safety problems. Even well-publicized and well-planned audits will still uncover poor management and operational practices. See Chapter 4 for more on safety audits.

PRACTICAL TIPS AND BEST PRACTICE

- When setting up a government oversight audit program, first define audit goals and objectives.
- Make sure that all the offices have appropriate input; you want to avoid auditing more than necessary.
- Use a target risk-based auditing process and use the last 12 months of industry data to determine issues.
- Make sure your staff are adequately trained in auditing, the audit scope, and audit etiquette.
- Give adequate notice to timing of the audit, ask for documents *before* the start of the audit, and make sure that they have them ready on day 1 of the audit.
- The audit should be based on document review, selected interviews, and observations on operations during site visits.
- In-brief the audited leadership and out-brief them on preliminary results on the last day of the audit.
- Quickly notify audited leadership of any immediate threats to life; brief potential suggested corrective actions at the out-brief.
- Be sure that final audited results are regularly tracked, trended, and shared with all relevant government departments for lessons learned.

This office also conducts safety investigations. As noted earlier, these investigations are punitive in nature and represent formal response to the accident and determine culpability. Fortunately, most accidents are relatively small and can be handled by this office. However, there are times that large-scale accidents (e.g., Deepwater Horizon, Waterfall rail accident, Space Shuttle Challenger) do occur that require an investigation by an independent investigation team.

12.3.1.5 Safety Accreditation Office

This office manages the safety accreditation process and will formally review the accreditation holder's SMS to ensure that they meet safety regulations. The safety accreditation package will be submitted to this office for review and approval, though the safety accreditation office will closely coordinate with the auditing, compliance, and enforcement office to ensure that the submitted data are verified and validated as adequate and appropriate. The safety accreditation package should center on the accreditation holder's SMS and include all relevant data of the entire system life cycle, including safety-critical supply chains. Typically, a formal accreditation is given and then is reviewed on a regular basis for reaccreditation or when there are major system changes, upgrades, or modifications. This office should not easily grant derogations, waivers, or exemptions to safety regulations or requirements except under extreme circumstances and only if an appropriate level of safety has been adequately demonstrated.

12.3.1.5.1   Safety Accreditation Process

The regulator is the ultimate approval authority and accredits the competency and capacity, and the systems are safe for the entire system life cycle, especially during operations. The accreditation package is a very large data file that should include all relevant safety information. Some of the typical data elements (but not comprehensive) include

- Description of the operational environment
- Complete description and evidence of the SMS
- System safety program plan and how it fits into the SMS
- Safety hazard log and/or risk register
- Employee safety capacity and training
- Internal accident, incident, near miss, and major defects' reports
- Memoranda of agreements and other organizational interface agreements with first responders such as fire and emergency medical services
- Engineering, safety, and risk analyses
- Safety impacts from supply chain
- Tests and operational performance reports
- Internal safety audit and corrective action program
- Evidence of compliance with safety regulations
- Quality assurance and management system
- Third-party inspections, audits, and testing, including independent safety assessors
- Other relevant data that impact safety

Figure 12.7 illustrates a notional representation of typical safety-relevant data and information flows as would be reviewed in the V-shape safety case process illustrated in Figure 12.4. Note that this is just showing some of the safety information exchanges that would occur within an SMS and is not how an SMS is audited. Of course, the SMS also includes all the management systems that manage these information flows.

The regulator will then review the data file and conduct an audit to verify that the SMS meets requirements. Once reviewed and approved, the regulator would issue a certificate of accreditation. Figure 12.8 illustrates the accreditation approval process.

12.3.1.6   Safety Risk Management Office

Experience has shown that many safety regulators and their corresponding suboffices create information islands; information can be in silos and not appropriately shared, evaluated, and aggregated to the benefit of ensuring safety of industry. The purpose of this office is to *connect the dots* of all the safety data within the safety oversight agency. In some countries, this office may be combined with the safety accreditation office. There are merits to housing them together, such as efficiency and linking both activities closely together. The reason that some wish to keep them separate is because of the emphasis on the SMS as a central unit in the accreditation process. The office focuses on best practices in SMS and how to best support industry in developing their own SMS so that it is not only viable but also sustainable.

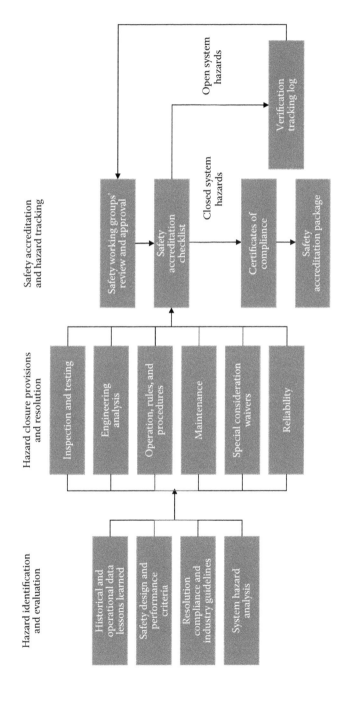

FIGURE 12.7 Notional representation of safety-relevant data and information flows.

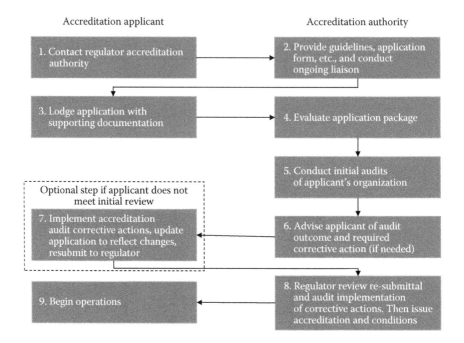

FIGURE 12.8 Accreditation approval process. (From National for Rail Safety Accreditation Package Guidelines, New South Wales, Australia, April 1999 with 2005 updates, Version 2.0, The Rail Safety Regulations Panel Australia.)

The other function of the office is to track, trend, and analyze all industry safety and risk data and maintain current risk profiles of industry organizations. The office is responsible for conducting industry-level risk assessments and determining the efficacy of current safety regulations and the accreditation process. Advanced research and new approaches to safety oversight or SMS implementation are also developed in this office. In other words, this office can be viewed as a continuous safety improvement office. Information from this office should feed back into the other safety offices reporting to the executive director.

12.3.1.7   Safety Promotion

Communicating safety information both internally within the safety oversight organization and to industry is paramount to safety improvements. Lessons learned need to be shared and both groups need to be educated to safety best practices. The two key activities of this office are to

1. Communicate safety awareness, lessons learned, and general safety information
2. Provide or coordinate safety training and education to industry

12.3.1.8   Independent Accident Investigation Board

As mentioned earlier and demonstrated by a lot of independent accident investigations, there is significant merit to creating an independent accident investigation board. It must have the authority and autonomy to fully investigate accidents down to the detailed causal level. It also must be able to protect witnesses so that honest information can be gleamed. This is critical for very large-scale accidents (e.g., commercial airliner crashes) can be investigated and the public assured that the root cause has been determined.

The board must be completely independent from the oversight authority. The Waterfall accident and the Upper Big Branch Mine explosion are important examples of why this is important. Both accidents had significant failures not just with the company that caused the accident but also on the regulatory side. The prime minister or president of the country should appoint the board. It should have full-time staff dedicated to the investigation and should not be seconded from the regulator or organization that had the accident. The final investigation report with suggested corrective actions should be reported directly to the country's chief executive and legislature. The purpose of the board is to

- Independently, and without outside influences or pressures, investigate significant, large-scale accidents
- Not assign blame and liability or carry out prosecutions

12.3.2 EXAMPLE PROCESS SAFETY OVERSIGHT IN THE UNITED STATES

One of the most complex and highest-hazard industries is the petrochemical industry. Though the Union Carbide Bhopal, India, gas leak in 1984 killed thousands and injured over 100,000 and is considered one of the worst industrial accidents in history, unfortunately, history does not end there. Though by some estimates process safety accidents have reduced almost 60% over the last 25 years, the twenty-first century is not immune to terrible large-scale industrial process accidents like a fertilizer plant explosion in Texas, killing 15 in 2013; Deepwater Horizon explosion and oil spill in the Gulf of Mexico in 2010, killing 11 and creating one of the worst environmental disasters ever; the 2005 Texas City Refinery hydrocarbon vapor cloud explosion; the LNG explosion in Skikda, Algeria, in 2004; and Atofina's Grande Paroisse fertilizer plant explosion in Toulouse, France, in 2001. Of course, there are many more and it seems like there is some sort of explosion or serious accident almost every day somewhere around the world. Chapter 6 described process safety analysis techniques that are very effective in managing hazards. But it is worth looking at an example of how industry and government have created good safety oversight and what lessons can be learned. The petrochemical process industry is one of the most hazardous and complex of all industries but also has very good and innovative methods to conduct safety oversight and partner with industry to prevent accidents and major disasters.

The Bhopal disaster was the trigger for the chemical industry to collaboratively develop better tools, technologies, and approaches to process safety management (PSM). Industry leaders through the American Institute of Chemical Engineers (AIChE) founded the Center for Chemical Process Safety (CCPS) in 1985 with 17 charter member companies. The CCPS quickly published the landmark *Guidelines for Hazard Evaluation Procedures* and now has over 100 books and publications, workshops, and training programs on the subject. Most of the world's leading petrochemical, pharmaceutical, and other process manufacturing companies are members.

CCPS promulgates process safety through formal university courses and has created the Safety in Chemical Engineering Education for university engineering programs. Their process safety incident database (PSID) collects, tracks, and shares process safety incidents with its members and serves as a mechanism to communicate lessons learned across their industry. The PSID has over 700 incidents and also tracks near misses, those incidents that never developed into accidents but almost occurred. It contains details of each incident including failure modes, consequences, and suggested design changes.

As described in Chapter 2, one thing that CCPS is not doing is standing still. In its Vision 2020, it has codified five tenets and four societal themes, which govern their focus for the near-term future.

Clearly, the chemical process industry is at the forefront of preventing accidents. They are not waiting for government to issue more regulations or oversight programs—they are working on solving their challenges themselves. But that is not to say that the government has not evolved and is an important partner in preventing accidents.

At about the same time that the CCPS started publishing their *Hazard Evaluation Guidelines*, the U.S. Occupational Health and Safety Administration started upgrading their oversight process as well. The most important is the *Process Safety Management of Highly Hazardous Chemicals* standard (29 CFR 1910.119). This PSM regulation incorporates many of the tools, techniques, and thought processes that the CCPS created. This is a good example of government and industry influencing each other in a very positive way. The PSM regulation focuses on 14 key system safety elements that are as follows (Occupational Safety and Health Administration, 2000):

1. Develop and maintain written safety information identifying workplace chemical and process hazards, equipment used in the processes, and technology used in the processes.
2. Perform a workplace hazard assessment, including, as appropriate, identification of potential sources of accidental releases, identification of any previous release within the facility that had a potential for catastrophic consequences in the workplace, estimation of workplace effects of a range of releases, and estimation of the health and safety effects of such a range on employees.
3. Consult with employees and their representatives on the development and conduct of hazard assessments and the development of chemical accident prevention plans, and provide access to these and other records required under the standard.
4. Establish a system to respond to the workplace hazard assessment findings, which shall address prevention, mitigation, and emergency responses.
5. Review periodically the workplace hazard assessment and response system.
6. Develop and implement written operating procedures for the chemical processes, including procedures for each operating phase, operating limitations, and safety and health considerations.
7. Provide written safety and operating information for employees and employee training in operating procedures, by emphasizing hazards and safe practices that must be developed and made available.

8. Ensure contractors and contract employees are provided with appropriate information and training.
9. Train and educate employees and contractors in emergency response procedures in a manner as comprehensive and effective as that required by the regulation promulgated pursuant to section 126(d) of the Superfund Amendments and Reauthorization Act.
10. Establish a quality assurance program to ensure that initial process-related equipment, maintenance materials, and spare parts are fabricated and installed consistent with design specifications.
11. Establish maintenance systems for critical process-related equipment, including written procedures, employee training, appropriate inspections, and testing of such equipment to ensure ongoing mechanical integrity.
12. Conduct prestartup safety reviews of all newly installed or modified equipment.
13. Establish and implement written procedures managing change to process chemicals, technology, equipment, and facilities.
14. Investigate every incident that results in or could have resulted in a major accident in the workplace, with any findings to be reviewed by operating personnel and modifications made, if appropriate.

Though this might appear as almost a *laundry* list of system safety concepts, they do fit well into a cogent unit as PSM plan, the same type of elements described in the system safety program plan and in general in a good SMS as described in Chapter 4.

In addition, in 1998, the U.S. Chemical Safety Board (CSB) was created. The CSB is designed similar to the National Transportation Safety Board that independently investigates transport accidents. It is an independent U.S. federal agency that is also nonpunitive and makes recommendations to government regulators (OSHA and Environmental Protection Agency in particular, but also others), the chemical process industry, labor unions, trade associations, and other groups. They continuously track and monitor closure of their recommendations—this is very important because it keeps significant pressure on groups to actually implement the recommendations and not ignore them. The Waterfall Special Commission of Inquiry in Australia successfully used a similar approach to ensure that their recommendations were acted upon.

Like the NTSB, the CSB focuses on understanding accident root causes. What is of particular interest is that the CSB sees most accident root causes as a failure in the SMS. This is very important because it recognizes that management (government, company, trade association, union, etc.) has the ultimate power and ability to prevent accidents. Even if a mechanical failure or act of nature was the root cause, a robust SMS should still prevent the accident or mitigate the consequences.

Of course, the great work being done in the U.S. chemical industry is not unique to the United States. Many other countries are doing the exact same thing, regulators and industry working together to forge better approaches to PSM. The Canadian Society for Chemical Engineering promulgates PSM. The National Occupational Health and Safety Commission in Australia do the same, and the UK Health and Safety Executive are equally focused on the issue. Also, we are seeing much more cooperation not just within countries but also transnationally. For example, the EU and OSHA culminated a risk prevention campaign at a summit in Bilbao, Spain, in November 2013. And regular meetings are occurring around the world with industry and government; the largest international gathering of process safety experts is through the Global Congress on Process Safety.

12.4 ALIGNING RESOURCE NEEDS TO THE OVERSIGHT ORGANIZATION: HOW TO DO MORE WITH LESS

Now the issue is how do I do more with less? To be effective, it is imperative to ensure that resources have real and positive impact. Here are some practical tips and best practices.

PRACTICAL TIPS AND BEST PRACTICE

- Resources are not only about the money. They are about being smart in how you align and allocate them. Align resources with priorities—focus on what matters most.
- Resources should be leadership driven. This should not be ad hoc but have clear goals and objectives that are driven directly from leadership. Organize resources around these goals and objectives.
- Of course, people are your most important resource; invest in them regularly and continuously work on upgrading their capabilities. Create networking relationship opportunities within government and industry.
- Conduct training needs assessments to ensure that resources are adequately prepared for their roles. Clearly define, measure, and enforce accountability.
- When aligning resources, think of how people, processes, and technology can work together and enable each other to get extra lift.
- Build support structures (processes and technology) around how people act and think. Do not force-fit people into unnatural situations; otherwise, they will go around the process to do things.
- Bundle, consolidate, and streamline resources where appropriate.
- Try to reduce as much administrative burden, bureaucracies, and system complexities as possible.
- Insight versus oversight. Be technically smart on the industry that you are regulating; you need to be just as smart and experienced as they are.
- Use metrics to identify resource risks, track improvements, and better understand resource impact. Develop and track key performance indicators. Set meaningful benchmarks.
- Practice selective abandonment. If programs do not work, dump them.
- Expect the unexpected. There will be surprises: budgetary, industry accidents, varying political pressures, new technology, etc. Try to view resources as flexible and fungible as appropriate.

12.5   CASE STUDY: US FEDERAL AVIATION ADMINISTRATION, IMPROVING OVERSIGHT THROUGH SYSTEM SAFETY

An interesting example to see how safety can be improved in a highly dynamic industry is to review the U.S. Federal Aviation Administration (FAA) Systems Approach to Safety Oversight (SASO) Program. Though the U.S. airspace is one of the safest in the world, its future is facing many strong pressures and changes:

- Air traffic demand continues to grow sharply, especially with regional jets.
- Aircraft are increasingly incorporating newer and more complex technologies.
- The FAA is upgrading the national airspace through a set of very complex technology programs called the Next Generation Air Transportation System (NextGen). NextGen is a satellite-based and digital technology suite that will significantly change how the national airspace is managed.
- NextGen has to incorporate new technologies *live* into the national airspace while still maintaining its current operating environment safe.
- Currently, the FAA inspects more than 2300 air operators and almost 4900 maintenance and repair facilities. But a large bulk of inspectors is set to retire in the coming years.
- The current safety inspection process is based on a very prescriptive regulatory model.

The FAA recognizes that all of these pressures can have a detrimental impact on air safety so they have decided to better incorporate system safety activities into the current inspection process and migrate from the prescriptive regulatory model toward a more proactive, risk-based self-certification model that uses an SMS as its core. The FAA created the SASO office in 2002 to focus resources on developing and transitioning from a prescriptive program to a more data-supported risk-based oversight program. However, it appears that pieces of the original prescriptive program will not go away. There still will be prescriptive regulations that industry will have to meet, albeit fewer and with more of an SMS approach.

One key area of safety oversight in general (and SMS) is the safety assurance function. The SASO Program (this is a long-term program that will take years to implement) is incorporating a system safety assurance approach into their safety assurance system (SAS). The new SAS emphasizes balancing people, process, and technology to ensure that system safety not only is maintained through all these industry changes but also is forward leaning and improves the aviation industry risk profile. The SAS focuses on these key areas (Federal Aviation Administration, 2013):

- *Business process reengineering*—This entailed mapping current *as-is* processes for conducting inspections as part of the oversight program. Once current processes were well documented and understood, SASO developed the *future state* of incorporating system safety techniques and processes into a *future* state of processes.

- *Systems alignment*—Creates technology tools and processes that support the *future* state of processes.
- *Enterprise architecture*—Integrates the technology and processes across the entire enterprise.
- *Change management and implementation*—Prepares inspectors, other staff, managers, and other offices for the process and technology changes forthcoming. Its focus is on training and education so that all parties understand and support the *future* state.

The SAS program stresses balancing people (change management), process (business process reengineering), and technology (system alignment and enterprise architecture). This enables the FAA to point resources appropriately at higher-risk areas, but still not forget about other areas that could be problematic but still not be considered high risk yet. It is developing into an interesting prescriptive (there still are and probably will remain some prescriptive regulations) management (SMS and incorporating system safety into the SAS uses a more risk-based management approach) model. This new approach will also help the FAA to better meet ICAO SMS requirements.

NOTES FROM NICK'S FILE

One of the key challenges that we faced on the SASO project was the changing demographics balanced with new technologies. We had to figure out how to balance the deep technical depth that much more experienced FAA inspectors had (also, many of them nearing retirement in the next few years) with the younger generation that had much less experience but were very smart with technology. We worked hard to blend the two together and try to get the benefit of both worlds of deep practical experience and using new technologies.

12.6 COMMON MISTAKES IN GOVERNMENT OVERSIGHT PROGRAMS

Of course, all programs, including government programs, can be improved. Experience shows that many government safety oversight programs sometimes have fallen into common pitfalls. Section 12.1 briefly discussed some of the typical gaps to avoid in safety oversight (see Figure 12.2). This section expands and discusses some of the common mistakes in oversight programs. The following section will suggest ways to fix them.

Figure 12.6 illustrates a suggested safety oversight office. Here, Table 12.3 shows a list of things to avoid while managing government safety oversight programs (there are many more than just these here).

Chapter 4 discussed SMSs in detail. A list of 29 key elements that should be included in a company's SMS is also discussed. Table 12.4 illustrates the 10 elements for government oversight of an industry SMS. Note that the elements for

TABLE 12.3
Common Mistakes in Government Safety Oversight Programs

Safety Oversight Office	Common Mistakes
Independent accident investigation board	• Punitive and not blame-free investigation. • Insufficiently independent in stature, funding, and ability to investigate. • Lack of appropriate resources. • Undue outside influences and pressures.
Safety regulator executive director	• Insufficient authority or unclear mandate to efficaciously manage oversight programs. • Inadequate funding and resources. • Does not report to the Cabinet Office (minister). • Lack of transparency across offices and toward industry.
Crisis management	• Unclear roles, tries to manage the crisis (e.g., manage first responders) instead of support resources. • Understaffed and inadequate technology to follow the crisis. • Poor planning and lack of practice exercises. • Inadequate or inappropriate crisis communication to leadership or the public.
Safety policy and regulations	• Legal framework governing regulations is overly complex and too detailed preventing future flexibility. • Safety and other (i.e., economic) regulations are mixed together and not clearly segregated. • Regulations are too broad, vague, and not measurable.
Audit, compliance, and enforcement	• Enforcement escalation is not clear and followed. • Auditors are not adequately trained nor understand the industry. • Audit results are not fed back into the safety accreditation and safety risk management process. • Does not focus adequately on evaluating the safety management system.
Safety accreditation	• Process is overly complex/simplistic. • Accreditation focuses too much on the process and not validating actual safety. • Inadequate surveillance program.
Safety risk management	• Incomplete definition of SMS. • Unclear application requirements. • Poor tracking and trending of safety data. • Industry and government oversight programs not fed back into the regulatory process.
Safety promotion	• Campaigns do not focus on making sure that safety is at the top of agenda for industry and its operations. • Programs are disjointed and do not focus on improving safety culture. • Inadequate understanding or input from industry to appropriately design promotion campaigns.

TABLE 12.4

Review Elements of Government Oversight

Review Elements for Evaluation of Government
Oversight of Industry SMS Programs

1	Regulatory independence
2	Regulatory mandate
3	Policy and objectives
4	Organization and function
5	Data analysis
6	Safety enforcement over industry operator
7	Government accident/incident investigation
8	Government audits
9	Safety accreditation
10	Partnership with industry

---

government oversight have been slightly modified to a generalized approach from one focused strictly on rail safety. This review template was taken from the 2003 Waterfall investigation (Bahr, 2005) and can be used to review government safety oversight programs. This template was very effective in determining some of the key deficiencies in the regulator's oversight of the railway industry at the time of the accident.

REFERENCES

Bahr, N. January 2005. Special commission of inquiry into the Waterfall Rail Accident, final report, Vol. 2, Appendix F. SMS Review Methodology Report. May 12, 2004. Sydney, New South Wales, Australia: Special Commission of Inquiry into the Waterfall Rail Accident.

Center for Chemical Process Safety. No date. *Vision 2020 AIChE, CCPS*. http://www.aiche. org/sites/default/files/docs/pages/vision2020.pdf, downloaded January 20, 2014.

Federal Aviation Administration. 2013. System Approach for Safety Oversight Program (SASO), http://www.faa.gov/about/initiatives/saso/, downloaded May 20, 2014.

Irish Aviation Authority. 2013. State Safety Program, https://www.iaa.ie/state-safety-programme, downloaded October 14, 2013.

McInerney, P. A. 2001. Special commission of inquiry into the Glenbrook Rail Accident, final report. Sydney, New South Wales, Australia: Government of New South Wales.

McInerney, P. A. January 2005. Special commission of inquiry into the Waterfall Rail Accident, final report. Vol. 2. Sydney, New South Wales, Australia: Government of New South Wales: p. xviii.

National Institutes for Occupational Safety and Health. 2012. An Independent Panel Assessment of an Internal Review of MSHA Enforcement Actions, requested by The Honorable Hilda L. Solis, Secretary of U.S. Department of Labor, March 22, 2012. U.S. Department of Labor, Mine Safety and Health Administration. http://www.msha.gov/ PerformanceCoal/NIOSH/Independent%20Assessment%20Panel%20Report%20w_ Errata.pdf, downloaded May 20, 2014.

National Rail Safety Accreditation Package, The Rail Safety Regulations Panel Guideline, Version 2.0, 1999, updated 2005, Rail Safety Regulators Panel.

National Transport Rail Safety Bill Regulations. 2006. Sydney, New South Wales, Australia Government.

Occupational Safety and Health Administration OSHA. 2012. *Process Safety Management.* OSHA 3132 (2000 reprinted) booklet, taken from the website: https://www.osha.gov/ Publications/osha3132.html, downloaded January 20, 2014.

Safe Work Australia. 2013. *How to Determine What Is Reasonably Practicable to Meet a Health and Safety Duty.* May 2013, p. 4, http://www.safeworkaustralia.gov.au/sites/ SWA/about/Publications/Documents/774/Guide-Reasonably-Practicable.pdf, down-loaded May 20, 2014.

Taylor, M. R. 2012. Deputy Commissioner for Foods and Veterinary Medicine, U.S. Food and Drug Administration, speech at *China International Food Safety and Quality Conference and Expo.* Shanghai, China, November 7, 2012.

United Kingdom. 2005. Statutory Instruments of 2005 No. 3117 Offshore Installations. *The Offshore Installations (Safety Case) Regulations 2005.* U.K.

FURTHER READING

Center for Chemical Process Safety. https://www.aiche.org/ccps.

Cooke, R., Ross, H., and Stern, A. January 2011. *Precursor Analysis for Offshore Oil and Gas Drilling: From Prescriptive to Risk-Informed Regulation.* RFF DP 10-62. Washington, DC: Resources for the Future.

International Civil Aviation Authority. 2006. *Safety Oversight Manual. Part A. The Establishment and Management of a State's Safety Oversight System.* Secretary General International Civil Aviation Authority.

Jakhu, R., Sgobba, T., and Dempsey, P. (eds.). 2012. *The Need for an Integrated Regulatory Regime for Aviation and Space: ICAO for Space?* Studies in Space Policy (Book 7). Vienna, Austria: Springer Publishers.

Kerwin, C. and Furlong, S. 2010. *Rulemaking How Government Agencies Write Law and Make Policy*, 4th edn. Washington, DC: CQ Press.

Pelton, J. and Jakhu, R. 2010. *Space Safety Regulations and Standards.* Oxford, U.K.: Butterworth-Heinemann.

Youngblood, R. and Kim, I. June 2005. Issues in formulating performance-based approaches to regulatory oversight of nuclear power plants. *Nuclear Engineering and Technology*, 37 (3):231–244.

13 Risk Assessment

We triumph without glory when we conquer without danger.

Le Cid
Tragic-comedy play first performed in 1637
Pierre Corneille

Absence of the evidence of risk is not evidence of the absence of risk.

U.K. Health and Safety Executive

It's a very sobering feeling to be up in space and realize that one's safety factor was determined by the lowest bidder on a government contract.

First American to travel in space and walked on the Moon as Commander of Apollo 14
Alan Shepard

Whether quantitative or qualitative, risk assessment is much like statistics; it can demonstrate almost any conclusion, depending on the assumptions and the frame of reference.

Risk Researcher
Peter Neumann

Risk is probably one subject we all feel that we understand yet admit that we know nothing about. We face risks every moment of our lives. If you choose to walk instead of driving to work because it is too risky, you still must face the risk of being hit by a car while crossing the street or even tripping over the curb and breaking an arm. And, of course, risk does not necessarily have to be related directly to safety. One example is the risk of your financial portfolio. But understanding the nature of risk, how to measure it, how to evaluate it, and how to respond to the results is very important to efficiently make our systems as safe as possible. The remaining chapters in this book discuss risk in the context of safety to systems, people, and the environment.

There are a great number of individuals studying risk. The current research is very interesting because in the last number of years, professionals from very diverse backgrounds have started to come together and redefine the meaning of risk. Engineers seeing risk from the technological point of view are mixing with sociologists who look at society's perception of risk. Psychologists are involved in studying how the cognitive processes analyze risk. Biologists, chemists, epidemiologists, and other medical doctors are studying how toxins from various sources (smoking, air pollution, or a chemical spill) affect the body and the mortality risk. In fact, not only have they started working on risk as part of interdisciplinary teams, it now seems to be the norm.

Environmentalists are studying how toxic releases affect the environment. Government regulatory agencies are involved with how the public reacts to risk information. Economists study how countries take risks and the economic effects of those risks. Politicians are asking why we cannot have zero risk. Philosophers debate the taxonomy of risk. Even international organizations are all looking at risk, for example, ISO 31000 is a family of risk management standards. And, of course, we cannot forget insurance companies; they were among the first organizations to take the concept of risk seriously and attempt to define and study it.

NOTES FROM NICK'S FILE

One thing I have found is be very careful with terminology, especially in an international context. I have worked around the world for many years and have found that I have to be careful how I use terms, taking into consideration their translations into other languages. For example, in many languages, safety and security are the same word. And many people tend to loosely use words like risk, hazard, safety, security, etc., as if they mean the same thing. For example, I was working in various South American countries and spent time making sure that Spanish, Portuguese, and English speakers all understood and agreed on the basic words and risk concepts before we started the project.

13.1 WHAT IS RISK?

Chapter 2 defined *risk as the severity of the consequences of an accident times the probability of its occurrence*. In reality, the concept of risk is much deeper and needs to be discussed. Many different factors go into defining risk. There are also many different kinds of risk. And strangest of all, there are many different perceptions of risk. One of the first to try to pull all these different ideas together was Chauncey Starr.

In 1969, Chauncey Starr published in *Science*, "Social Benefit versus Technological Risk." His seminal article brought forth the idea that the traditional engineering management method of cost–benefit risk analysis—equating risk to monetary return—was not sufficient to accurately determine the risk of technological systems. Starr argued that it is possible to quantify, to some extent, societal perceptions of risk and that these too should be included in the risk evaluation equation.

Starr said that people judge risk based on whether it is a risk they voluntarily take or one that is thrust upon them. For example, many people mistakenly believe that riding in a car is less risky than flying on a commercial airliner. In fact, the risk is the same under certain conditions (see Table 13.1). But because people feel that they are controlling their destinies in a car, they feel that they can avert a disaster, whereas on an airliner, it would be impossible for a passenger to prevent a crash.

As engineers, it is counterintuitive that calculating the risk and presenting the data are not sufficient. Many feel that an increased risk of death of one in a million is acceptable to the population. In Table 13.1, Wilson (1979) calculates and compares a variety of activities that increase the risk of death in any year by one in a million. All activities carry approximately the same amount of risk.

TABLE 13.1
Risk Comparison Chart

Risks that increase chance of death by 0.000001 (or 1 part in 1 million)

Smoking 1.4 cigarettes	Cancer, heart disease
Drinking 1/2 L of wine	Cirrhosis of the liver
Spending 1 h in a coal mine	Black lung disease
Spending 3 h in a coal mine	Accident
Traveling 300 miles by car	Accident
Traveling 10 miles by bicycle	Accident
Flying 1000 miles by jet	Accident
Flying 6000 miles by jet	Cancer caused by cosmic radiation
Living 2 months in Denver on vacation from New York	Cancer caused by cosmic radiation
Living 2 months in average stone or brick building	Cancer caused by natural radioactivity
Living 2 months with a cigarette smoker	Cancer, heart disease
Eating 40 tablespoons of peanut butter	Liver cancer caused by aflatoxin B
Living 150 years within 20 miles of a nuclear power plant	Cancer caused by radiation
Eating 100 charcoal-broiled steaks	Cancer from benzopyrene
Risk of accident by living within 5 miles of a nuclear reactor for 50 years	Cancer caused by radiation

Source: Wilson, R., *Technol. Rev.*, 81(4), 41–46, February 1979. Abridged chart reprinted with permission from *Technology Review*, Copyright 1979.

If we take the first entry in the table, smoking, we can calculate the risk of death in 1 year. Of course the risk of smoking has changed since Wilson published with fewer people smoking and more people living longer. Just the same, his calculation is illustrative. Using 1975 data, Wilson quickly calculates

675 billion cigarettes produced per year = 3000 cigarettes per person

(including children)

Fifteen percent of the American population dies of smoking-related disease per year (which equals an average lifetime risk = 0.15). Taking the 0.15 annual risk to a typical 70-year life span gives 0.002 (or 0.15/70), and again, dividing the number of cigarettes per person, per year, gives an annual risk of dying from smoking of

0.7×10^{-6}(deaths/year/cigarette) or risk of death per cigarette

The inverse of this number gives the number of cigarettes smoked to increase the risk of death by one in a million or

1.4 cigarettes / 1×10^6

For the average engineer, this may seem to be an esoteric exercise, but Starr's work opened the door for engineers to take this information and better design their systems. Remember the products we build are sold to the public; if we do not understand the public's concerns, we will have difficulty convincing people to buy what we build.

A clear example is the commercial nuclear power industry in the United States. In the 1960s and early 1970s, nuclear power looked like a very promising career and an incredible growth industry. Nuclear power was to be our cheap form of energy. But the industry was stymied by the American public's perception that nuclear power is an unwarranted risk. And the 1979 Three Mile Island nuclear accident sealed its fate. Now commercial nuclear power in the United States is a relatively small industry in comparison to other forms of generating power. And yet, in France, 40% of the electric power is generated by nuclear reactors.

The Fukushima Daiichi nuclear disaster has reopened the debate, just when it looked like commercial nuclear power was *coming back online*. In the United States, the Energy Policy Act of 2005 spurred 25 license applications in a few short years. But immediately after Fukushima, Germany announced that it would shut down all its nuclear reactors by 2022. And, in Japan where the Fukushima accident occurred, the government vacillates between reopening its nuclear power plants (all were taken off-line for safety testing immediately after the Fukushima accident) and being nuclear free. The public has reacted negatively due to their concern (and perception) of the safety risk. It is clear how we perceive risk greatly impacts how we accept risk.

A similar debate is raging over the siting of hazardous waste incineration facilities, though this time it seems that the engineers are more sensitive to the public's perception of the risk to their health and safety. All industries must confront this issue, from chemical plants to mass transit, aerospace, and manufacturing plants.

Rowe (1988) defines *risk as* "the potential for realization of unwanted, negative consequences of an event. Risk aversion is action taken to control risk." This sounds a lot like a definition for safety. What is relatively difficult to notice is that risk does not mean the same thing to everyone. So in defining the potential for the realization of unwanted negative consequences, that potential may vary from person to person. This is one point engineers must understand: Because you think the risk of an accident is 1×10^{-9}, the public may not agree that this is acceptable or even understand what that means. That assumption is exactly what led to the demise of the commercial nuclear industry in the United States. Even the black swan event in Japan, something that Germany will not face (the same black swan event, earthquake and tsunami), impacts Germany's perception and acceptance of risk too. This means that you must define what risk you are willing to accept before you start to work.

To comprehend risk, we must understand that we can have an unsafe event and the risk could still be low. Remember if you lower the probability of occurrence to a very small number, then even a catastrophic event (e.g., radioactive fallout from a nuclear power plant) can still be a low risk to the public. That brings the real challenge to engineers, which is to decide how safe is safe enough.

One of the problems of safety engineering is that safety analysis many times identifies the worst-case scenario. The risk analysis process is a way to optimize a safety analysis and look more closely at the problem. Probabilistic risk analysis gives us a way to quantify the process. Kaplan and Garrick (1981) succinctly define risk as three questions:

1. What can go wrong that could lead to an exposure to a hazard?
2. How likely is this to happen?
3. If it happens, what are the consequences?

Or we could further define *risk* as the triplet: *event scenario, probability of occurrence*, and *consequence*. *Event scenario* is a description of the event under study. In safety parlance, what are the hazards? *Probability of occurrence* is our quantification of the likelihood that the event will occur. *Consequence* is defined as the severity of the event.

As engineers, going from the esoteric back to the practical, we want to know how to manage the actual risks in our system. How do we allocate resources? The risk assessment process addresses this issue. Before this can be done, however, here are a few thoughts on risk perception.

13.2 RISK PERCEPTION

Montooth (1984) stated in *Public Epidemiology*: "The Tylenol killings claimed seven lives … a one-time event, over a 1-week period. The public was outraged. The pharmaceutical industry has … re-tooled at [a cost of almost] $225 million. Annually, 50,000 people die on our highways. About half of those deaths are preventable by inexpensive means, which are already at hand, but complacency prevails. Risk perceptions … are reactions, are not rational."

Much research has been conducted in trying to understand people's concept of risk. What makes one individual (or country) take more risks? Why do people feel that the risk of dying in a car crash is lower than the risk of flying in a commercial jetliner? Slovic et al. (1979) have defined a number of factors that affect the way people perceive a risk. These factors have already become standard bearers for risk perception. Not everyone may agree with the risk perception factors. For some, involuntary risks are not important; for others, they may carry much weight.

One factor is whether the risk is voluntary or not. The siting of a hazardous waste facility in someone's town would be considered involuntary and therefore could be perceived to carry more risk. People feel that risks faced voluntarily are well known and controlled. For example, the risk of skydiving could be considered low to the individual who does it because he chooses to.

Another important factor in perceiving risk is whether the consequences of an accident are chronic or catastrophic. Perceived catastrophic consequences raise the risk concern. Some people feel that flying in an airplane is riskier than driving in a car because more people die at one time when there is an airplane crash.

If the risk is dreaded as opposed to common, then many times it is viewed as worse. Radioactive fallout from a nuclear power plant in the event of an accident is dreaded much more than if a coal-fired plant had an accident (which is perceived to be more common).

Another factor is whether the fate is certain to be fatal or not certain to be fatal. Handling a gun is considered very risky because an error will most likely result in death, but people feel that the risk of dying of food poisoning is lower because death is not certain.

If the people exposed know the risk, it is ranked lower than if people feel that they are being exposed to a risk without their knowledge. The alar used on apples years ago is a good example. People were outraged because they were exposed to a danger without their knowledge.

If the danger is immediate and not delayed, many people feel that the risk is greater. That is one way in which people rationalize smoking. They know that they are facing a much higher risk of dying from smoking-related diseases, but the effects will not be felt until much later. Interestingly, one of the reasons that there is a significant decrease in smoking is that the public has become aware (through media) of the dangers and risks to secondhand smoke. This risk education to the public pushed politicians to greatly control where and how people can smoke.

If the technology is controllable, then people feel that the risk is lower. If it is difficult to control the technology or event, then people feel that they are in more danger. The controllability factor pertains to whether an individual feels that he or she can personally control the risk (e.g., driving a car).

The last factor is whether the risk is new or old. The newer it is, the riskier it is perceived to be. People feel a certain comfort level with the old. People feel that chemical plants pose a smaller risk to their health than nuclear plants, yet the number of accidents at chemical plants is much greater than at nuclear plants.

13.3   RISK ASSESSMENT METHODOLOGY

As a reaction to a string of commercial nuclear reactor accidents, the U.S. Nuclear Regulatory Commission published a reactor safety study, popularly known as WASH-1400 (U.S. Nuclear Regulatory Commission, 1975). The study laid formal groundwork for conducting probabilistic risk assessments in the commercial nuclear industry. The assessment technique has been used in numerous industries including automobiles, food safety, environmental, petrochemicals, and aerospace.

Risk assessment is the formal process of calculating the risk of an event and making a decision on how to react to that risk. This risk assessment process (Bahr, 1993) is a useful technique for conducting a probabilistic risk assessment consistent with the U.S. Nuclear Regulatory Commission WASH-1400 reactor study.

There are many risk assessment concepts in existence. Some look at security risk assessments—how adversaries, with intent and ability, can create risk. Others look at risks in a very quantitative fashion, such as human health and toxic chemical exposure. The UK rail sector uses a quantitative approach to defining a tolerable level of risk to the public from a rail accident mixed with a qualitative safety risk assessment based on hazard analyses. Some people are looking at risk from a business perspective and will focus on financial risk (credit and liquidity risk in financial institutions). And even others will look at the entire risk to an enterprise or a business and evaluate enterprise risk—risk across the entire business enterprise. This approach will be very comprehensive and not just consider financial or technical (engineering)

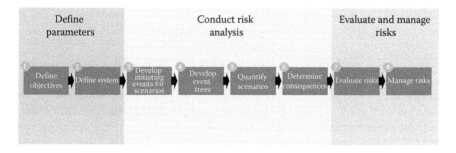

FIGURE 13.1 Risk assessment methodology.

risk but also reputational (to the company), brand, and legal. Chapter 14 illustrates a quantitative or probabilistic risk assessment example.

As Figure 13.1 illustrates, the risk assessment methodology is straightforward and very similar to the method shown in Figure 2.2 for the system safety process. This is appropriate, because the system safety process is part of the overall risk management system. Here, we are going to look just at the aspect of risk assessment. The risk assessment methodology uses many of the system safety techniques discussed in Chapter 5 to identify and analyze the risks to the system. Those risks are quantified and their consequences determined. Then the risk is evaluated, and from that information, the risk manager is able to decide whether it is justifiable to accept the risk.

NOTES FROM NICK'S FILE

One of the most frequent and biggest mistakes that I have seen in risk assessments is that many people start in the middle. They do not spend enough time defining objectives and defining the system. They just jump to event scenarios. This is very risky (no pun intended) because without defining these parameters first, it becomes very easy to forget or leave out key risks.

The first step in the risk assessment methodology is to define the objectives of the assessment. This always seems to be a trivial matter, yet many engineers do a poor job of it and waste precious time, resources, sometimes even getting inconclusive results. It is not uncommon to find that many completely skip this step—do not do that.

In defining the assessment objectives, you should clearly state the objectives, scope, purpose, and damage states that are of interest. This is the bounding of the problem. If the objectives and purpose are unclear, then the assessment will be unclear. It is already difficult enough to get non-engineers to read technical reports and agree with engineering assessments; you should do everything possible not to make the process more difficult. Even if the report is going to other engineers, if it is not clear, it will be difficult to convince them of its merits.

This step is also where risk perception can be taken into account. It will also appear again later in the risk management process, but it is impossible to manage something if you do not know first what you wish to manage.

The typical damage states that would be of interest are catastrophic, critical, minor, and negligible. Tables 5.2 through 5.4 define these terms in risk assessment terminology.

Step 2 in risk assessment is to define the system of interest. Of course, it is important to look at the total system. This does not necessarily mean that you must study the entire plant, but if your concern is one section of the plant, then it should be viewed from a system's point of view. In other words, understand the human–machine interfaces, the environmental conditions, the operating conditions, the mechanical and electrical designs, and any other important aspects. Also, do not forget the organizational issues.

PRACTICAL TIPS AND BEST PRACTICE

A very good thumbnail way to remember to consider the entire system as discussed in Chapters 2 and 5 is to think of these four considerations:

- *People*—What kind of people issues or human characteristics do you need to consider, organizationally, governance, decision rights, human factors, training, awareness, etc.?
- *Process*—What are the processes, procedures, work activities, regulations, and business processes in place that impact the system?
- *Technology*—What are the technologies involved, IT, instrumentation, etc.?
- *Infrastructure*—What are the physical facilities, layout, plant location, etc., and how do they impact the system?

The next step is to identify initiating events for scenarios. These are the events of interest. They are identified and studied by any one, or a combination, of the system safety analysis techniques discussed in Chapters 5 through 9. For example, a hazard analysis can identify the events that present the hazards of most concern. A fault tree can further refine how the event could occur, and failure mode and effect analysis (FMEA) can give specific failure information about particular components that led to that event.

From this, functional event trees are developed that describe these hazard event scenarios. The event trees show the relationship between initiation of the hazard event and the safeguards (or hazard controls) that must be challenged.

The scenarios are quantified by calculating the failure probability of each event. Having quantified the system fault tree becomes very useful at this point.

Step 6 determines the consequences. This is really a two-part step: a qualitative determination and a quantitative calculation.

The damage states are analyzed in a qualitative fashion, as discussed in Section 5.1. The same damage states can be quantified in numerous ways, depending on the kinds of results you seek. A typical cost–benefit analysis can be conducted in which an accident is equated with dollars lost. Or you could look at number of people killed or injured, measurable damage to the environment, or even percent loss of productivity.

In the risk evaluation, the risks are calculated, and risk profiles—used to compare the different risks—are generated. The purpose of the risk profiles is to understand better which scenarios are relatively riskier. This will be very important for the risk management.

Again, any risk perception concerns identified in step 1 are applied to the risk management. This is one point many engineers do not want to face. As much as someone may dislike the politics of work, or the consequences of the product in the public's eye, it is very important. These issues can help a product sell well, or it can drive it out of the market. Risk management is the systematic decision-making processing of deciding how to disposition the risks that have been identified in the risk assessment.

13.4 IDENTIFYING RISK IN A SYSTEM

Chapters 5 through 9 are the best sources for tools to identity risk in the system. Once the system is defined, developing a quick preliminary hazard list will detect the gross hazards of concern to the system. The hazard analysis further refines the hazard list and clearly recognizes which hazards are of greatest concern. Also, the Hazard Risk Index is a good qualitative tool with which to note some of the qualitative risks that are required in step 6 of the risk assessment methodology.

HAZOP and what-if/safety checklists perform the same function as the hazard analysis in the risk assessment process.

Having done this, the fault tree is extremely useful, especially if it has been quantified. Fault trees are typical inputs to event trees in the nuclear power industry. There is no reason that other industries cannot do the same.

Later in the risk assessment process, once the scenarios of particular interest are clear, FMEA is a very powerful tool for focusing on which component is the trigger in the event and how to make it more robust in the system. The same is true of sneak circuit analysis, cause–consequence analysis, or dispersion modeling.

Chapter 5 discusses tracking hazards to closure. A sample facility hazard analysis template is shown. Many safety and risk experts will use the term risk register. It is essentially the same thing. Table 13.2 illustrates an example risk register.

13.5 RISK COMMUNICATION

Rayner and Cantor (1987) conducted a very interesting pilot study in risk perception. They looked at how three different constituencies—the utilities, state public utility commissions, and public-interest groups critical of nuclear power—view the risk of siting a new nuclear power technology. Rayner and Cantor found that the various groups' concerns about the risks were fundamentally different. The utilities viewed risk as the rate of return on their investment. The utilities felt that the engineers were competent in managing the technology. The state public utility commissions, on the other hand, were more concerned about how the technology would fare economically over the years. They felt that the public was concerned more about the cost of the power than about the safety of the delivery system. They felt it was safe by virtue of the licensing process. The public-interest groups, however, were concerned

TABLE 13.2

Risk Register

					Risk Register Tool								
		Hazard Identification			**Hazard Rating (Preresolution)**			**Hazard Mitigations**		**Hazard Rating (Postresolution)**			
Seq. No. (1)	Control Number (2)	Potential Hazard (3)	Potential Cause (4)	Effects (5)	Hazard Category (6)	Freq. (7)	Hazard Risk Index (8)	Current Safety Assurance Provisions (9)	Recommendations for Safety Controls (10)	Hazard Category (11)	Freq. (12)	Hazard Risk Index (13)	Mitigation Status (Open/Closed) (14)
1. Permanent way													
		Train derailment or collision between trains	Loss of points setting capability due to adverse environmental conditions, especially debris on points	Loss of life or severe injury. Service interruption/cessation.	1	1	1	Daily inspection of tracks to identify debris buildup.	Consider remote surveillance capability to verify points are clean.				
2. Rolling stock													
		Gap between coaches	Inadequate provisions to prevent person from falling between coaches	Person falls between coaches causing severe injury or death.	1	2	2	No controls were identified.	No current recommendations. However, it is noted that rolling stock is now fitted with curtains between cars that provide some protection against this hazard.	I	D	2	

3. Train operation, signaling, and block working

Possible collision or derailment	Train speed control failure, braking distance envelop encroached	Collision between two trains. Severe casualties.	1	2	2	Driver can apply emergency break.	I	E	3

4. Telecommunications

Inadequate audio announcement	Speaker or electronic failure	Patrons unable to understand announcements in station resulting in confusion or inefficient egress/ingress.	1	2	2	Preexisting approved design utilized tested and reliable components.	III	D	3

primarily with the safety and health aspects of the new technology. They felt that the technology was being imposed on the community and that the risk was unacceptable.

The utilities felt that technical managers would make good decisions and that the technical risk of an accident was minimal. The public utility commissions felt that the process was acceptable and would work properly to control any technical risks. The public-interest groups felt that only a collective agreement from the community was acceptable with regard to accepting the risk.

What is fascinating about this study is that all three groups were looking at the same phenomenon. Many lived in the same area. How you communicate the results of the risk assessment to engineering management, the public, or another company is crucial to winning acceptance of the project. As the public becomes more politically aware of technology and is inundated with conflicting information, the need for appropriate methods to communicate risk information becomes paramount. And it is virtually guaranteed that your public project will be discussed, dissected, criticized, debated, and opined on the Internet.

Accepting the premise that people view risk differently, and therefore behave differently, the question is how you can communicate with the public in ways the public can understand, without compromising the scientific method.

The first step is to take to heart the information presented in Section 13.2. Many people become engineers precisely because they do not want to deal with these types of issues, but the reality is that they must if they wish to succeed.

Sandman (1985) makes two points that will facilitate communicating risk assessment information to the public. The first point is to accept and publicly acknowledge that the community is important and does have significant power. Many times, companies do not recognize the will of the community. Repeatedly this has created problems and stopped the project cold. This is important for a chemical plant or for someone selling airbags to auto manufacturers. Politicians have felt the public's ire on Election Day when the city or state did not respond quickly enough to a recent disaster.

The second point is to be very careful not to imply that any opposition from the community to the technology is irrational or just ignorant. No one likes to be called stupid. The public's distrust of science is not always unfounded. Repeated scandals and the callous manipulation of data by some have given us a very cynical body politic.

Sandman goes on to suggest a few methods that are very good:

- Participate with the community to study, measure, and discuss the issue.
- Use the appropriate communication with the appropriate group (as in the case of the pilot project mentioned previously).
- Make consultation with the community required as part of the process.
- Involve the community in direct negotiations to meet its concerns.
- Be as open with your information as possible but understand that the community will want independent assessment.

Going back to the risk assessment process in Figure 13.1, it does make sense to try to communicate with your stakeholders during each stage of the risk assessment process.

Because of the deep public reaction to nuclear power, the commercial nuclear power industry has been very adept at reaching out to the public for every step in the process.

PRACTICAL TIPS AND BEST PRACTICE

Look at all your stakeholders in your community:

- Reach to each group of stakeholders and reach to each one for each step of the risk assessment process.
- Try to get true buy-in for each step so that they feel that they are not just part of the process but also own it with you.
- Actually listen to what they say and explain and document why you cannot implement their recommendations (if you cannot). Do not leave stakeholders hanging not knowing if you even considered their thoughts on the matter.
- Spend time explaining the costs and benefits and resultant risk reduction of their options. They need to understand that what they may be recommending is cost-prohibitive even if it makes the system safer.
- Be open and honest and clearly state what you know and do not know.
- Look at Chapters 4 and 11 for more ideas.

NOTES FROM NICK'S FILE

I was working a project in which we were rolling out a new risk management regulation to the water sector. This particular approach had never been done before and was brand new to the industry. We were very successful in getting industry buy-in because we had constant and persistent outreach to all stakeholders from the very beginning. We did this through group workshops and separate private meetings. They gave us feedback, and we incorporated what made sense and told them why we could not accommodate other suggestions. They even recommended a different title for the regulation—which we adopted. It was a great success and served us as a model for other regulations.

One last and very important point to remember is that the risk assessment process must be part of the safety management system.

REFERENCES

Bahr, N. J. 1993. Risk assessment of a cryogenic subsystem for a NASA shuttle payload. *American Society of Mechanical Engineers Winter Annual Meeting*, New York, SERA-Vol. 1, pp. 1–20.

Kaplan, S. and Garrick, J. 1981. On the quantitative definition of risk. *Risk Analysis*, 1 (1): 11–27.

Montooth, R. L. 1984. Quotes. As quoted in *Hazard Prevention*, September/October p. 13. Reproduced from *Public Epidemiology*, 18 (12).

Rayner, S. and Cantor, R. 1987. How fair is safe enough? The cultural approach to societal technology choice. *Risk Analysis*, 7 (1):3–9.

Rowe, W. D. 1988. *An Anatomy of Risk*. Malabar, FL: Robert E. Krieger Publishing, 24.

Sandman, P. M. 1985. Getting to maybe: Some communications aspects of siting hazardous waste facilities. *Seton Hall Legislative Journal*, 9:442–465.

Slovic, P., Fischhoff, B., and Lichtenstein, S. 1979. Rating the risks. *Environment*, 21 (3):14–20, 36–39.

Starr, C. 1969. Social Benefit versus Technological Risk. *Science*, 165:1232–1238.

U.S. Nuclear Regulatory Commission. 1975. *Reactor Safety Study–An Assessment of Accident Risks in U.S. Commercial Nuclear Power Plants*. WASH-1400. Washington, DC: U.S. Nuclear Regulatory Commission.

Wilson, R. 1979. Analyzing the daily risks of life. *Technology Review*, 81 (4):41–46.

FURTHER READING

Beck, U. 1992. *Risk Society: Towards a New Modernity*. London, U.K.: SAGE Publications.

Bernstein, P. 1998. *Against the Gods: The Remarkable Story of Risk*. New York: Wiley.

Fischhoff, B. and Kadvany, J. 2011. *Risk: A Very Short Introduction*. Oxford, U.K.: Oxford University Press.

Fischhoff, B., Slovic, S., Derby, S. L., and Keeney, R. L. 1984. *Acceptable Risk*. New York: Cambridge University Press.

Fischhoff, B., Watson, S. R., and Hope, C. 1984. Defining risk. *Policy Sciences*, 17:123–139.

Frank, M. 2008. *Choosing Safety: A Guide to Using Probabilistic Risk Assessment and Decision Analysis in Complex, High-Consequence Systems*. Washington, DC: Routledge.

Graham, J. D., Green, L. C., and Roberts, M. J. 1991. *In Search of Safety: Chemicals and Cancer Risk*. Cambridge, MA: Harvard University Press.

Henley, E. J. and Kumamoto, H. 2000. *Probabilistic Risk Assessment and Management for Engineers and Scientists*. New York: Wiley-IEEE Press.

McCormick, N. J. 1981. *Reliability and Risk Analysis: Methods and Nuclear Power Applications*. London, U.K.: Academic Press.

Ostrom, L. and Wilhelmsen, C. 2012. *Risk Assessment: Tools, Techniques, and Their Applications*. Hoboken, NJ: Wiley.

Otway, H. J., Phaner, P. D., and Linnerooth, J. 1975. *Social Values in Risk Acceptance*. Laxenburg, Austria: International Institute for Applied Systems Analysis.

Rowe, W. D. 1988. *An Anatomy of Risk*. Malabar, FL: Robert E. Krieger Publishing.

14 Risk Evaluation

When you can measure what you are speaking about and express it in numbers, you know something about it.

Lecture on "Electrical Units of Measurement" 1883
Lord Kelvin

A witty statesman said, you might prove anything by figures.

Essay on Chartism, 1839
Thomas Carlyle

The future is no more uncertain than the present.

Song of the Broad-Axe, from *Leaves of Grass*, 1855
Walt Whitman

Nothing from nothing ever yet was born.

On the Nature of Things
Lucretius, c 58 BCE

The method of nature: who could ever analyze it?

Nature: Addresses and Lectures, 1841
Ralph Waldo Emerson

It is not certain that everything is uncertain.

Pensées, Section VI, The Philosophers, 1669
Pascal

Chapter 13 discussed the concept of risk and risk assessment. This chapter will focus on how to conduct a risk assessment, especially emphasizing the risk evaluation process. As demonstrated in Chapter 13, risk assessment has grown immensely in the last few years. For a long time, most engineers outside the narrow field of nuclear energy have had very little contact with risk assessment; this has now changed. Engineers are seeing risk assessment applied to almost every industry.

As we all know, international trade has increased incredibly in the last few decades with the advent of the European common market (especially the Maastricht Treaty), the North American Free Trade Agreement, various regional trade agreements in Asia, the common market in South America (Mercosur), and various other smaller agreements. With the large cross border traffic and even the internationalization of the engineering profession, risk assessment becomes critical to assuring a safe and cost-effective operation.

Now, so many accidents no longer affect only one country or community, but many. To be sure that our systems are designing safety into the process, the application of risk assessments becomes paramount. And the risk assessment process is a very good structure in which to put the safety analysis tools discussed in Chapters 5 through 9. To complete the process, the safety analysis results need to be put into a context that engineers can use to make good design and operations decisions.

The risk assessment process assists engineers in determining how to allocate resources to fix the problems that the safety analysis has discovered. With risk assessments, engineers are much more capable of rating and ranking hazards and using that information to decide whether it makes sense to make changes and what those changes should be.

Risk assessment is the entire process. *Risk evaluation* is taking that information and using it to make decisions. Because so many people use the term *risk evaluation* loosely to mean the studying and the decision-making process, this chapter takes the generic term *risk evaluation* to indicate steps 2 through 8 (see Chapter 13) of the risk assessment process.

14.1   PROBABILISTIC APPROACH

The basis of quantitative risk assessments and risk evaluations is probability. Probabilities of equipment failures and human errors are fed into the risk assessment process. Quantitative risk assessment (called probabilistic safety assessment in Europe) depends very heavily on this.

Of course, the problem is the danger of quantifying information that is not easily quantifiable. When data are too ambiguous or nebulous to be of utility, you can use a *pseudo-quantification* approach such as the hazard risk index. Risk assessment is based on the numerical comparison between risk events. Part of the risk assessment process is to compare, contrast, and evaluate risks. This risk evaluation becomes a ranking and comparison process and uses probability differences (matched with other factors such as severity of consequences) to do it.

Before applying the probability numbers and calculating a risk, review the information presented in Chapter 10. It is crucial that you fully understand the data you are using, their sources, and their limitations. Most data you find will not be in the appropriate form you need to do a risk calculation. By studying where the data came from, you can adjust the numbers up or down, depending on other factors (environmental, operational, component application, etc.).

You should follow all the rules of probability theory in manipulating the data. Do not forget that you can use expert judgment and apply Bayesian updating to give you more accurate numbers.

14.2   RISK ANALYSIS MODEL

In conducting a risk evaluation, we must first develop a risk analysis model. This model is a representation of the system, its protective systems (to prevent an accident), the accident scenarios we are worried about, various events that could lead to a dangerous situation, and finally the various potential outcomes of each event possibility.

This model should look at all aspects of the system, its design, failure history, and current operations and then identify scenarios of concern. These scenarios of concern typically are the hazards identified in the system.

PRACTICAL TIPS AND BEST PRACTICE

Estimates should be as realistic as possible, not superconservative. If you over- or underestimate, you will not be able to optimize the risk. Obviously, a risk assessment is very different from a worst-case analysis.

With scenarios developed, the model looks at how the system responds to the various scenarios. It focuses on how well the system can respond to the scenarios. And with those system responses are the resulting consequences.

The model then puts the information together into a format that can be evaluated. These risk profiles help to give an understanding of how the risks compare to each other. This is a very important step because it is what will be used to determine whether the hazard is worth controlling. For example, even if the hazard is catastrophic, such as an airliner crash, giving everyone a parachute may not significantly lower the risk of death. This is because it would be almost impossible to get everyone out of the plane before it hits the ground. In order to arrive at that conclusion, however, you must go through the process of evaluating the different risk scenarios. And only through evaluating those different risks in detail will you be able to manage them.

14.2.1 DEVELOPING ACCIDENT SCENARIOS AND INITIATING EVENTS

First, you need to identify the accident scenarios of concern and the initiating events that lead to that accident scenario. An initiating event is an event that triggers departure from the normal (preplanned) operations and desired mission.

Accident scenarios are identified through the safety analysis tools presented in Chapters 5 through 9. It is particularly important that this step be performed well. If inappropriate accident scenarios are developed, then the risk assessment is pointless since you will not have first identified what is important.

Use a preliminary hazard list as part of the brainstorming process. Appendices A and B are very good jumping-off points. With this information, a hazard analysis can be performed on the system. Any hazards of particular concern to the public or workers should be noted. These hazards are the event scenarios that we wish to manage. Some typical hazards are inadvertent release of chlorine to the environment, train collision, loss of in-flight aircraft control, explosion and fire from boiler rupture, or breach of oil tanker cargo hold, releasing large amounts of oil to the environment.

A HAZOP can be used in lieu of a hazard analysis. It is important to apply a safety technique that is fairly comprehensive in identifying hazards and their causes.

Once the hazard scenarios have been identified, you will need to understand how those hazards came about. If the hazard analysis is not sufficient to understand the underlying causes of the hazard scenario, then further analysis will be necessary. A fault tree is particularly useful because it gives a good sequence of events that lead to the hazard scenario. Failure modes and effects analyses (FMEAs) are also very good for identifying the failure causes.

You will need to understand which particular events led to the event scenario. For example, it is important to identify the sequence of events or failures that could lead to a loss of in-flight aircraft control. Or, for example, the event could be that a blocked valve causes system overpressure. If humans are in the process, then some sort of human factors safety analysis will help find human error causes.

This information must be quantified, even if the estimates are rough. The probability of occurrence and the severity of the scenario need to be determined. This together is called the triggering or initiating event.

Initiating events create the hazard scenario (assuming all the safety systems have been compromised, which will be studied in the event trees) and have a probability of occurrence attached to them. Remember that an initiating event itself is not always negative; it is the consequences of the initiating event that are hazardous.

14.2.2 EVENT TREES

Each of the initiating events is studied. Review system information to determine what barriers exist that could prevent the initiating event from occurring or that might mitigate some of its effects. An example is a relief valve. If system overpressure is the hazard scenario, one initiating event could be an overspeed pump motor. A relief valve downstream of the pump is a barrier to system overpressurization. Back *emf* on the motor, preventing overspeed, is another barrier to system failure. An additional barrier to system failure is pressure-sensing devices downstream of the pump, shutting down the motor if a certain pressure increase is noted.

The *hazard analysis controls* column is very useful in identifying the kinds of barriers that exist to prevent the hazard or mitigate the results. This part of the analysis is also one of the most difficult. The probability of breaching the barrier must be estimated. If one of the barriers is proper operation of relief valve, then you need to determine the probability that the relief valve will not operate. Various failure probabilities could be tied to this one barrier. Obviously, you should choose the most likely failure probability.

The event tree also illustrates each scenario's final damage states. The typical damage states are as follows: I = catastrophic, II = critical, III = minor, and IV = negligible. Of course, you will need to define what those terms really mean. Catastrophic could mean that people are killed or over 100 lb of ammonia is discharged to the environment or even the plane flies with only one engine. The hazard risk indices used in Chapter 5 are good damage state denominations.

The consequences part of the event tree is a more quantitative estimation of the final damage states. You can use the damage states in a qualitative way (even using quantitative data to develop the qualitative damage state) resulting from the accident scenario. Consequences can be divided into two parts: dollar value and risk expectation value. The *dollar value* is the amount of dollars lost if the damage state is reached. For example, if the ship sinks, over $20 million is lost. The *risk expectation value* is the probability of the damage state occurring times the dollar value of the damage state.

Figure 14.1 is an example of what one event tree might look like. The initiating events, barriers, damage states, and consequences are shown at the top of the figure.

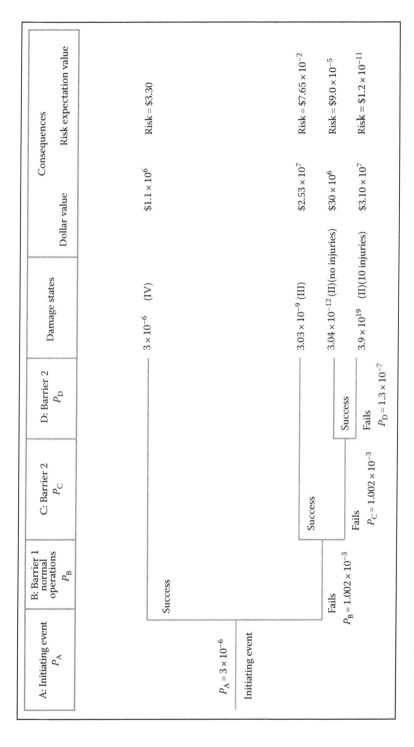

FIGURE 14.1 Sample event tree.

This particular initiating event has three barriers to prevent a hazardous outcome. The first barrier is just the normal operation of the system. In other words, the system operating normally can handle this initiating event with a particular kind of outcome or consequence. Whether barriers 2 and 3 are in the system and whether they function or not determines the different severities of consequences.

NOTES FROM NICK'S FILE

There are many standards on how to conduct risk assessments. Some are very complicated and some very, very simple. When I am working with a client, I like to spend time with them to discuss what a risk assessment is and how to pick a method that is useful for them. One of my biggest challenges has been to get people to stop looking at risks from a purely asset focus and move more toward risk scenarios. If I look only at assets and their vulnerabilities (security risk assessments typically do this), then I may not identify all my *risky scenarios*. But if I start with developing scenarios based on my data, irrespective of the asset viewed in isolation, I am much more likely to more comprehensively identify *risky scenarios*.

P_A is the probability of the initiating event occurring. If a fault tree is used as input to this event tree, then the top-event probability in the fault tree is the initiating event. Each accident scenario could have various initiating events. And each initiating event has its own event tree. It quickly becomes apparent why it is important to understand the hazard scenarios clearly.

After the initiating event occurs, its completion is either successful or not. This means that if nothing stops the initiating event from occurring, a certain damage state (IV in this case) is reached. If it is successful, such that barrier 1, normal operations (e.g., normal vent path to scrubber in hazardous process is open) occurs, then the probability that the normal vent path can handle the initiating event is 1 (success probability of 1) times the probability that the initiating event occurred to begin with.

However, if the normal barrier does not work, then a failure occurs. That failure probability is taken from a fault tree or FMEA that indicates the failure probability of that component or subsystem operating.

The initiating event continues traveling through the system and reaches the next barrier—barrier 2. P_c is the probability that barrier C will fail. Again, if it is successful, then the success probability is 1 and the resulting probability of reaching the damage state is P_c.

The last barrier that the initiating event encounters is barrier 3. The failure of barrier 3, P_D, is the probability that barrier 3 cannot mitigate the hazard consequences.

Starting at the top of the figure, the damage state IV indicates that if the system operates normally, the resulting damage to the system or environment or person is negligible. This means that normal operation of the system does not present a hazard. This is very important because, as mentioned earlier in the book, in some cases normal operations can result in catastrophic events. Dumping hazardous waste in a nonsanctioned dump would have a damage state of I, even if the system operated normally. That is because the normal system operation is already a hazard.

The last two damage states differentiate equipment hazards and personnel hazards. In this case, the first indicates that critical hardware damage has occurred, but there were no injuries to operators. The second illustrates the same hardware damage but with 10 operators injured. An example of the latter is that the damage state was a fire and 10 workers suffered from smoke inhalation.

The dollar value on the figure is the calculated *total* dollars lost if each of the damage states occurs. It is important to be cognizant of all dollars lost. If personnel are involved, then it is not just equipment destroyed or damaged but also workers' compensation claims, medical bills, or lawsuits.

The risk expectation value is a very simple calculation. This number tells you immediately how risky this scenario is. If you have 10 different scenarios, then by comparing the risk expectation values, you can determine which of the scenarios is riskiest. You can also rank the risks and list the greatest to lowest system risks. The total risk of a scenario is the summation of all risk dollars (risk expectation value).

14.2.3 Consequences Determination

Part of the event tree analysis is determining the consequences of the events. Again, the information developed from Chapters 5 through 9 safety analyses is extremely useful. Additional engineering analyses may have to be conducted to get a clearer understanding of the resulting consequences. For example, if the hazard scenario is the release of a toxic gas cloud, then gas dispersion modeling will need to be performed to determine the consequences to the surrounding communities. Different releases will yield different damage states and different dollar value consequences.

It is appropriate to use both qualitative and quantitative expressions of the scenario consequences. Alternatively, the I–IV damage states could be defined as potential political or community opposition to a particular event. Or they could be defined as different levels of system availability or productivity.

As the risk assessment progresses, you will see that quite a number of event trees are generated. To help alleviate this overabundance of data, it becomes useful to truncate some of the less important information. If the event tree has very small risk expectation values, such as the last two branches on the tree in Figure 14.1, then the tree can be pruned or dropped. Rare-event approximation can be used to further reduce the number of branches of the event tree.

Of course, using only minimal cut sets of the fault tree as input to the event tree is important and will help keep the number of initiating events down.

Another way to prune the tree is to determine what percent of the total risk each consequence represents. The previous section stated that the total risk is the summation of all the individual risk expectation values. If you assume that the total risk percentage is 100%, then it is a simple calculation to determine what percent of the total risk value each consequence value includes. Most systems tend to have only a handful of events that have significant impact on the entire system risk.

Why even generate all these event trees? Why not just prune the input probabilities? Table 14.1 shows what this would do. It looks like we could drop cut sets 1, 4, and 6 without thinking twice. And maybe we could even drop set 2. This is very

TABLE 14.1
Cut Set Probabilities

Cut Set	Probability of Failure	Percent Total Failure
1	1.7×10^{-7}	0.004
2	3.7×10^{-4}	7.74
3	2.9×10^{-3}	60.67
4	9.8×10^{-6}	0.205
5	1.5×10^{-3}	31.38
6	1.5×10^{-6}	0.031
Total	4.8×10^{-3}	100

misleading, however. If cut set 1 is combined with a very high-consequence state, then the risk expectation value could be higher than that of cut set 3.

For example, if cut set 4 has a dollars-at-risk value of $40 million and cut set 3 has a dollars-at-risk value of $10,000, then

$$9.8 \times 10^{-6} \times 10 \times 10^{6} = \$98.0 \quad \text{risk expectation value}$$

$$2.9 \times 10^{-3} \times 10 \times 10^{3} = \$29.0 \quad \text{risk expectation value}$$

It becomes clear that cut set 4 carries almost three times more risk than cut set 3. If we drop cut set 4, because it is such a small percentage of the total system failure, then we are also dropping a significant risk factor along with it. It is much more advantageous to do the pruning at the event tree level.

NOTES FROM NICK'S FILE

Though this chapter focuses on quantitative risk assessments and risk evaluation, in reality, the other safety analysis tools listed in Chapters 5 through 9 are risk (safety risk) assessment tools. My biggest challenge in communicating quantitative risk assessments to nontechnical people is to make sure that they do not focus on the numbers—which they always do. It is important for them to understand the numbers are useful when we want to compare different risks relatively and not just look at the absolute risk number. There is too much uncertainty in our analysis to say definitively that there is x amount of risk. But I can say that this risk is much bigger (e.g., 2× bigger) than the other risk. In that sense, I can rank my risks and figure out how best to mitigate.

14.2.4 UNCERTAINTY

As the example demonstrates, we must be very careful how we manipulate the data. An important point to remember is that none of these data are absolute. There always is some error in the probability estimations. Uncertainty enters the analysis in many places: Are all the hazards modeled? Are the hazards modeled correctly? How reliable are the data? Are we using the correct probability density functions?

As mentioned earlier, Bayesian updating with expert judgment can add some clarity to the parameters and put them in better context. Of course, expert judgment also carries uncertainty.

The two principal methods of considering uncertainty are through classical statistics and through probabilistic methods. Two of the classical statistical methods are Taylor series and the system reduction method.

Probabilistic methods are also very useful. Monte Carlo simulation is probably one of the most often used tools. Methods of moments and discrete probability distributions are also popular tools.

14.2.5   RISK EVALUATION: THE USE OF RISK PROFILES

The calculation of risk expectation values is extremely useful in determining which risk scenarios are greater, but this is not the last step. In the risk assessment process, we still must have a systematic process for studying the risk and deciding whether to accept the risk, reject it, or modify the system to mitigate the effects of the hazard. This process is called risk evaluation, and it uses a set of risk profiles to compare the various risks. After risk profiles are generated, you can decide how to proceed. The risk profiles help identify which risks are worth modifying and which offer little payback for modifying them.

The information taken from the risk profiles is then put into a risk management matrix. A *risk management matrix* is a decision-making tool that indicates which risks the company is willing to accept. It is very important that the risk management matrix is developed before any analysis is started.

PRACTICAL TIPS AND BEST PRACTICE

Never look at the risk expectation value as an absolute final number. It is only to be used to compare risks. The risk number is a relative number, relative to other risks. The risk number should never stand alone. If you say the risk of a fire in your plant is 1.5×10^{-6}, you will be in an endless battle trying to prove it is 10^{-6} and not 10^{-5}, and you will never win. A better approach is to use risk profiles and compare *your* plant fire risk with other known, accepted risks.

NOTES FROM NICK'S FILE

I was working with an operator of one of the oldest subways in the world. They wanted to bring the design up to meet more modern fire codes. But because the system was so old and so large, it would cost billions and billions of dollars. We did a probabilistic risk assessment using fault trees to identify and quantify how fires can start in the system. We then used a quantitative risk assessment as part of a cost–benefit analysis of countermeasures to reduce the chance of either a fire starting or a fire developing out of control. For example, one countermeasure was using sweeper trains (trains that clean the trash off the track bed) more frequently significantly reduced the risk of fire. Of course, we had other controls put in place. Our quantitative risk assessment was detailed enough that it convinced the regulators that we did not have to spend billions of dollars to upgrade the system because we could show how we adequately controlled the hazards.

Many times, it is difficult for the engineer to decide which risk scenario is worse. Is a 10% difference between risk expectation values important? Even if the risk of one scenario is higher than another, does that mean I will fix the first and not the second?

Risk profiles are visual representations of the event tree consequence determination. Two risk profiles are developed for each tree. The first profile is the probability of occurrence of the scenario (scenario mean frequency) versus the qualitative consequences (damage state). The second profile is the probability of occurrence of the scenario versus dollars-at-risk or risk expectation value.

Figure 14.2 illustrates the power of risk profiles. You can immediately see that scenarios 3 and 4 are the only ones that are important (within the probability range given). Scenario 3 consequences are minor, but somewhat likely to occur. Depending on how *minor* was defined, this risk profile tells us that we probably do not need to make any changes to the system for scenario 3. If the scenario is risk of a toxic cloud release and minor is defined as the U.S. Environmental Protection Agency–approved release rate, then nothing needs to be changed in the system.

Scenario 4 is a different story. It is obvious that the likelihood is fairly high and the results are from critical to catastrophic. If we use the toxic cloud release example, this indicates that scenario 4 is a problem. It cannot stand as is: the system needs to be modified in some fashion to lower the risk profile. If the failure scenario of that particular scenario is *motor fails on*, then the fix may be fairly easy. Various fail-safe controls could probably be put in place without much expense to mitigate the consequences.

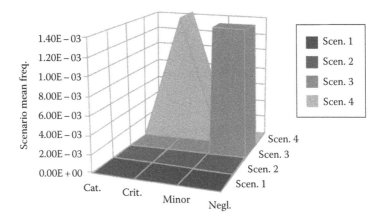

FIGURE 14.2   **(See color insert.)** Sample failure consequences risk profile.

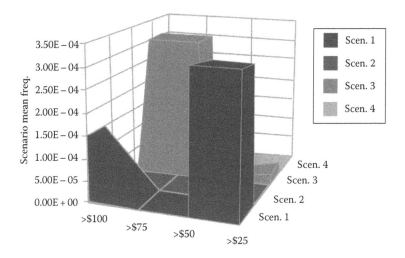

FIGURE 14.3 **(See color insert.)** Sample risk expectation profile.

Figure 14.3 is read in exactly the same way. It is immediately apparent that only scenarios 1 and 3 are of interest. Of the two, scenario 3 is the most significant. The risk expectation profile demonstrates the power of risk assessment and the danger.

The advantage is obvious: You can identify which scenarios to modify (and therefore where to make your change in the system). You can also see how people can get the wrong impression if they view the numbers as absolute and not relative. Again, it is very important to remember that risk assessment numbers should be used to *compare* risks, not as a presentation of an absolute risk.

14.3   CALCULATING SAFETY COSTS

There are numerous ways to calculate how much safety is going to cost. The risk assessment process, especially the risk expectation model or dollars at risk, is one method this chapter has explored. Before describing the various methods available to calculate safety costs, it is important to note that there are other, less tangible activities that can affect safety costs. Their costs can be calculated using normal engineering economics methods.

For example, Kletz (1991) states that from 5% to 10% of the capital costs of a new plant could be reduced if it were possible to reduce inventories of hazardous materials, thus lowering the cost of safety protective systems (fire protection sprinklers, etc.). Kletz goes on to point out that using inherently more efficient processes, such as smaller reactors, pressure, storage vessels, etc., will reduce the size of the process units and therefore lower the overall cost of the plant and also make an inherently safer plant.

Of course, the best way to reduce the cost of safety in a system or process is to design it in. The beginning of Chapter 4 documents numerous examples of how having a good safety management system and designing safety from the start is the most cost-effective method.

The most typical safety measurement tool is to track the number of accidents. The types of data needed to do this are costs of system downtime (and lost productivity and product or market share), equipment damaged during an accident, accident cleanup, equipment replacement, and, of course, personnel injuries and death (including medical costs, workman's compensation, and potential lawsuits). These data can be easily trended and tracked on a monthly or quarterly basis. You can then compare your statistics to national averages. The U.S. National Safety Council (many countries publish comparable information at the federal governmental level) publishes accident costs across all industries. These costs include estimates of lost wages, medical expenses, insurance administration costs, and uninsured costs.

Many argue that safety is too expensive for developing countries and that they just cannot afford it. However, the same reasons developed countries should use system safety and risk assessment methods apply to developing countries. Pe Benito Claudio (1988) states that even though quantitative risk assessment is not practiced in developing countries as much as in the developed world, qualitative risk assessments are used.

Much of this is also changing with international lending organizations such as the World Bank and the Interamerican Development Bank requiring environmental risk assessments to be conducted as part of the development project. Many countries now have their own local risk assessment experts and are not as dependent on outside expertise as they formerly were. One thing is sure: safety does save money. Throughout this book, you can see numerous examples and sources of cost–benefit to safety. Businesses attribute this to better preventative safety management systems.

One common method of calculating the cost of safety is the use of *expected value*. This is a fairly typical method taken from economics management theory. It states:

$$E(V_i) = \sum_i P_i V_i$$

where
  E is the expected value function
  V_i is the utility value function of the ith accident
  P_i is the probability density function of the ith accident

The most popular method of calculating safety is the cost–benefit analysis approach. It takes the present value of the costs of injury and death to people involved in the accident with costs of equipment damage. Accident rates of similar systems or industries are estimated using a regression model, and marginal probabilities are determined. All of this information is combined into the cost–benefit model.

14.4 BRIEF EXAMPLE: RISK ASSESSMENT OF LAUNCHING A SPACE SHUTTLE PAYLOAD

This example demonstrates how to perform a risk assessment and how it can be used for assessing the safety of a technological system. One of the problems that engineers constantly face is which design alternative should be chosen. Many make the choice based purely on economics. As the adage goes, however, what starts off cheap may end up being very expensive. Risk assessment is a very useful tool for assisting engineers in deciding which design change yields the best results for the best price.

The example (Bahr, 1993) is a hypothetical case of assessing the risk of launching a payload on the Space Shuttle. Though the Space Shuttle has been retired since this case was put together, it is still instructive. A true assessment is fairly complex and involves things such as calculating ballistic characteristics and dispersion rates of debris if an in-flight accident did occur. Because the risk assessment fills various volumes of analyses, this example will focus only on the part of the assessment that would be most applicable to other engineering fields.

We are presented with a payload that will be testing new technology to be used on the International Space Station. One problem in space is pumping liquids in microgravity. NASA would like to test a new liquid helium storage system that will eventually be used to cool space telescopes to a few kelvin.

Payloads are very complex (involving electrical, mechanical, and other subsystems); this assessment will study just the cryogenic handling system. Payload safety analysis also entails studying the mission from launch to landing. To keep the focus on what is most interesting—the actual risk assessment of the design and design decisions—we will look only at the initial launch phase of the mission. The question to be answered is "What is the risk of launching this payload?"

The first step is to define the objectives and scope. The objective is to calculate the risk of launching the payload. Is it sufficiently risky to ground the system and not launch? If so, what can be done to make the risk acceptable and go forward with a launch?

The scope is limited to assessing the risk of the cryogenic system overpressurizing and rupturing the system during ground operations before launch. If a rupture does occur, then ground personnel will be injured and even possibly killed. Some of the hazards of an explosive rupture are displacement of air (causing asphyxiation), freezing of equipment and personnel from cryogenic liquids, and shrapnel. For simplicity, epidemiological effects are not included in this study.

The damage states used for studying the consequences are

- *Catastrophic*—May cause personnel death or loss of Space Shuttle flight opportunity, Space Shuttle hardware, or high-value payload equipment
- *Critical*—May cause severe personnel injury, reschedule of Shuttle flight, or payload equipment damage
- *Minor*—May cause loss of payload mission, but not loss of Shuttle flight, and minor payload equipment damage
- *Negligible*—Will not result in injury but may result in reduced payload mission equipment capability

It is assumed that the payload is designed and built well. This means that valves and lines are sized appropriately and are fluid compatible and no components are in the wear-out phase.

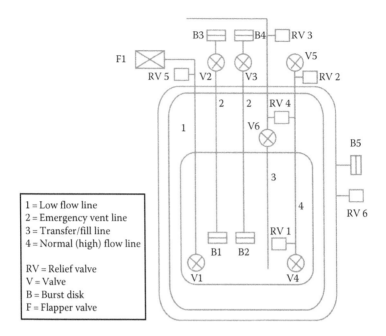

FIGURE 14.4   Cryogenic system payload.

The next phase is to define the system. Figure 14.4 is a simplified schematic of the cryogenic system. The system is a cryostat mounted inside a dewar (two shells with a vacuum between them). The dewar main shell has a vacuum pump-out port and a burst disk. Both the pump-out port and burst disk can fully relieve any pressure buildup in the main shell.

The schematic shows the major components of the system. Some of the principal components are the cryotank, a flapper valve, fluid lines, valves, relief valves, burst disks, electric switches, and bayonet couplers.

The dewar must be kept at liquid helium temperatures during the launch operations, 88 h before launch. After the system is set for launch, there is no way to monitor the status of the system while it sits on the launch pad, so the engineers must be fairly confident that the risks have been well understood and appropriately controlled.

The next step in the risk assessment process is to identify accident scenarios and develop the initiating events for those scenarios. A hazard analysis was performed and various hazards were identified. Of the hazards identified, the most significant were related to the uncontrolled release of cryogenic fluid or gas. With that information, a fault tree was constructed for the system with the top event designated as *uncontrolled cryogenic release*. An FMEA was performed on those components that were determined to be critical to the fault tree.

Numerous hazard scenarios could lead to a dangerous situation. Initiating events were developed from the safety analyses. Four initiating events were identified for further study.

The first initiating event is rapid pressure rise from a leak of the outer shell of the dewar and cryostat. If a leak occurs, heat will be introduced into the system and quickly vaporize the helium, causing the internal pressure to rise. The identified barriers to prevent damage propagation are the normal high-rate vent path, the emergency vent line, and the cryotank. As Figure 14.4 illustrates, all three subassemblies have pressure-relief valves and/or burst disks. You might wish to ask (and the risk assessment will answer), are all these safety systems necessary?

The next initiating event is failure of the high-flow vent line. There are various leakage points in the subassembly, such as leakage through motor valve V5 or through the bayonet couplings. If there is air leakage into the system (remember that the payload is sitting on the launch pad), then the moisture in the air will condense (from the cold helium) and freeze and form an ice plug, compromising the safety relief system. Leaving V5 in the open position is considered human error.

The third initiating event is failure in the low-flow vent line flapper valve. Of course, there always is some heat input into a system and that does cause the helium to vaporize. A flapper valve burps the excess pressure, closing in time to prevent air ingestion. The failure is the flapper valve failing open. Again, the hazard is from air ingestion and an ice plug forming, compromising the safety system.

The last initiating event is air ingestion through the emergency vent line pump-out port, burst disks, or relief valves. Similarly, an ice plug could form and compromise the safety system.

An interesting point to note is that it is possible to have any of these failures and still have an operable system. One of the difficult decisions an engineer needs to make is to decide when it is sufficiently unsafe to stop all activities and fix them or continue forward. The risk assessment will help answer this question.

The fourth step in the risk assessment process is to develop the event trees. Table 14.2 lists the component failure probabilities in the cryo system. There were no data for these components in the specific environment, so some Bayesian updating was needed to adjust the numbers. Also, in some cases, no data were available, and that is a good example of where you must just use your best judgment and continue forward.

A fault tree was used as input to each initiating event. Table 14.2 failure frequency probability distribution is lognormal.

Before you can determine the extent of the consequences of each scenario, it is very important to develop a consequence matrix. In this case, the matrix in Table 14.3 illustrates the damage states from negligible to catastrophic. It indicates both the qualitative and quantitative consequences of various kinds of consequences of an uncontrolled cryo leak.

The mission status category indicates how each event termination will affect the ability to launch the Space Shuttle. Because the cryo payload is one of a number of other payloads, it is possible to fly even if the cryo payload is not functioning. Although there are a number of barriers, all result in some cryogen release, thus differing mission results.

Loss of the Shuttle flight opportunity means that the Shuttle is grounded for 6–12 months. Launch delay is a 30-day hold.

Dollar amounts that could be lost are based on both injuries to ground crews and hardware damage.

TABLE 14.2
Component Failure Probabilities

Component	Probability Fails	Source
Cryotank	1×10^{-8}	(1)
Burst disk fails to operate	1×10^{-6}	(3)
Burst disk leaks	0.85×10^{-6}	(2)
Flapper valve fails closed	1×10^{-5}	(3)
Flapper valve fails open	0.13	(4)
Flapper valve leaks	0.85×10^{-5}	(2)
Relief valve fails to operate	1×10^{-5}	(1)
Relief valve leaks	0.85×10^{-6}	(2)
Motor valve fails to operate	1×10^{-3}	(1)
Motor valve leaks	0.85×10^{-6}	(2)
Bayonet couplers leak	1×10^{-8}	(3)
Pressure lines leak	3×10^{-6}	(1)
Pressure lines fail (rupture)	1×10^{-8}	(1)
Pump-out port leaks	0.85×10^{-6}	(2)
Porous plug blocked	1×10^{-6}	(3)
Human operator error	1×10^{-3}	(3)

Notes: (1) Henley and Kumamoto (1992); (2) Ploe and Skewis (1990); (3) author's engineering judgment, based on experience with similar equipment; (4) actual failure data from a similar system.

The risk calculation is as follows:

$$\text{Probability of occurrence} \times \text{Dollar value} = \text{Dollars at risk}\,(\text{expected risk})$$

Two event trees are shown to give an indication of the kind of information available to the engineer. Figures 14.5 and 14.6 are the event trees for operator error of leaving valve 5 open and the flapper valve, respectively.

Figure 14.5 indicates that there is a 1 in 1000 chance that this valve will be left open. If it is, then the dollar value is high and the dollars at risk are also very high. The likelihood is fairly high and the consequences of forgetting the valve open could cost the payload mission. The other damage states in this scenario are not very important in comparison. If this scenario occurs, it would not necessarily stop the launch but would ruin this payload's mission.

Figure 14.6 illustrates the results of a very high failure rate in the flapper valve. With the valve stuck open, an ice plug forms in the vent line, resulting in a very high risk at $143,000. The various other barriers lower the risk significantly.

The event trees also show the risk of injuring or killing personnel. Other event trees gave similar data as these two. Now the engineer has a much better understanding

TABLE 14.3

Consequence Matrix

	Normal Vent Path (High-Flow Vent Line Release)	Emergency Vent Release	Cryotank Assembly Failure	Outer Shell Failure
Event termination	Reduced mission capability; No equipment damage (but may have reduced capability)	Loss of mission (but not loss of flight opportunity); Minor equipment damage	Loss of Shuttle flight opportunity (launch delay); Major equipment damage; Personnel injury	Loss of Shuttle flight opportunity; Significant equipment damage (including Shuttle damage); Personnel death
Dollars lost	Reduced mission capability, $1M[a]; Other equipment capability reduction, $100K[a]	Loss of mission, $25M; Minor payload equipment damage, $250K	Reschedule of flight opportunity, $30M[a]; Payload equipment damage, $800K[a]; Personnel injury (this includes med. and lawsuits), $23K/person[b]	Loss of flight opportunity, $212M[c]; Shuttle equipment damage, $25M[c]; Payload equipment damage, $10M[a]; Personnel death (includes total death compensation), $730K/person[b]
Severity classification	Negligible	Minor	Critical	Catastrophic
Classification number	IV	III	II	I

[a] Author's estimate based on experience of a *typical* loss.
[b] U.S. National Safety Council, (1991). These figures include estimates of wage losses, medical expenses, insurance administration costs, and uninsured costs.
[c] National Aeronautics and Space Administration. Actual 1992 costs include flight operations, launch and landing operations, and associated administrative costs.

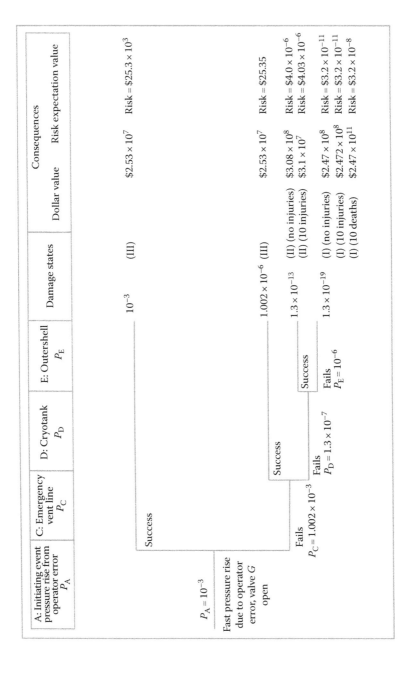

FIGURE 14.5   Operator error (valve 5) event tree.

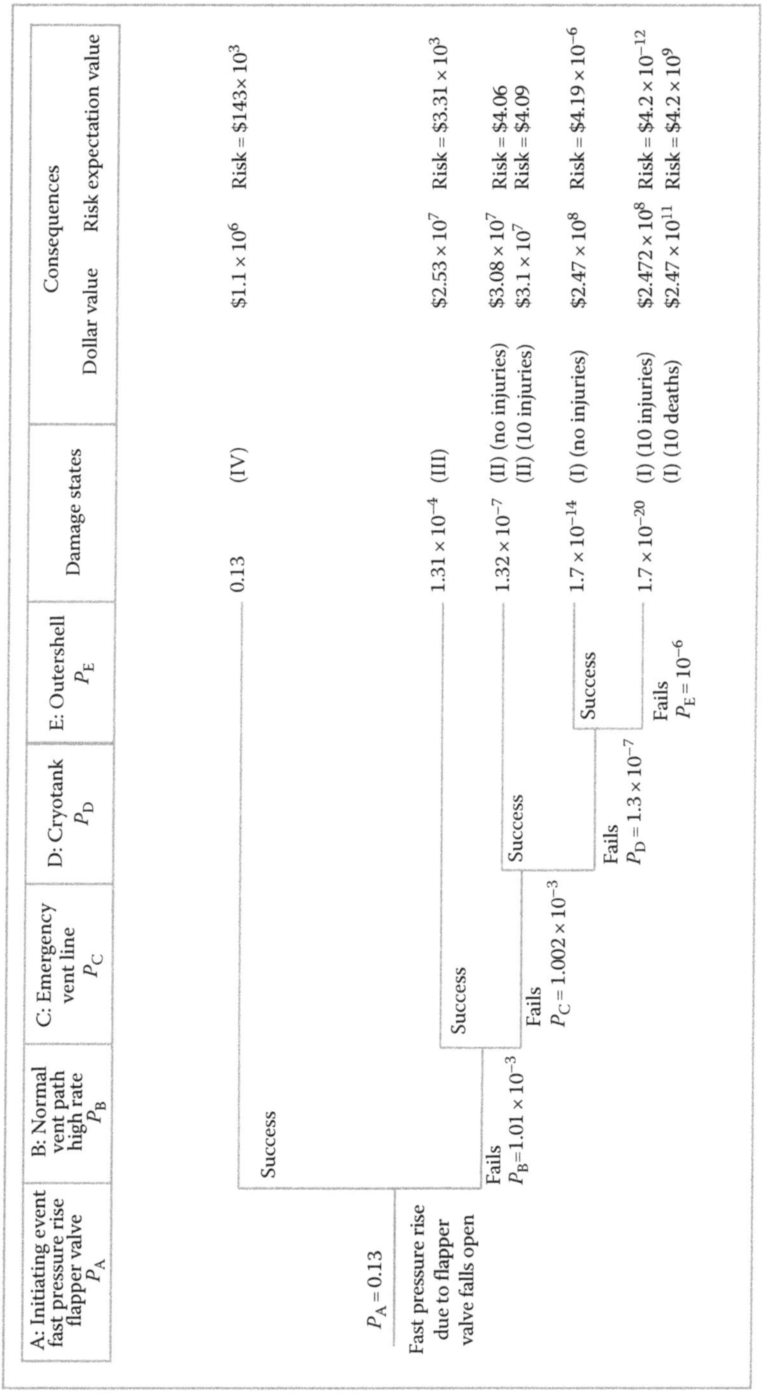

FIGURE 14.6 Flapper valve failure event tree.

of which scenarios are important, but it is still difficult to decide where to allocate resources to solve the problem. It is obvious that the top event in each tree is significant, but what about the others? Here is where the risk evaluation really starts.

Remember that risk evaluation is the systematic approach to accepting, rejecting, or modifying the risk. Two risk profiles—severity of consequences and dollars at risk versus probability of occurrence—are developed for each event tree. Figures 14.7 and 14.8 are risk profiles of the operator error event tree. The two profiles indicate dramatically which scenarios in each event tree are important. This is very important, especially if the risk expectation dollars appear close to each other on the event tree. The profiles answer the question from the event tree whether scenario 2, with $25M at risk, is significant or not.

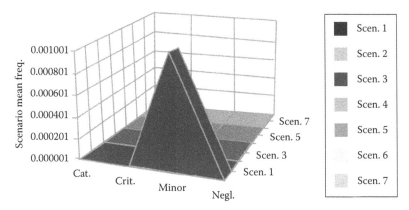

FIGURE 14.7 **(See color insert.)** Operator error (valve 5 open) severity classification risk profile.

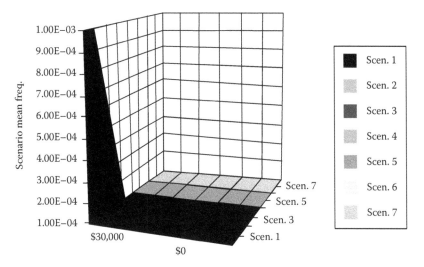

FIGURE 14.8 **(See color insert.)** Operator error (valve 5 open) dollars-at-risk profile.

Remember that the risk profiles are used to help the engineer visualize which risks are important and which are not.

None of these event trees or risk profiles indicates the risk of losing the Space Shuttle completely. As you can see from the event trees and profiles, the risk is not even on the chart (the risk of the payload causing loss of the Space Shuttle). In fact, reviewing the drawing of the cryo system, it is obvious that the system has a lot of redundancies in the safety systems. A few scenarios, however, still present a significant amount of risk, in spite of the safety systems. This is a good illustration of the power of risk assessments.

The last step in the risk assessment is to manage the risk of the system. We still must determine if these risks are enough to ground the payload and not fly, and to do that we use a set of risk decision matrices.

NOTES FROM NICK'S FILE

It is very important to spend time thinking about (before you start the analysis) how to define the risk decision matrices. They will be different for each company. I was working with one manufacturing company, and we argued a lot about how to define them and then socialized the risk decision matrices to all levels of management to get consensus—then we started the analysis. That was very important because we were going to actually make decisions of when we will shut down line operations based on the risk assessment, and that means lost revenue.

Table 14.4 is very similar to the qualitative risk management (hazard risk index) in Chapter 5. Note the risk management criteria at the bottom of the table. It is very important to define when action must be taken and when it can wait. This is really just prioritizing the risk.

TABLE 14.4
Launch Risk Management Matrix

Probability of Occurrence	Severity Classification	Catastrophic	Critical	Minor	Negligible
	Classification Number	I	II	III	IV
High probability	$>10^{-4}$	A	A	A	B
Medium probability	$10^{-7}–10^{-4}$	B	B	B	C
Low probability	$<10^{-7}$	C	C	D	D

Note: Risk management criteria—A, prompt action required. This hazard must be resolved immediately; B, need to develop a plan to mitigate hazard. Hazard still cannot remain, but there is more time to develop mitigation. Note that the hazard must be controlled before launching; C, acceptable, but with management oversight (this includes both the mission manager and the launch control manager); D, no action necessary.

TABLE 14.5
Launch Risk Decision Matrix

Hazard Risk Index	Launch Commit Criterion
IA, IIA, IIIA	Unacceptable, delay launch to resolve issue
IB, IIB, IIIB, IVB	Undesirable, upper NASA management decision to accept or reject risk
IC, IIC, IVC	Acceptable, with review by launch management authority
IVD, IIID	Acceptable, without launch management review

Table 14.5 uses the hazard risk index from the previous table and develops what NASA calls a *launch commit criterion*. This is the actual decision of whether they wish to launch or not.

Table 14.6 is one table of the results from the analysis. In this case, it is immediately apparent that scenario 1 of the valve 5 operator error risk is unacceptable and cannot stand as is. In fact, the risk is so important that the entire launch will be delayed until this risk is resolved. The other scenario risks are either undesirable (needs upper management approval) or are acceptable with review.

Taking together all of this information, it is now very easy to answer the question of whether it is safe to fly: not until some of these risks have been dealt with. Also, we know exactly what part of the system must be changed. In this example, the valve must be closed at launch. A very good and inexpensive solution is to verify that the valve is closed at a mandatory inspection point and have the valve wired closed so that it cannot be opened again by mistake. If this is done, the probability of occurrence falls to 10^{-8} and thus reduces the risk to a very low number.

Though the flapper valve has not been discussed much, it too has a high failure rate and risk expectation number. In this case, the only courses of action are to either redesign the flapper valve to be more reliable or replace it with a more reliable valve.

Similar information can be taken from all the event trees, risk profiles, and risk matrix results developed in the course of a risk assessment. The technique becomes very powerful when you wish to weigh the advantages of one change

TABLE 14.6
Operator Error (Valve 5) Risk Matrix Results

Scenario Number	Hazard Risk Index	Launch Commit Criterion
Scenario 1	IIIA	Unacceptable, delay launch to resolve issue
Scenario 2	IIIB	Undesirable, upper NASA management decision to accept or reject risk
Scenario 3	IIC	Acceptable, with review by launch management authority
Scenario 4	IIC	Acceptable, with review by launch management authority
Scenario 5	IC	Acceptable, with review by launch management authority
Scenario 6	IC	Acceptable, with review by launch management authority
Scenario 7	IC	Acceptable, with review by launch management authority

versus another. It also is a very systematic method of deciding which risks to take and which risks not to accept.

This risk management system did not take into consideration the political implications of the launch. However, that would not be that difficult to add. Say, for example, that this payload was going to be launched with another satellite that had a very small launch window. Many interplanetary probes must be launched during certain times, and the chance may not come again for years. The risk of missing that launch date could be very great. Then it may be worth the risk to launch the cryo payload without waiting to make any design changes (since the Shuttle is still not at risk).

REFERENCES

Bahr, N. J. 1993. Risk assessment of a cryogenic subsystem for a NASA shuttle payload. *American Society of Mechanical Engineers Winter Annual Meeting 1993*, New York, SERA-Vol. 1, pp. 1–20.

Kletz, T. A. 1991. *Plant Design for Safety: A User-Friendly Approach*. New York: Hemisphere, p. 16.

Pe Benito Claudio, C. 1988. Risk analysis in developing countries. *Risk Analysis*, 8 (4):475–478.

FURTHER READING

Center for Chemical Process Safety. 1999. *Guidelines for Chemical Process Quantitative Risk Analysis*, 2nd edn. New York: Wiley-AICHE.

Frank, M. 2008. *Choosing Safety: A Guide to Using Probabilistic Risk Assessment and Decision Analysis in Complex, High-Consequence Systems*. New York: Routledge.

Henley, E. J. and Kumamoto, H. 1992. *Probabilistic Risk Assessment*. New York: IEEE Press.

Henley, E. J. and Kumamoto, H. 2000. *Probabilistic Risk Assessment and Management for Engineers and Scientists*. New York: Wiley-IEEE Press.

Iman, R. L. and Hora, S. C. 1989. Bayesian methods for modeling recovery times with an application to the loss of off-site power at nuclear power plants. *Risk Analysis*, 9 (1):25–36.

Keeney, R. L., Kulkarni, R. B., and Nair, K. October 1978. Assessing the risk of an LNG terminal. *Technology Review*, 81 (1):64–72.

Leveson, N. 2012. *Engineering a Safer World: Systems Thinking Applied to Safety*. Cambridge, MA: MIT Press.

McCormick, N. J. 1981. *Reliability and Risk Analysis: Methods and Nuclear Power Applications*. London, U.K.: Academic Press.

National Aeronautics and Space Administration, *Budget Estimates Fiscal Year 1994*, Vol. 1, Agency Summary. Washington, DC: National Aeronautics and Space Administration.

Office of Commercial Space Transportation Licensing Programs Division. U.S. Department of Transportation. 1995. *Hazard Analysis of Commercial Space Transportation*. Washington, DC: U.S. Department of Transportation.

Pelton, J. and Jakhu, R. 2010. *Space Safety Regulations and Standards*. Oxford, U.K.: Butterworth-Heinemann.

Ploe, R. J. and Skewis, W. H. May 1990. *Handbook of Reliability Prediction Procedures for Mechanical Equipment*, DTRC-90/010. Bethesda, MD: United States Navy—David Taylor Research Center.

Seixas de Oliveira, L. F. 1987. Cost-effectiveness of risk-reduction measures from a national viewpoint: A case study of the Angora nuclear plant in Brazil. *Risk Analysis*, 7 (3):321–328.

Sgobba, T. (Chief Ed.). 2013. *Safety Design for Space Operations*. Oxford, U.K.: Butterworth-Heinemann.

Smith, C. 2012. *Probabilistic Risk Assessment Procedures Guide for NASA Managers and Practitioners: NASA/SP-2011-3421*. Washington, DC: CreateSpace Independent Publishing Platform.

Suokas, J. 1988. Evaluation of the quality of safety and risk analysis in the chemical industry. *Risk Analysis*, 8 (4):581–591.

U.S. National Safety Council. 1991. *Accident Facts*, Chicago, IL: National Safety Council.

Appendix A: Typical Energy Sources

Note: Do not forget that potential energy sources can be just as dangerous as kinetic energy sources. Also, most systems will have a combination of energy sources. Do not look at energy sources in isolation; look at their interaction and how they influence each other in the creation of hazards.

- Acoustic and other noise producers
- Actuating devices
- Boilers and other heated pressure systems
- Catapulted objects
- Charged electrical capacitors
- Chemical reaction sources
- Combustion systems
- Compression devices
- Cooling devices
- Cryogenic and refrigerant systems and storage vessels
- Displacement systems
- Electrical generators and delivery systems
- Electric static discharge
- Electromagnetic devices (e.g., radio-frequency sources)
- Etiologic (viral, bacterial, fungal)
- Explosive charges and devices
- External sources (i.e., earthquake, flood, landslide, and weather conditions)
- Falling objects
- Friction devices
- Flammable materials
- Fuels and propellants
- Fluid devices
- Gas generators
- Hazardous material flow systems
- Heating devices
- Human interaction and power
- Illumination devices
- Impact and shock test devices
- Initiators/igniters
- Ionizing radiation sources
- Lifting equipment
- Magnetic devices and sources
- Material-handling devices

- Material-mixing devices
- Nonionizing radiation sources (lasers, ultraviolet, infrared, etc.)
- Nuclear systems
- Pressure containers, systems, and devices
- Pumps, blowers, and fans
- Rotating machinery
- Spring-loaded devices
- Storage batteries
- Suspension systems
- Tensor systems
- Vacuum systems and devices
- Vibration devices

Appendix B: Generic Hazard Checklist

This generic hazard checklist can be used to help identify hazards and hazard sources in any industry. Of course, every industry has unique hazards that will not be on this list. Some of the hazards on this list are unique to certain hazard scenarios; others are common cause factors that will cross all subsystem boundaries. Many of the hazards are repeated in different categories. As can be seen, the generic hazard checklist has many of the same entries as the energy source checklist. That is because energy sources make up the vast majority of hazards.

Note: As for all checklists, it is impossible to be all inclusive. This list should be viewed as a starting point. As you gain more experience, you may wish to add to this list and keep it for future reference. And as the book says many times, a checklist is *not* a substitute for a safety analysis. What it is good for is helping you to make sure that you are including a long list of potential hazards. By definition, it is nonexhaustive.

Acceleration/Deceleration

- Acceleration/deceleration
- Falling objects
- Fragments/projectiles
- Impacts
- Inadvertent motion
- Sloshing liquids

Contamination/Corrosion

- Chemical dissociation
- Chemical replacement/combination
- Corrosion due to electrolysis
- Hydrogen embrittlement
- Moisture
- Oxidation
- Organic (fungal/bacterial, etc.)
- Particulate
- Stress corrosion

Control Systems

- Inappropriate control system operation
- Inappropriate software operation
- Interference to control system
- Sneak circuit

Electrical

- Arcing
- Bent pins
- Breakdown of dielectric
- Burns
- Corona
- Distribution feedback
- Electrical noise
- Electrical surges
- Electromagnetic interference
- Excessive solder
- Grounding
- Ignition of combustibles
- Improper electrical connections (mismating) and wiring
- Inadequate heat dissipation
- Inadvertent activation
- Incorrect voltage, current, cycle, etc.
- Inductive or capacitive coupling
- Lightning strike
- Magnetic surge
- Mismating of power connectors
- Polarity
- Poor insulation
- Power outage
- Shock
- Short, open circuit
- Static discharge
- Stray currents/sparks

Environmental/Weather

- Fog
- Foreign matter contamination
- Fungal/bacterial
- Humidity
- Lightning

- Outside versus inside environment
- Precipitation (fog, rain, snow, ice, sleet, hail)
- Radiation
- Salt
- Sand/dust
- Temperature extremes (and variations)
- Vacuum
- Wind

Ergonomic

- Fatigue
- Faulty/inadequate control/readout labeling
- Faulty workstation design
- Glare
- Heating/ventilation and air conditioning
- Inaccessibility
- Inadequate control/readout differentiation
- Inadequate/improper illumination
- Inappropriate control/readout location

Explosives

- Chemical contamination
- Dust explosion
- Electrostatic discharge
- Explosive liquids, gases, or vapors present
- Friction
- Heat/cold
- Humidity levels
- Impact/shock
- Lightning
- Normally nonflammable material in finely powdered form (dust, aluminum, magnesium, etc.)
- Pyrophoric
- Welding
- Vibration

Fire

- Chemical change (exothermic/endothermic)
- Combustible material and combustible atmosphere
- Fuel and oxidizer in the presence of pressure and ignition source
- Pressure release
- High-heat source

Human Factors

- Failure to operate
- Inadvertent operation
- Operation too brief/too long
- Operation early/late
- Operation out of sequence
- Operator error
- Right operation/wrong control

Leaks/Spills

- Dusts
- Flooding
- Gases/vapors
- Liquids
- Porosity
- Radioactive leaks
- Runoff
- Solids

Life Cycle

- Maintenance
- Start-up
- Steady-state operation
- Stressed operation
- Shutdown (standard, emergency, unexpected)

Materials

- Bad protective paint
- Chemical combinations
- Compressible/incompressible fluids
- Combustible material
- Dissimilar materials
- Exo-/endothermic reactions
- Halogens and other oxidizing agents
- Lack of resiliency
- Lubrication
- Incompatible materials or media
- Polymerization
- Solvent residues

Mechanical

- Crushing surfaces
- Ejected parts/fragments
- Fatigue/cyclic stresses
- Flexure
- Friction surfaces
- Hysteresis
- Lifting
- Misalignment
- Pinch points
- Rotating equipment
- Sharp edges
- Stability/toppling potential
- Torquing (over/under)
- Vibration

Physiological

- Allergens
- Asphyxiates
- Baropressure extremes
- Carcinogens
- Fatigue
- Irritants
- Lifted weights
- Mutagens
- Noise
- Nuisance dusts/odors
- Pathogens
- Radiation
- Temperature extremes
- Vibration

Pneumatic/Hydraulic Pressure/Vacuum

- Backflow/siphon effect
- Blown objects
- Blast
- Cavitation
- Dynamic pressure loading
- Hydraulic hammer
- Implosion
- Inadequate pressure/flow capacity relief

- Inadvertent release of material
- Over-/underpressurization
- Pipe/vessel rupture
- Pipe/hose whip
- Pressure/fluid entrapment in system
- Rapid pressure change

Radiation

- Ionizing (alpha, beta, gamma, x-ray)
- Nonionizing (infrared, laser, microwave, ultraviolet)
- Thermal radiation

Structural

- Accelerations (high/low)
- Aerodynamic and acoustic loads
- Bad welds
- Brittleness/ductility of materials
- Cracks
- Fatigue/cyclic stresses
- Load- and non-load-bearing paths
- Stress concentrations
- Vibration/noise

Temperature

- Altered structural properties
- Burns (hot/cold)
- Compressive heating
- Cryogenic properties
- Elevated flammability
- Elevated gas/liquid pressure
- Elevated reactivity
- Elevated volatility
- Freezing
- Heat source/heat sink
- Hot/cold surfaces
- Humidity/moisture
- Joule–Thomson cooling
- Solar effects

Appendix C: Generic Facility Safety Checklist

This generic checklist can help identify hazards and hazard sources in a facility. As you can see, this list is very similar to the two previous appendixes, mixing both functional areas and specific devices. Also, some of the areas overlap, just as they do in many industrial plants. Because there are a plethora of different kinds of facilities around the world and in different industries, this list is only something to help get you started on a facility hazard analysis.

Software by itself cannot be a hazard. What is a hazard is what the software control will enable to occur. Integrated control systems are ubiquitous in our plants. Software, computer control, and Internet-enabling and monitoring systems perform many of the following functions listed. Most of these items are controlled by software. So, when you look at hazards in the following list, you also need to look at the systems that control them to be complete in your analysis.

General Plant Layout

- Location of hazardous operations
- Location of shop processes
- Location of laboratories and testing facilities
- Office locations
- Location, handling, and storage of hazardous materials
- Emergency systems
- Compatibility of operations
- Storage areas
- Staging areas
- Disposal areas
- Fire control and exclusion zones
- Public access areas
- Plant modifications, upgrades, and retrofits

Building Materials

- Compatibility of materials
- Flammability
- Structural integrity (especially roof, floor, and wall loading)
- Useful life of materials
- Appropriate use of materials
- Construction

Access/Egress

- Life Safety Code requirements
- Emergency (evacuation, emergency response team)
- Restricted
- Ease of maintenance
- Operations
- Persons with disabilities
- Stairs/railings
- Loading/unloading people, materials
- Traffic

Facility Utilities

- Control, monitoring, shutoff of local utilities
- Automatic utilities
- Electrical power supply
- Potable water supply
- Utility water supply
- Sanitary, sewage, and waste water/material disposal
- Other utility supplies (bulk natural gas, petroleum, coal, regenerated and cogeneration supplies)
- Transmission lines and grid
- Emergency services

Fire Protection

- Fire/smoke detection
- Alarms/annunciation
- Automatic fire suppression
- Fire-resistant design
- Extinguisher selection and location
- Adequateness of fire protection system
- Fire protection during loss of utility services

Ventilation

- Heating
- Ventilation (recirculation and reentrainment)
- Air conditioning
- Humidity
- Hazardous materials and gases
- Ventilation during emergencies
- Airborne particles
- Toxicity
- Explosive environments

Lighting

- Ambient
- Emergency
- Special lighting
- Light sources
- Use of color lights
- Heat generation from lighting

Sound

- Sound production systems
- Plant noise levels
- Noise from machinery and other plant processes (i.e., gas flow systems)
- Emergency warning systems
- Ultrasonic sound

Electrical

- Lockout/tagout
- Grounding/bonding
- Switch gear
- Insulation
- Electrical shock
- High voltage/low voltage
- Power surges
- Sparking sources
- Electrostatic discharge
- Electromagnetic compatibility
- Wiring and fusing
- Isolation capacitors, rheostats, power resistors, rectifiers, shock boards, contactors, and relays
- Shutoffs/breakers
- Power tools
- Inadvertent operation
- Maintenance
- Lightening protection
- Emergency power
- Emergency shutdown
- Explosion-proof components in explosive environments
- Electrical motors, generators, amplifiers, bus bars, and other equipment
- Electrical distribution system
- Batteries, battery packs, and charging and dc distribution systems
- Electrical substations and transformers
- Electronic systems

Mechanical

- Machine guards
- Rotating machinery
- Lifting equipment, including cranes, dollies, forklifts, etc.
- Machine tools
- Material handling and transportation
- Vibration
- Mechanisms
- Fans
- Gearing, fastenings, bearings, packings and seals, and other machine elements
- Gas turbines
- Steam turbines
- Heat exchangers, condensers, and cooling towers
- Nuclear power
- Air injectors
- Internal combustion engines
- Maintenance operations

Pressure Systems

- Hydraulics
- Pneumatic/air system
- Compressed gases, bottles, tanks
- Pressure systems, including relief valves, valves, quick disconnects, and other pressure components
- Boilers
- Pumps
- Compressors
- Vacuum pumps and vacuum systems
- Emergency relief
- Thermal controls
- Monitoring and control

Refrigeration and Cryogenics

- Deep refrigeration
- Ice making
- Thermal expansion
- Compatibility of materials
- Gas liquefaction
- Refrigerants and gases
- System control and monitoring
- Vapor-compression circuits
- Absorption systems

- Thermoelectric cooling
- Direct expansion systems
- Brine systems
- Insulation
- Asphyxiants

Material Handling

- Hoists, skips, and holding mechanisms
- Cranes
- Elevators, dumbwaiters, and escalators
- Winches
- Slings, chains, and wire ropes
- Lifting magnets
- Industrial cars
- Dozers and draglines
- Car-unloading machinery
- Containers and containerization
- Earth-moving and off-highway equipment
- Lift trucks and palletized loads
- Above-/below-surface handling
- Overhead conveyors
- Flight conveyors
- Carrying conveyors
- Bucket and belt conveyors
- Pneumatic conveyors
- Automatic metering
- Spill control and containment
- Exhaust and ventilation

Radiation

- Ionizing radiation systems (alpha particles, beta particles, neutrons, x-rays, and gamma rays)
- Ionizing radiation detection systems
- Radioactive isotope control systems and management
- Laboratory equipment with radioactive isotopes
- Nuclear reactors and fuel systems
- Nonionizing radiation sources (laser, radar, ultraviolet and infrared light, microwave, electromagnetic interference, radio-frequency [RF] waves, and high-frequency equipment)

Hazardous Materials

- Flammable/combustibles systems and storage
- Explosive and pyrophoric handling systems and storage areas

- Toxic substances handling, storage, and disposal systems
- Use of corrosives
- Use of oxidizers
- Water-reactive mixing
- Unstable substances handling and storage
- Use of irritants, asphyxiants, carcinogens, and pathogens
- Radioactive material handling, monitoring, storage, and disposal

Confined Space

- Utility tunnels
- Storage tanks, bins, boilers, piping, and other sealed chambers
- Raised floors
- Vacuum and pressure chambers

Laboratories

- Space utilization
- Benches and work surfaces
- Chemical and hazardous material storage
- Drainage systems
- Exhaust and ventilation systems
- Spills, containment, and cleanup
- Utilities
- Material compatibility
- Personnel protection
- Pressure and material handling systems
- Leak detection and warning
- Emergency protection systems
- Waste generation and disposal systems

Shop Processes

- Molding processes and machinery
- Die-casting machines
- Vibrators
- Sand slinger
- Squeezing and jolt molding machines
- Casting alloys
- Melting furnaces
- Sand-blasting machinery
- Cleaning materials, solvents, and equipment
- Nondestructive inspection (dye penetrant, x-ray, ultrasound, Magnaflux, etc.)
- Destructive testing machinery
- Metalworking and metal-cutting operations
- Hot- and cold-working operations

- Rolling operations
- Protective coating operations
- Power-press operations
- Hydraulic presses
- Drop hammers
- Steam and pneumatic hammers
- Plate-bending machines
- Arc and gas welding
- Resistance welding
- Thermal metal-cutting machinery
- Electroslag welding
- Laser-beam welding
- Lathes
- Boring machines
- Drilling, reaming, threading, and milling machines
- Grinding and polishing machinery
- Wood cutting tools and machines

Fuels and Furnaces

- Fuels
- Combustion furnaces
- Incineration
- Electric furnaces and ovens
- Exhaust and material disposal

Exhaust Systems

- General
- Local
- Fume hoods
- Emergency exhaust
- Discharge systems
- Scrubber and filtration systems
- Recirculation, migration, and reentrainment systems
- Regeneration and cogeneration systems

Natural Phenomena

- Rain
- Droughts
- Flooding and mud slides
- Tornadoes, hurricanes, and earthquakes
- Snow, ice, and blizzards
- High winds
- Extreme temperatures

Process Monitoring and Integrated Control Systems

- Process monitoring
- Integrated control systems monitoring
- Utility services monitoring
- Pressure, temperature, flow, voltage, current, and vibration levels
- Environmental (air quality, temperature, humidity)
- Hazardous material discharge to the environment
- Human health levels
- Fire and smoke detection
- Hazardous gas and vapor detection
- Oxygen-level detection
- Leak detection
- Safety-critical subsystem monitoring
- Mass, weight, and volumetric monitoring
- Use of consumables
- Chemical and physical property measurements
- Ionizing radiation levels
- Nonionizing radiation levels
- Overriding automatic controls
- Large-scale computer control systems
- Microprocessor control systems
- Programmable logic devices
- Internet-enabled controls, monitoring, and health
- Software controls
- Alarms, annunciation, and other warning systems

Communications

- Public address systems
- Emergency communication systems
- Public affairs
- Right-to-know
- Machine–human interface
- Management–employee relations
- Written and verbal procedures
- Emergency operations, procedures
- Emergency response team

Operations

- Normal operations
- Emergency operations
- Training
- Shift work
- Maintenance (planned and emergency)
- Testing

Personnel Safety

- Personal protective equipment (gloves, gowns, eye, face, and ear protection, respirators)
- Eyewashes and showers
- Exposure control systems
- First aid
- Alarm and annunciator systems

Documentation

- Material safety data sheets
- Training plans
- Emergency management plans
- System safety program
- Operating procedures
- Maintenance procedures
- Accident investigation reports and tracking
- Test procedures
- Chemical hygiene plan
- Radiation control plan
- Hardware and facility configuration control plan

Appendix D: Internet Sources

It is impossible to list all the useful safety websites around. Every country has its own as well. Here are just a few:

American Chemical Society: gopher://acsinfo.acs.org.
American Society of Mechanical Engineers: https://www.asme.org/.
American Society of Safety Engineers National Home Page (future): http://www.asse.org.
Board of Certified Safety Professionals: http://www.bcsp.org/.
Canadian Centre for Occupational Health and Safety: http://www.ccohs.ca.
Canadian Society of Safety Engineers: http://www.csse.org/.
Center for Chemical Process Safety: https://www.aiche.org/ccps.
Consumer Product Safety Commission: gopher://cpsc.gov.
Defense Systems Information Analysis Center (formerly Reliability Information Analysis Center): http://www.theriac.org/.
Environmental Health Center/National Safety Council: http://envirolink.org.
European Agency for Safety and Health at Work: https://osha.europa.eu/en.
European Strategic Safety Initiative: http://easa.europa.eu/essi/index.html.
Federal and State Regulations: http://www.gate.net/~gwarbis/solutions.
Federal Emergency Management Agency: http://www.fema.gov.
Fire Information: http://life.anu.edu.au/firenet/firenet.html.
Hazardous Materials Information Database: http://atsdrl.cdc.gov: 8080/atsdrhome.html.
Institution of Occupational Safety and Health: http://www.iosh.co.uk/.
International Association for the Advancement of Space Safety: http://iaass.space-safety.org/.
International Association of Personal Protection Specialists: http://www.mps.ohio-state.edu/cgi-bin/hpp/lapps_home.html.
International Civil Aviation Organization: http://www.icao.int/safety/Pages/default.aspx.
International Institute of Risk and Safety Management: http://www.iirsm.org/.
International Labor Organization: http://www.ilo.org/global/lang—en/index.htm.
International Network of Safety and Health Practitioner Organizations: http://www.inshpo.org/.
International Standards Organization: http://www.iso.org/iso/home.html.
International System Safety Society: http://www.system-safety.org/.
Index of Occupational Health and Safety Internet Resources: http://turva.me.tut.fi/~tuusital/oshalinks.html.
Martindale's Health Science Guide '95: http://sss-sci.lib.uci.edu/~martindale/HSGuide.html.
Nancy Leveson home page: http://sunnyday.mit.edu/.
National (U.S.) Fire Protection Association: http://www.nfpa.org/.
National (U.S.) Institute for Occupational Safety and Health: http://www.cdc.gov/niosh/.
National (U.S.) Safety Council: http://www.nsc.org/pages/home.aspx.
Natural Toxins Database (at National University of Singapore): http://biomed.nus.sg.

NIOSH: http://www/.cdc.gov.niosh/homepage.htm.

Nuclear Regulatory Commission: http://www.nrc.gov/about-nrc/organization.html.

OSHA: http://www.osha.gov.

OSHA regulations: http://www.osha-sic.gov.

OSHA Salt Lake Technical Center: http://www.osha-slc.gov.

RISKWeb: http://riskweb.bus.utexas.edu/riskweb.html.

Safety Online: http://www.safetyonline.net.

Safety Groups UK: http://www.safetygroupsuk.org.uk/.

Safety Technology Institute (Italy): ftp://willow.sti.frc.it.

Safe Work Australia: http://www.safeworkaustralia.gov.au/sites/SWA.

Society of Risk Analysis: http://www.sra.org/.

University of Illinois Environmental Health and Safety: http://romulus.ehs.uiuc.edu/ DEHS/dehs.html.

UK Health and Safety Executive: http://www.hse.gov.uk/.

U.S. Centers for Disease Control and Prevention: http://www.cdc.gov.

U.S. Dept. of Energy—Office of Environment, Safety and Health: http://130.20.92. 130:8001/esh/home.html.

U.S. EPA: http://www.epa.gov.

U.S. EPA Center for Exposure Assessment Modeling: http://ftp.epa.gov/epa_ceam/ wwwhtml/ceam_home.html.

U.S. FEMA: http://www.fema.gov.

U.S. Food and Drug Administration: http://vm.cfsan.fda.gov/index.html.

Workplace Safety and Health Council (Singapore): https://www.wshc.sg/.

World Health Organization: http://www.who.ch.

World Safety Organization: http://www.worldsafety.org/.

SOME HEALTH AND SAFETY MAILING LISTS

To subscribe, send this message to the listserv: *subscribe [name of list] [your real name]*

Name of List	Listserv Address	General Subject Area
BIOSAFTY	listserv@mitvma.mit.edu	Safe handling of biohazards
CHEMED-L	listserv@uwf.cc.uwf.edu	Chemical safety
CMTS-L	listserv@cornell.edu	Chemical management and tracking
FIRENET	listserv@life.anu.edu.au	Firefighting and emergency response
LEPC	listproc@moose.uvm.edu	Hazardous materials and emergency response
RISKNET	listproc@mcfeeley.ccutexas.edu	Safety and health
SYSTEM-SAFETY	listserv@listserv.gsfc.nasa.gov	System safety

To subscribe, send this message: *subscribe [name of list] [your e-mail address]*

EMC-PSTC	majordomo@ieee.org	Product safety
HAZMATMED	listserv@mediccom.norden1.com	Hazardous material emergency response

Index

Milton Keynes UK
Ingram Content Group UK Ltd.
UKHW030900141024
449569UK00025B/1302

9 781138 893368